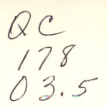

ASTRONOMICAL CONSTANTS

Sun:
mass $\qquad M_\odot = 1.99 \times 10^{33}$ g
radius $\qquad R_\odot = 6.96 \times 10^{10}$ cm
surface gravity $\qquad g_\odot = 2.74 \times 10^4$ cm/sec^2
luminosity $\qquad \mathscr{L}_\odot = 3.9 \times 10^{33}$ erg/sec

Earth:
mass $\qquad M_\oplus = 5.98 \times 10^{27}$ g
equatorial radius $\qquad R_\oplus = 6.38 \times 10^8$ cm
polar radius $\qquad R'_\oplus = R_\oplus - 2.15 \times 10^6$ cm
surface gravity $\qquad g = 9.81 \times 10^2$ cm/sec^2
moment of inertia:
  about polar axis $\qquad I^{33} = 0.331\ M_\oplus R_\oplus{}^2$
  about equatorial axis $\qquad I^{22} = I^{11} = 0.329\ M_\oplus R_\oplus{}^2$
period of rotation $\qquad$ 1 sidereal day $= 8.62 \times 10^4$ sec
mean distance to sun $\qquad$ 1 A.U. $= 1.50 \times 10^{13}$ cm
orbital period $\qquad$ 1 sidereal year $= 3.16 \times 10^7$ sec
orbital velocity $\qquad$ 29.8 km/sec

Moon:
mass $\qquad M_\mathrm{C} = 7.35 \times 10^{25}$ g
radius $\qquad R_\mathrm{C} = 1.74 \times 10^8$ cm
mean distance from earth $\qquad 3.84 \times 10^{10}$ cm
orbital period $\qquad$ 1 sidereal month $= 27.3$ days

Planetary orbits:

	Period	Perihelion Distance	Eccentricity
Mercury	0.241 years	45.9×10^6 km	0.206
Venus	0.615	107	0.00682
Earth	1.00	147	0.0167
Mars	1.88	207	0.0933
Jupiter	11.9	741	0.0483
Saturn	29.5	1350	0.0559
Uranus	84.0	2730	0.0471
Neptune	165	4460	0.0085
Pluto	248	4420	0.249

GRAVITATION
AND
SPACETIME

GRAVITATION
AND
SPACETIME

Hans C. Ohanian
RENSSELAER POLYTECHNIC INSTITUTE

W · W · NORTON & COMPANY

New York · London

GRAVITATION AND SPACETIME

Excerpt from "Cosmic Gall" by John Updike from *Telephone Poles and Other Poems*
© 1963, Alfred A. Knopf, Inc. Reprinted by permission of the publisher.

Library of Congress Cataloging in Publication Data

Ohanian, Hans C
  Gravitation and spacetime.

  Bibliography: p.
  Includes index.
  1. Gravitation. 2. Space and time. I. Title.
QC178.035   530.1'1   75-38991
ISBN 0-393-09198-8
Printed in the United States of America
  2 3 4 5 6 7 8 9 0

To Helmut and Emmi, with thanks

CONTENTS

* See Preface.

4.   Gravitational Waves   140

5.   The Measurement of Spacetime   182

6.   Riemannian Geometry   221

7.   Geometrodynamics   250

8.   The Schwarzschild Solution   280

9.   Black Holes and Gravitational
     Collapse   306

PREFACE

Einstein discovered his theory of gravitation in 1916. By rights this theory should not have been discovered until 20 years later, when physicists acquired a complete understanding of relativistic field equations and of gauge invariance. Einstein's profound and premature insights into the nature of gravitation had more to do with intuition than with logic. In contrast to the admirably precise and clear operational foundation on which he based his theory of special relativity, the foundations on which he based general relativity were vague and obscure. As has been emphasized by Synge and Fock, even the very name of the theory indicates a misconception: there is no such thing as "general relativity." But, whatever murky roads he may have taken, in the end Einstein's intuition did lead him to create a theory of dazzling beauty. If, using Arthur Koestler's image, one regards Copernicus, Kepler, Galileo, and Newton as sleepwalkers who knew where they wanted to go and managed to get there without quite knowing how, then Einstein was the greatest sleepwalker of them all.

It will be my objective to develop gravitational theory in the most logical and straightforward way — in the way it probably would have developed without Einstein's intervention. This means that I will begin with the linear approximation and regard gravitation as a field theory, entirely analogous to electrodynamics. The geometrical interpretation and the nonlinear Einstein equations gradually emerge as one attempts to understand and improve the equations of the linear approximation. This approach is not new; Gupta, Feynman, Thirring, and Weinberg have presented it from somewhat different points of view and with varying amounts of detail. One advantage of this approach is that it gives a clearer understanding of how and why gravitation is geometry. Another advantage is that the linear theory permits one to get right on with the physics: light deflection, retardation, redshift, and gravitational radiation can be directly treated in the context of the linear approximation without any lengthy preliminary digressions on the mathematics of Riemannian spacetime.

Since my approach is completely divorced from the historical development of the subject, it would have been awkward to attempt to trace the history of the theoretical concepts and ideas I have borrowed and used. Only in the description of experimental results have I tried to give some historical perspective; experiments become convincing only through repetition and therefore the lists of experimental results are as complete as I could make them.

This book is intended to serve as an introduction to Einstein's theory of gravitation at an undergraduate level. The book is based on material that I have used for several years in an undergraduate course at Rensselaer Polytechnic Institute. Although the course is aimed mainly at seniors, some graduate students and juniors also participate. The only essential prerequisite is an acquaintance with the physical basis of special relativity at the level of, say, Taylor and Wheeler's *Spacetime Physics* (my Chapter 2, on special relativity, is self-contained but is mostly concerned with the mathematical formalism; the physical motivation for special relativity must be sought elsewhere). In addition, some acquaintance with Maxwell's equations is helpful.

As a concession to the brevity of a one-semester course, it may be necessary to compromise between a broad and a deep view of the subject. In the table of contents I have indicated by asterisks some of the sections that may be omitted without destroying the balance and logical development of the presentation.

The exercises that are scattered throughout the chapters are an integral part of the text; they amplify the discussion, supply proofs, and are intended to be done as the book is read. Only a fanatic will find the time to do them all; the reader is invited to look upon these exercises as challenges which should not always be refused. The problems at the end of the chapters are for the traditional homework assignments. I have tried to make them interesting and informative, yet not too hard.

Thanks are due to many students who pointed out flaws in the earlier drafts of this book. Matthew Alexander carried out the calculations on which the light cone diagrams of Chapter 9 are based. My colleagues at RPI, Alan Meltzer and Richard Smith, patiently answered my naive questions about matters astronomical. William Press of Princeton University and Richard Price of the University of Utah provided valuable critical commentary. It is also a pleasure to thank Walter Knight of the University of California, Berkeley, for his careful review of the manuscript and for many suggestions on how to improve it. Finally, I would like to express my deep gratitude to John Wheeler, my teacher, who showed the way.

H. C. O.
Princeton, New Jersey, November 1975

I am indebted to many readers for their kindness in sending me lists of misprints and mistakes found in the first printing, and especially to Harvey Picker of Trinity College for many detailed comments and to Peter van Nieuwenhuizen of SUNY, Stony Brook for pointing out a mistake in the derivation of the formula for geodetic precession.

H. C. O.
December 1979

A Note on Notation

The components of a three-dimensional vector **A** with respect to the three-dimensional rectangular coordinates will be indicated by a *superscript* with the values 1, 2, 3:

$$A^1 = A_x \qquad A^2 = A_y \qquad A^3 = A_z$$

In particular, the rectangular coordinates are

$$x^1 = x \qquad x^2 = y \qquad x^3 = z$$

The symbol A^k, where the *Latin* superscript k takes on the values $k = 1$, 2, 3, then stands for the kth component of the vector. This symbol will also stand for the set (A^1, A^2, A^3) of all the components taken together; in the latter case A^k represents the entire vector **A**.

The Einstein summation convention applies: when a repeated Latin index appears, a summation is to be carried out over the values 1, 2, 3 of that index, e.g.,

$$A^n B^n \equiv \sum_{n=1}^{3} A^n B^n$$

The Kronecker delta will be written as

$$\delta_k{}^n = \begin{cases} 1 \text{ if } k = n \\ 0 \text{ if } k \neq n \end{cases}$$

and integration over a three-dimensional volume as

$$\int f(\mathbf{x}) d^3 x \equiv \iiint f(\mathbf{x}) dx\, dy\, dz$$

The components of a four-dimensional vector will be indicated by a superscript with the values 0, 1, 2, 3. In the special case of flat four-dimensional spacetime,

$$A^0 = A_t \qquad A^1 = A_x \qquad A^2 = A_y \qquad A^3 = A_z$$

and

$$x^0 = t \qquad x^1 = x \qquad x^2 = y \qquad x^3 = z$$

The symbol A^μ, where the *Greek* superscript takes on the value $\mu = 0, 1,$ 2, 3, stands for the μth component of the vector; it also stands for the set (A^0, A^1, A^2, A^3) and in the latter case represents the entire four-dimensional vector.

The definition of the four-dimensional Kronecker delta is the same as in the three-dimensional case,

$$\delta_\mu{}^\nu = \begin{cases} 1 \text{ if } \mu = \nu \\ 0 \text{ if } \mu \neq \nu \end{cases}$$

The Minkowski metric of flat four-dimensional spacetime is taken as

$$\eta_{\mu\nu} = \begin{pmatrix} \eta_{00} & \eta_{01} & \eta_{02} & \eta_{03} \\ \eta_{10} & \eta_{11} & \eta_{12} & \eta_{13} \\ \eta_{20} & \eta_{21} & \eta_{22} & \eta_{23} \\ \eta_{30} & \eta_{31} & \eta_{32} & \eta_{33} \end{pmatrix} = \begin{pmatrix} 1 & 0 & 0 & 0 \\ 0 & -1 & 0 & 0 \\ 0 & 0 & -1 & 0 \\ 0 & 0 & 0 & -1 \end{pmatrix}$$

The proper time interval of flat spacetime is

$$d\tau^2 = + \sum_{\mu=0}^{3} \sum_{\nu=0}^{3} \eta_{\mu\nu} dx^\mu dx^\nu$$
$$= (dt)^2 - (dx)^2 - (dy)^2 - (dz)^2$$

The proper time interval of curved spacetime is

$$d\tau^2 = + \sum_{\mu=0}^{3} \sum_{\nu=0}^{3} g_{\mu\nu} dx^\mu dx^\nu$$

If x^0 is the time coordinate, then $g_{00} > 0$ (timelike sign convention).

When a repeated Greek index appears in a term of an equation, a summation is to be carried out over the values 0, 1, 2, 3 of that index, e.g.,

$$\eta_{\mu\nu} A^\nu \equiv \sum_{\nu=0}^{3} \eta_{\mu\nu} A^\nu$$

In general, indices are raised and lowered with the metric tensor $g_{\mu\nu}$ of curved spacetime, e.g.,

$$A_\mu = g_{\mu\nu} A^\nu$$

However, in all equations which are written *in the linear approximation*, the indices are raised and lowered with the Minkowski metric $\eta_{\mu\nu}$.

Partial derivatives are indicated by a comma or by the differential operator ∂:

$$\frac{\partial f}{\partial x^\mu} = f_{,\mu} = \partial_\mu f$$

A dot over a variable indicates a derivative with respect to time (e.g., $\dot{z} = dz/dt$ in Chapter 4 and Appendix 1), or else a derivative with respect to proper time (e.g., $\dot{r} = dr/d\tau$ in Chapter 8), or else a derivative with respect to a "time parameter" (e.g., $\dot{a} = da/d\eta$ in Chapter 10).

GRAVITATION
AND
SPACETIME

1. NEWTON'S GRAVITATIONAL THEORY

It was occasioned by the fall of an apple, as he sat in a contemplative mood . . .

William Stukeley, *Memoirs of Sir Isaac Newton's Life*

Few theories can compare in the accuracy of their predictions with Newton's theory of universal gravitation. The predictions of celestial mechanics for the positions of the major planets agree with observation to within a few seconds of arc over time intervals of many years. The discovery of Neptune, and the rediscovery of Ceres, are among the spectacular successes that testify to the accuracy of the theory. But Newton's theory is not perfect: the predicted motions of the perihelia for the inner planets deviate somewhat from the observed values. In the case of Mercury the excess perihelion precession amounts to 43 seconds of arc per century. This small deviation was discovered through calculations by LeVerrier in 1845 and recalculated by Newcomb in 1882. The explanation of the precession was one of the early successes of Einstein's relativistic theory of gravitation.

Optical observations of planetary (angular) positions stretching over hundreds of years are needed to detect the excess perihelion precession. However, with the recent development of radar astronomy it has become possible to measure the distances to the (inner) planets directly and very accurately by means of the travel time of a radio signal sent

1

from the earth to the planet and reflected back. Also, Doppler radar makes it possible to measure the instantaneous velocity of planets. When the radar observations of distances are combined with optical observations of angular positions, the small deviations from Newton's theory can be detected after just a few years of observation.

Although Newton's theory is not perfect, it is an excellent approximation in the limiting case of motion at low velocity in a weak gravitational field. Any relativistic theory of gravitation ought to agree with Newton's theory in this limiting case. We therefore begin with a brief exposition of some aspects of Newton's theory.

1.1   THE LAW OF UNIVERSAL GRAVITATION

According to Newton the law governing gravitational interactions is "that there is a power of gravity pertaining to all bodies, proportional to the several quantities of matter which they contain. . . . The force of gravity towards the several equal parts of any body is inversely as the square of the distance of places from the particles."[1] If one mass is at the origin and the other at a radial distance r, then the force equation takes the mathematical form

$$F = -\frac{Gmm'}{r^2} \qquad\qquad [1]$$

where $G = 6.67 \times 10^{-8}$ dyne cm^2/g^2.

Strictly speaking, the mass that enters the force law, (1), is the *gravitational mass,* which is to gravitation what the electric charge is to electromagnetism. We will worry about the distinction between gravitational mass and inertial mass later on. In the following discussion of gravitational fields and potentials (Sections 1.1–1.3), the masses are always gravitational.

According to the above law, gravitation is action-at-distance: a mass at one point acts directly and instantaneously on another mass even though the other mass is not in contact with it. Newton had serious misgivings about such a ghostly tug-of-war of distant masses and suggested that the interaction should be conveyed by some material medium. The modern view is that gravitation acts locally through fields: a mass at one point produces a field, and this field acts on whatever masses it comes in contact with. The gravitational field may be regarded as the material medium sought by Newton; the field is material because it does possess an energy density. The description of interactions by means of local fields has the further advantage of leading us to a relativistic theory in which gravitational effects propagate at finite velocity. Instantaneous action-at-a-distance makes no sense as a relativistic theory because of the lack of an absolute time: what is instantaneous propagation in one reference frame need not be instantaneous in another. Of course, in the case of static or quasistatic mass distributions, retardation effects are insignificant and there is no physical distinction between local interaction and action-at-a-distance.

In the solar system, Newton's theory is an excellent approximation. The force (1) can be derived from a potential

$$V(r) = -\frac{Gmm'}{r} \qquad [2]$$

In general, we can say that relativistic effects will be small provided that the potential energy is much less than the rest mass energy and that the velocities are much less than the velocity of light. For a mass m moving with velocity v around a central mass m' we can express these conditions as

$$|V| << mc^2 \qquad \text{and} \qquad v << c$$

where c is the velocity of light. Note that the former condition is equivalent to $r >> Gm'/c^2$. Hence the deviations from Newton's theory are expected to be very small if the distance from the central mass is sufficiently large and the velocity sufficiently low. For the sun, with a mass $m' = M_\odot \cong 2 \times 10^{33}$ g, we have $Gm'/c^2 \cong 2$ km and the condition $r >> 2$ km is obviously very well satisfied, even for comets with a perihelion close to the surface of the sun.

Is it possible that there are deviations from the inverse square law at large distances? By "large distances" we mean distances of up to 10^4 or 10^5 light years; such distances are large compared to the dimensions of the solar system, but small compared to the typical dimensions of the universe.* Could the force law be some other function of distance? There are some general properties of relativistic field theory that put very tight restrictions on a possible alternative to the inverse square force law. It is best to express these restrictions in terms of potentials. The inverse square law has the special potential given by Eq. (2). The general potential consistent with field theory turns out to be

$$V(r) = -Gmm' \frac{e^{-\kappa r}}{r} \qquad [3]$$

where κ is a constant. This is called a *Yukawa potential;* obviously (2) is a special Yukawa potential with $\kappa = 0$. Besides (3), the only other possibility is some combination of *several* Yukawa potentials, which would mean that we are dealing with *several* gravitational fields. It is

* At *very* large distances (more than 10^7 light years), there may be cosmological deviations from the $1/r^2$ force (see Section 7.4). These deviations are not our concern in the present context.

hard to disprove this experimentally (if the combination has cleverly chosen coefficients), but we can appeal to Newton's first Rule for Reasoning in Philosophy*: "We are to admit no more causes of natural things than such as are both true and *sufficient* to explain their appearances."[2] To the best of our knowledge, one gravitational field is sufficient to explain all gravitational effects, and more fields would therefore be a wretched excess.

If the gravitational potential is as given by (3), then what can we say about κ? We can say that κ is extremely small, because the Yukawa potential, and the force it produces, become negligible when r appreciably exceeds κ^{-1}; one says that the *range* of the potential is κ^{-1}. The range of the gravitational force is known to be large. We know that our galaxy is held together by gravitation and this implies that the gravitational potential does not deviate much from $1/r$ out to distances of $r \simeq$ (galactic radius) $\simeq 10^{22}$ cm. Hence we conclude that

$$\kappa < 10^{-22} \text{ cm} \qquad\qquad\qquad [4]$$

Since κ is certainly very small, and since a value $\kappa = 0$ is consistent with all our experimental information, we will assume from now on that $\kappa = 0$.

Incidentally, the value of κ is related to the mass of the graviton**,

$$m_\Gamma = \kappa \hbar / c$$

The inequality (4) implies that

$$m_\Gamma < 10^{-59} \text{ g}$$

It is interesting to note that by setting similar limits on possible deviations from Coulomb's law, one finds that the mass of the photon is bounded by

$$m_\gamma < 10^{-48} \text{ g}$$

Even though the gravitational interaction is much weaker than the electromagnetic interaction, we can place better limits on m_Γ than on m_γ! The gravitational constant G that appears in Eq. (1) is not known very precisely. Thus, while e, $\hbar$, and c are known to six or eight significant figures, the value of G is only known to three figures. Measurements of G are difficult because of the extreme weakness of the gravitational force between masses of laboratory size. The gravitational force between masses of planetary size is not so weak, but this is of no help in determin-

* Also known as "Occam's razor."

** A particle of spin two (and mass believed to be zero) which is to gravitation what the photon is to electromagnetism.

ing G because only the combination GM (where M is the mass of the attracting body) appears in the equations of motion of bodies with purely gravitational interactions; hence, planetary observations cannot determine the separate values of G and M. Table 1.1 summarizes results of laboratory measurements of G.

Fig. 1.1 shows the torsion balance used by Cavendish. A beam

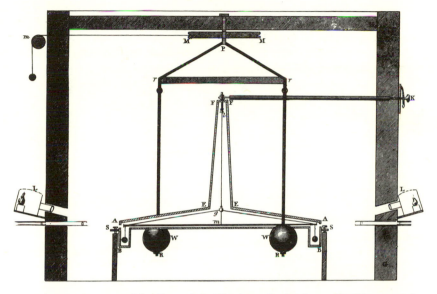

Fig. 1.1 *The apparatus used by Cavendish. The large lead spheres* (W, W) *attract the small spheres* (B, B) *which are attached to the beam of the torsion balance.* (*From Cavendish,* Phil. Trans. Roy. Soc. London *18, 388 (1798).*)

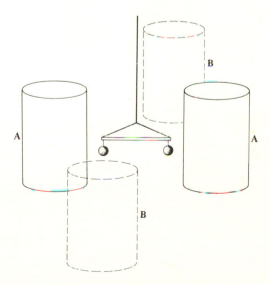

Fig. 1.2 *The method of Heyl. The large masses may be placed either in the "near" position* (A, A) *or else in the "far" position* (B, B).

with two small masses (B, B) is suspended from a thin fiber. The small masses are gravitationally attracted by the two large lead spheres (W, W), and this results in a measurable deflection of the beam of the balance through some angle around the vertical. From the known torsional constant of the fiber and the geometry of the apparatus the gravitational constant can then be calculated.

The last two measurements listed in Table 1.1 have given the most precise values for G. The torsion-balance method of Heyl uses an arrangement similar to that of a Cavendish balance, but instead of measuring the deflection produced by the presence of the large masses, one measures the period of oscillation, first when the large masses are placed

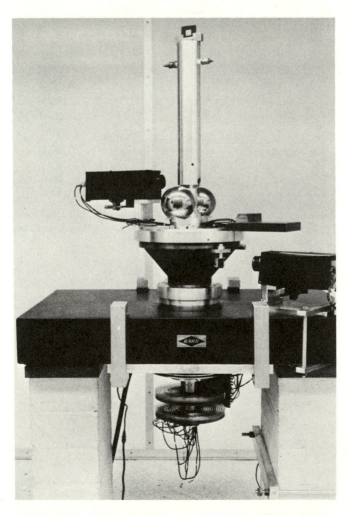

Fig. 1.3 *The apparatus of Beams et al. (From* Physics Today, *May 1971.)*

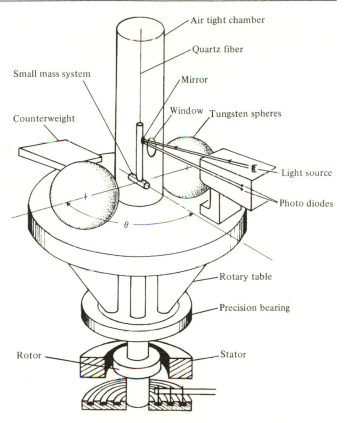

Fig. 1.4 *Schematic diagram of the apparatus. The small mass system consists of a single horizontal cylinder; the gravitational attraction between the cylinder and the spheres tends to twist the torsional pendulum. During a typical experimental run the angle θ is held steady at 45° by the angular acceleration of the rotary table. (From Rose, Parker, Lowry, and Beams,* Phys. Rev. Lett. **23,** 655 *(1969).)*

near the equilibrium position of the small masses and again when the large masses are placed at 90° to this equilibrium position (see Fig. 1.2). In the former position the period is shortened since the large masses produce an extra restoring force; in the latter position the period is lengthened since the large masses tend to pull the small ones away from the equilibrium position; the change in period determines G. The advantage of the method is that the period can be measured more precisely than the deflection. The error in G is probably 0.5%.

The apparatus of Beams et al. consists of a Cavendish balance mounted on a table that can rotate about the vertical axis (see Figs. 1.3 and 1.4). As the large masses begin to twist the torsional pendulum, a servomechanism begins to rotate the table at exactly the rate needed to keep the torsional pendulum from ever getting any nearer the large masses. This means that throughout each experimental run ($\sim$ 2 hr) the rotary table must be continuously accelerated; to keep the angle θ fixed at $\sim$ 45°, an angular acceleration of $\sim$ 10^{-6} radians/sec^2 is required. Measurement of this acceleration yields the value of G to a precision of about 0.1%.

TABLE 1.1   DETERMINATIONS OF G[3]

Experimenter(s)	Year	Method	G(dyne cm^2/g^2)
Cavendish	1798	torsion-balance deflection	6.754×10^{-8}
Reich	1838	torsion-balance deflection	6.70
Wilsing	1887	metronome balance	6.596
von Jolly	1881	common balance	6.465
Poynting	1891	common balance	6.698
Boys	1895	torsion-balance deflection	6.658
Braun	1896	torsion-balance deflection	6.658
Eötvös	1896	torsion-balance deflection	6.65
Richarz and Krigar-Menzel	1898	common balance	6.685
Burgess	1901	torsion-balance deflection	6.64
Crémieu[4]	1909	torsion-balance deflection	6.67
Heyl[5]	1930	torsion-balance period	6.670
Zahradnicek[6]	1933	torsion-balance resonance	6.659
Heyl and Chrzanowski[7]	1942	torsion-balance period	6.673
Beams et al.[8]	1969	rotating torsion-balance	6.674

1.2   THE GRAVITATIONAL POTENTIAL

The gravitational force exerted by N particles at positions $\mathbf{x}_1, \mathbf{x}_2, \mathbf{x}_3, \ldots,$ $\mathbf{x}_N$ on a particle of mass m located at $\mathbf{x}$ is given by*

$$\mathbf{F}(\mathbf{x}) = -Gm \sum_{i=1}^{N} \frac{m_i}{|\mathbf{x} - \mathbf{x}_i|^3} (\mathbf{x} - \mathbf{x}_i)$$

To this force there corresponds a potential energy

$$V(\mathbf{x}) = -Gm \sum_i \frac{m_i}{|\mathbf{x} - \mathbf{x}_i|} \qquad [5]$$

with

$$\mathbf{F}(\mathbf{x}) = -\nabla V(\mathbf{x}) \qquad [6]$$

---

* $\mathbf{x}$ is the position vector, $\mathbf{x} = x\hat{x} + y\hat{y} + z\hat{z}$. Instead of x, y, z we will often use x^1, x^2, x^3. In this notation the components F_x, F_y, F_z of an arbitrary vector will be written F^1, F^2, F^3. The reason for the use of numerical *superscripts*, rather than subscripts, will become clear later.

In component notation, Eq. (6) may be written as

$$F^k(\mathbf{x}) = -\frac{\partial V}{\partial x^k} \qquad k = 1, 2, 3 \tag{7}$$

The *gravitational field* that we regard as the carrier of the interaction is defined as force per unit mass,

$$\mathbf{g}(\mathbf{x}) \equiv \frac{1}{m}\,\mathbf{F}(\mathbf{x})$$

The corresponding *gravitational potential* is defined as

$$\Phi(\mathbf{x}) \equiv \frac{1}{m}\,V(\mathbf{x}) = -\sum_i \frac{Gm_i}{|\mathbf{x} - \mathbf{x}_i|} \tag{8}$$

This definition makes the potential negative, as expected for an attractive force. The gravitational potential is sometimes defined with a sign opposite to that of Eq. (8), but we prefer to choose our signs by analogy with electrostatics. For a continuous mass distribution we have

$$\Phi(\mathbf{x}) = -\int \frac{G\rho(\mathbf{x}')}{|\mathbf{x} - \mathbf{x}'|}\,d^3x' \tag{9}$$

where $\rho(x)$ is the mass density. Eq. (9) implies that Φ obeys the Poisson equation

$$\nabla^2 \Phi(\mathbf{x}) = +4\pi G\rho(\mathbf{x}) \tag{10}$$

Exercise 1. Prove this.

Problems in the theory of the Newtonian gravitational potential involve exactly the same mathematics as in the theory of the electrostatic potential. Because of this we will not spell out all of the details of the following derivations. One important difference between gravitation and electrostatics is that the mass density, as opposed to the charge density, can never be negative. This implies that there can be no shielding of gravitational fields corresponding to the shielding of electric fields by conductors.

By ingenious geometrical arguments Newton proved that a spherically symmetric mass distribution behaves in the same way as a point particle located at its center, in that (1) it produces the same field in its exterior and (2) it responds in the same way to fields produced by external sources. These results, and some others, can be obtained very directly by appealing to the uniqueness theorem and the mean value theorem of potential theory.

EXERCISE 2. Use the uniqueness theorem for the potential function to show that the potential due to a spherically symmetric mass distribution is $-GM/r$ in the exterior region.

EXERCISE 3. The potential energy of an object placed in a given gravitational potential $\Phi(\mathbf{x})$ (produced by given external sources) is

$$V = \int \rho(\mathbf{x})\Phi(\mathbf{x})d^3x$$

where $\rho(\mathbf{x})$ is the mass density of the object. Use the mean value theorem for the potential to show that a spherically symmetric mass distribution, placed in a potential due to external sources, has a potential energy $M\Phi(\mathbf{x}_0)$ where $\mathbf{x}_0$ is the coordinate of the center of the sphere and $M = \int \rho d^3x$ is the mass.

EXERCISE 4. Use the uniqueness theorem for the potential function to show that the potential produced by a uniform spherical mass shell is constant in the interior of the shell.

The gravitational self-energy of a continuous mass distribution is given by

$$-\tfrac{1}{2}\iint G\,\frac{\rho(\mathbf{x})\rho(\mathbf{x}')}{|\mathbf{x}-\mathbf{x}'|}\,d^3x d^3x' \qquad [11]$$

This can also be expressed as

$$\int \frac{1}{8\pi G}(\nabla\Phi)^2 d^3x + \int \rho\Phi d^3x \qquad [12]$$

This equation has an interesting interpretation: the quantity $(\nabla\Phi)^2/8\pi G$ may be regarded as the energy density of the gravitational field and $\rho\Phi$ may be regarded as an interaction energy density of field and matter. In the case of electromagnetism, all of the electric energy may be regarded as field energy; in the gravitational case this is not possible because field energy is positive and hence something negative must be added to it in order to get a negative total energy as given by (11).

EXERCISE 5. Show that the expressions (11) and (12) are equivalent.

1.3 GRAVITATIONAL MULTIPOLES

The gravitational potential has the simple form $-GM/r$ only in the space surrounding a mass distribution with spherical symmetry. An arbitrary mass distribution produces a potential

$$\Phi(\mathbf{x}) = -G \int \frac{\rho(\mathbf{x}')d^3x'}{|\mathbf{x} - \mathbf{x}'|}$$ [13]

If $\mathbf{x}$ is far away from the region that contains the mass (see Fig. 1.5), then we can construct a multipole expansion for the potential by using the Taylor series expansion of $1/|\mathbf{x} - \mathbf{x}'|$ about $\mathbf{x}' = 0$:

$$\frac{1}{|\mathbf{x} - \mathbf{x}'|} \equiv \frac{1}{[(x - x')^2 + (y - y')^2 + (z - z')^2]^{1/2}}$$

$$= \frac{1}{r} + \sum_k \frac{x^k x'^k}{r^3} + \tfrac{1}{2} \sum_k \sum_l (3x'^k x'^l - r'^2\delta_k{}^l) \frac{x^k x^l}{r^5} + \ldots$$ [14]

where $r^2 = x^2 + y^2 + z^2$.

Fig. 1.5

EXERCISE 6. Derive Eq. (14).

The integral (13) can therefore be written as

$$\Phi(\mathbf{x}) = -\frac{GM}{r} - \frac{G}{r^3} \mathbf{x} \cdot \mathbf{D} - \frac{G}{2} \sum_{kl} Q^{kl} \frac{x^k x^l}{r^5} + \ldots$$ [15]

where

$$M = \int \rho(\mathbf{x}')d^3x'$$

$$\mathbf{D} = \int \mathbf{x}'\rho(\mathbf{x}')d^3x'$$ [16]

$$Q^{kl} = \int (3x'^k x'^l - r'^2\delta_k{}^l)\rho(\mathbf{x}')d^3x'$$ [17]

The quantity $\mathbf{D}$ is the mass dipole moment. If the origin of coordinates is chosen to coincide with the center of mass, then $\mathbf{D} = 0$; we will usually assume that this is so (no mass dipole). The quantity Q^{kl} is the mass quadrupole tensor. Eq. (15) shows that whenever the quadrupole is non-zero, the potential will contain a term $\sim 1/r^3$ and hence the force will deviate from the inverse square law by a term $\sim 1/r^4$. Note that the quadrupole moment of a mass distribution with spherical symmetry is zero.

EXERCISE 7. Show this.

The earth's polar and equatorial diameters differ by about 3 parts in 10^3. The deviation from spherical shape produces a quadrupole term in the gravitational potential which causes perturbations in the Kepler orbits of satellites. In fact, the observed perturbations of the orbits are used to calculate the multipole moments and the mass distribution in the earth.

The sun is almost a perfect sphere. However, even a quite small difference between the polar and equatorial radii could have important consequences for gravitational theory. The reason is that a quadrupole field would cause a precession of the perihelion of planetary orbits, i.e., the Kepler ellipse would slowly rotate about the sun. Fig. 1.6 shows an orbit with precession; both the eccentricity of the orbit and the precession rate are shown very exaggerated. Einstein's theory of gravitation predicts a separate perihelion precession effect and in order to test the theory it is necessary to determine how much of the total observed precession is accounted for by a solar quadrupole moment.

In the case of the planet Mercury, the total observed perihelion precession amounts to ~ 500 seconds of arc per century (relative to the fixed stars). This large precession is not surprising since Mercury's orbit is perturbed by other planets. But if all the planetary perturbations are calculated from Newtonian theory, then the observed precession exceeds the calculated one by 43 seconds of arc per century. As will be shown later, this excess agrees exactly with the effect calculated from Einstein's theory of gravitation on the assumption that the sun has *no appreciable quadrupole moment.* If the sun does have a quadrupole moment, then it will contribute an extra precession and there will be a corresponding disagreement between observation and Einstein's theory; the remarkable agreement obtained on the assumption of no quadrupole moment would then have to be regarded as an accidental coincidence.

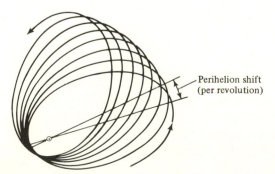

Perihelion shift
(per revolution)

Fig. 1.6 *Planetary   orbit   with (large) precession.*

Table 1.2 gives the results of determinations of the sun's shape. The quantity $\Delta_\odot$ is the difference between the equatorial and polar diameters in seconds of arc; a positive value of this quantity indicates that the equatorial diameter is larger.

TABLE 1.2 SHAPE OF THE SUN[9]

Observer	Year	Method	$\Delta_\odot$(sec)
Auwers	1874, 1882	heliometers	-0.038 ± 0.023
Schur	1891–1901	heliometer	0.007 ± 0.015
Ambronn	1891–1902	heliometer	-0.002 ± 0.009
Hayn, Schaub	1912	long exposure photographs	-0.02 to -0.04
Dicke and Goldenberg[10]	1966	scanning with photocell	0.0866 ± 0.0066
Hill and Stebbins[11]	1973	scanning with photocell	0.0184 ± 0.0125

Since the sun rotates, we expect it to have a small equatorial bulge. The observed rotation period at the surface is 25 days and under the assumption that the interior rotates at the same rate, we expect $\Delta_\odot = 0.016$ seconds of arc. Most of the measurements agree with this value, within one or two standard deviations. The important exception is the measurement of Dicke and Goldenberg, which indicates that the sun is oblate, having an equatorial bulge about five times as large as that expected from uniform rotation. Since the diameter of the sun is 1,920 seconds, the corresponding fractional difference between the equatorial and polar diameters would be $(4.51 \pm 0.34) \times 10^{-5}$. Such an oblateness would give the sun an appreciable quadrupole moment. Let us define a dimensionless parameter J_2,

$$J_2 \equiv -Q^{33}/(2M_\odot R_\odot^2)$$

where $M_\odot$ and $R_\odot$ are the mass and radius of the sun and the z-axis is taken to coincide with the polar axis. This parameter is a convenient measure of the quadrupole moment. The above oblateness would give

$$J_2 = (2.47 \pm 0.23) \times 10^{-5}$$

The corresponding quadrupole moment would produce an extra perihelion precession of 3 seconds of arc per century for Mercury. Einstein's theory of gravitation would then disagree with the observed precession by 3 seconds of arc per century. Brans and Dicke[12] have proposed a "scalar-tensor" theory of gravitation which would avoid any conflict with the observed precession. In this theory, the gravitational "constant" is a function of space and time.

However, the results of Dicke and Goldenberg are contradicted by the other measurements. The discrepancy with the measurement of Hill and Stebbins is particularly serious because both experiments use quite similar techniques which avoid the dependence on human judgment that entered into the earlier measurements. In both experiments a moving slit with a photocell behind it is used to measure the intensity of the light in the solar image as a function of position.

A possible explanation of the discrepancy may be that the equatorial region of the sun tends to be somewhat brighter; there are bright spots (faculae) on the surface of the sun, and they tend to concentrate in the equatorial region. This would make the sun look bigger along the equator. There is some disagreement however as to whether this effect can account for all of the oblateness found by Dicke and Goldenberg. The experiment of Hill and Stebbins should be free of this particular trouble because observations were conducted during a period of a few weeks when the excess equatorial brightness was smaller than usual.

A different approach to the problem has become possible with the development of radar astronomy. Since the quadrupole field produces perturbations in the orbits, the quadrupole moment can be determined quite directly by precise observations of planetary positions as a function of time. The very high precision required can be achieved by combining optical and radar data for planets with tracking data for Mariner spacecraft. Reliable results should soon be available.

To sum it up: the preponderance of the experimental evidence is consistent with a spherical, or very nearly spherical, sun. It seems likely that the quadrupole moment of the sun is quite small and does not make any appreciable contribution to the perihelion precession of Mercury.

Let us now briefly consider time-dependent multipoles. If we have a mass distribution that changes in time, then the multipoles will also change. Time-dependent quadrupole moments will be of great importance in the relativistic theory of gravitation because they act as sources of gravitational radiation. Vibrating and rotating dumbbells (see Fig. 1.7) are simple examples of time-dependent quadrupoles.

In general, the values of the elements of the quadrupole tensor

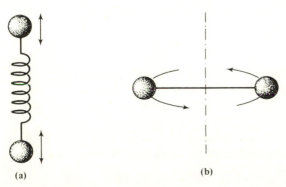

(a)   (b)

Fig. 1.7 *Oscillating   quadrupoles.*

depend on the choice of the origin of coordinates. Suppose that the coordinates x^k have their origin at the center of mass. If we shift the origin so that we obtain new coordinates

$$x'^k = x^k + b^k$$

where b^k is a constant, then the quadrupole tensors in the old and new coordinates are related by

$$Q'^{kl} = Q^{kl} + (3b^k b^l - \delta_k{}^l b^2)M \tag{18}$$

EXERCISE 8. Derive Eq. (18).

Eq. (18) shows that a change of origin only modifies Q^{kl} by an additive constant. Thus, the *time-dependent part of Q^{kl}* is independent of the choice of origin. This result will be useful in our study of gravitational radiation (Chapter 4).

1.4 THE EQUIVALENCE OF INERTIAL AND GRAVITATIONAL MASS

The equation of motion of a pointlike particle in a given gravitational potential is

$$m \frac{d^2 x^k}{dt^2} = -m \frac{\partial \Phi}{\partial x^k} \tag{19}$$

If we cancel the masses, we obtain

$$\frac{d^2 x^k}{dt^2} = -\frac{\partial \Phi}{\partial x^k} \tag{20}$$

This says that *in a given gravitational field, all pointlike particles fall with the same acceleration.* For historical reasons, we will call this statement *Galileo's principle of equivalence.* This statement implies that to some extent gravitational forces behave in the same way as the pseudo-forces that result from the use of a noninertial reference frame. In an accelerated reference frame, free particles appear to accelerate spontaneously and, at a given point, *all particles have exactly the same acceleration.** This equivalence between inertial and gravitational

---

* To be precise, we should also specify that all the particles are put at the given point with the *same initial velocity.* This is necessary because some pseudo-forces (e.g., Coriolis force) are velocity dependent. This restriction on the velocity is also needed for Galileo's principle, since it turns out that the "gravitational force" given by general relativity is velocity dependent.

effects can make it rather hard to distinguish between the two. Thus an astronaut in a freely falling spaceship will have a hard time deciding whether he is in free fall in a gravitational field, or in unaccelerated motion in a region far away from all fields (if he only performs experiments in the interior of his spaceship).

We must now ask: is it correct to cancel the masses in Eq. (19)? Are the m's on both sides really the same? The mass that appears on the left side of Eq. (19) is the *inertial mass*. For any arbitrary body, the inertial mass is defined in the usual way: we take the body and let it interact, somehow, with the standard kilogram. Both the body and the standard will accelerate towards or away from one another. Designating the acceleration of the body and of the standard by a and a_s, respectively, we can define the inertial mass by

$$m_I \equiv \frac{a_s}{a} \qquad [21]$$

This gives the inertial mass of the body in kilograms.

The gravitational mass can be defined in a similar manner. We take a standard object and *define* its gravitational mass to be one unit; for convenience we will use the standard kilogram as a standard for both inertial and gravitational mass (perhaps an atomic standard would be more relevant). We now place the body at some distance r and let it interact *gravitationally* with the standard. The body will accelerate towards the standard. We define the gravitational mass of the body in terms of this acceleration and the distance between the objects:

$$m_G \equiv \lim_{r \to \infty} \left(\frac{a m_I r^2}{G} \right) \qquad [22]$$

Alternatively, we can use (21) to write this as

$$m_G \equiv \lim_{r \to \infty} \left(\frac{a_s r^2}{G} \right) \qquad [23]$$

This gives the gravitational mass in kilograms. The limiting procedure $r \to \infty$ is needed in (22) and (23) in order to eliminate the effects of multipole fields, which depend on the mass distribution of the two bodies. Also, proceeding to the limit $r \to \infty$ eliminates the effect of short-range forces (nuclear force, Van der Waals force, etc.). At large distances only the gravitational and electrostatic forces will remain and the latter can be eliminated by taking the precaution of keeping the standard body neutral.*

---

* Occasionally a distinction between *active* and *passive* gravitational mass has been made. Thus the force exerted *by m' on m* would be written as $G m_{GP} m'_{GA}/r^2$ and the force exerted

If we take two identical copies of the standard of mass and let them fall towards each other, the acceleration of each serves to define the constant G:

$$G \equiv \lim_{r \to \infty} (a_{ss}r^2) \qquad [24]$$

With these precise definitions of m_I and m_G it is clear that the gravitational force between two particles is

$$F = -G \frac{m_G m'_G}{r^2} \qquad [25]$$

and that the equation of motion is

$$m_I \frac{d^2 x^2}{dt^2} = -m_G \frac{\partial \Phi}{\partial x^k} \qquad [26]$$

Whether all particles fall with the same acceleration depends on whether all particles have the same value of m_I/m_G. If this ratio is a universal constant, it must have the value one (the standard body has this value by definition). The question is then, is the equation

$$m_I = m_G \qquad [27]$$

satisfied for all bodies? We will call the equality (27) *Newton's principle of equivalence of inertial and gravitational mass.*

Before we turn to the experimental evidence, we remark that for a body of appreciable size the acceleration of the center of mass is given by

$$m_I \frac{d^2 x^k}{dt^2} = -\int \rho_G(\mathbf{x}') \frac{\partial \Phi}{\partial x'^k}(\mathbf{x}') d^3 x' \qquad [28]$$

where the integral is over the volume of the body and $\rho_G(\mathbf{x})$ is the gravitational mass density. Extended bodies will fall at the same rate as point particles only if $\partial \Phi/\partial x^k$ is approximately constant over the volume of the body; in that case we can factor $\partial \Phi/\partial x^k$ out from the integral and we then obtain (26).* Hence Newton's principle of equivalence implies equal

---

* The case of extended bodies with a spherically symmetric mass distribution is exceptional; such bodies behave as point particles regardless of their size.

---

by m on m' as $Gm_{GA}m'_{GP}/r^2$; then (22) and (23) would define the passive and the active mass, respectively. (One might likewise be tempted to introduce active and passive charges in Coulomb's law.) But this is nonsense. The equality of active and passive mass is required by the equality of action and reaction; an inequality would imply a violation of momentum conservation.

acceleration only for bodies of sufficiently small size placed in sufficiently homogeneous gravitational fields. Bodies of different size and shape placed in an inhomogeneous field will fall at different rates.

EXERCISE 9. Show that, as a consequence of Newton's principle of equivalence applied to the mass densities (i.e., $\rho_I = \rho_G$), two bodies of the same shape and size and similar (i.e., proportional) mass distribution will have the same linear and angular acceleration when placed in a given gravitational field.

The derivation given at the beginning of this section shows that for point particles the Galileo principle implies the Newton principle and vice versa. This derivation hinges on the validity of Newtonian mechanics; that is, the gravitational fields must be weak ($GM/rc^2 << 1$) and the velocities low ($v << c$). If these assumptions are not satisfied, then the Galileo and Newton principles must be regarded as *complementary* rather than equivalent: in general, the Galileo principle is a statement about a special class of particles (point particles) moving with arbitrary velocities in arbitrary fields, while the Newton principle is a statement about arbitrary systems moving at low velocity in weak fields. The experiments to be described next constitute evidence for the Newton principle; since they were performed with particles of low velocity in weak fields, they only give limited support to the Galileo principle.

1.5   THE EMPIRICAL EVIDENCE FOR THE EQUALITY OF m_I AND m_G

An experiment performed by Galileo gives some early evidence in favor of the equality of inertial and gravitational mass:

> . . . I took two balls, one of lead and one of cork, the former being more than a hundred times as heavy as the latter, and suspended them from two equal thin strings, each four or five bracchia long. Pulling each ball aside from the vertical, I released them at the same instant, and they, falling along the circumferences of the circles having the strings as radii, passed thru the vertical and returned along the same path. This free oscillation, repeated more than a hundred times, showed clearly that the heavy body kept time with the light body so well that neither in a hundred oscillations, nor in a thousand, will the former anticipate the latter by even an instant, so perfectly do they keep step.[13]

We can interpret Galileo's observation to mean that the ratio m_I/m_G is the same for lead as it is for cork, to within a few parts in a thousand.

EXERCISE 10. What is the equation for the period of a simple pendulum if the bob has $m_I \neq m_G$? Galileo's claim that the two pendulums showed no deviation after a thousand oscillations is an exaggeration (on pendulums of different weight, the effects of friction will be quite noticeable); his claim for a hundred oscillations is credible. Suppose that after 100 oscillations the pendulums deviate by no more than $\frac{1}{10}$ of a cycle. What limit does this set on the difference between the m_I/m_G values for lead and cork?

The first careful experiments specifically designed to test the equality of inertial and gravitational mass are due to Newton:

> ... I tried the thing in gold, silver, lead, glass, sand, common salt, wood, water, and wheat. I provided two equal wooden boxes. I filled the one with wood, and suspended an equal weight of gold (as exactly as I could) in the centre of oscillation of the other. The boxes, hung by equal threads of 11 feet, made a couple of pendulums perfectly equal in weight and figure, and equally exposed to the resistance of air: and, placing the one by the other, I observed them to play together forwards and backwards for a long while, with equal vibrations. And therefore (by Cor. I and VI, Prop. XXIV, Book II) the quantity of matter in the gold was to the quantity of matter in the wood as the action of the motive force upon all the gold to the action of the same upon all the wood; that is, as the weight of the one to the weight of the other.
>
> And by these experiments, in bodies of the same weight, one could have discovered a difference of matter less than the thousandth part of the whole.[14]

Table 1.3 summarizes the experimental evidence that has accumulated in favor of the equality of inertial and gravitational mass. These experiments were performed with a large variety of substances. In setting upper limits on $|m_I - m_G|/m_I$, one of the substances used by the experimenter has been taken as a standard for which $m_I \equiv m_G$. For example, Eötvös* used platinum as standard and compared other substances with it.

TABLE 1.3   EQUALITY OF m_I AND m_G

Experimenter(s)	Year	Method	$|m_I - m_G|/m_I$
Galileo[13]	~1610	pendulum	$<2 \times 10^{-3}$
Newton[14]	~1680	pendulum	$<10^{-3}$
Bessel[15]	1827	pendulum	$<2 \times 10^{-5}$
Eötvös[16]	1890	torsion-balance	$<5 \times 10^{-8}$
Eötvös et al.[17]	1905	torsion-balance	$<3 \times 10^{-9}$
Southerns[18]	1910	pendulum	$<5 \times 10^{-6}$
Zeeman[19]	1917	torsion-balance	$<3 \times 10^{-8}$
Potter[20]	1923	pendulum	$<3 \times 10^{-6}$
Renner[21]	1935	torsion-balance	$<2 \times 10^{-10}$
Dicke et al.[22]	1964	torsion-balance; sun	$<3 \times 10^{-11}$
Braginsky et al.[23]	1971	torsion-balance; sun	$<9 \times 10^{-13}$

---

* Pronounced ut-vush (u as in turn).

The most precise results have been obtained with torsion balances. The apparatus of Eötvös is shown in Figs. 1.8 and 1.9. Two pieces of matter, labeled "weight," are attached to the arms of a torsion balance. These weights are made of different substances, e.g., platinum and copper. If m_I/m_G has a different value for the two substances, then a torque will be exerted on the balance. To see how this comes about, consider the forces that act on the masses. Fig. 1.10 shows these forces. In this figure both masses are shown placed directly on the beam of the balance; the asymmetric suspension used in the apparatus of Eötvös is of no relevance to the present experiment (the suspension was originally designed for a different purpose). The forces are of two kinds: there is the gravitational force $m\mathbf{g}$ exerted by the earth and the *centrifugal* pseudo-force $m\mathbf{a}$ produced by the rotation of the earth; the quantity $\mathbf{g}$ is

Fig. 1.8 *Instrument used by Eötvös and his collaborators for the first gravity survey at Lake Balathon (1902). The instrument used to measure m_I/m_G was of the same type. (Courtesy of Hungarian State Geophysical Institute Eötvös Loránd, Budapest.)*

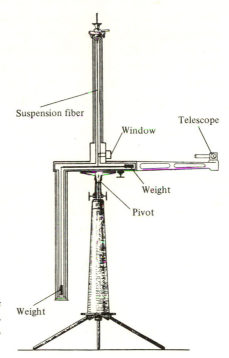

Fig. 1.9 *Schematic diagram of the Eötvös apparatus. (After Eötvös, Pekar, and Fekete,* **Ann. d. Physik** *68, 11 (1922).)*

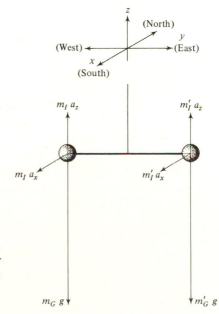

Fig. 1.10 *Torsional pendulum suspended near the surface of the earth at a latitude of about 45° N. Note that the centrifugal force vectors* ($m_I a$) *are shown exaggerated by a factor of ~ 100.*

the acceleration of gravity (*without* centrifugal effects*) and $\mathbf{a}$ is the centrifugal acceleration due to earth rotation at the locality of the experiment. In Fig. 1.10 it has been assumed that the beam of the balance points in the east-west direction; the centrifugal force then has a vertical (opposite to $\mathbf{g}$) component $m_I a_z$ and a horizontal component $m_I a_x$.

The torque about the z-axis is obviously

$$\tau = m_I a_x l - m_I' a_x l' \qquad [29]$$

We can eliminate l' by using the equilibrium condition for rotation about the x-axis,

$$(m_G g - m_I a_z)l = (m_G' g - m_I' a_z)l' \qquad [30]$$

and get

$$\tau = m_I a_x l \left[1 - \frac{m_I'}{m_I} \frac{m_G g - m_I a_z}{m_G' g - m_I' a_z} \right]$$

$$= m_I a_x l \frac{g \left[\frac{m_G'}{m_I'} - \frac{m_G}{m_I} \right]}{g m_G'/m_I' - a_z} \qquad [31]$$

From this it is obvious that a torque exists if and only if $m_G/m_I \neq m_G'/m_I'$. In equilibrium, the gravitational torque (31) is compensated by a torque produced by the suspension fiber. The presence of the gravitational torque can be detected by rotating the entire apparatus by exactly 180° about the vertical axis (see pivot, Fig. 1.9). If the equilibrium position of the beam was exactly along the east-west direction before this rotation, then it will be slightly off after the rotation; this change in equilibrium position occurs because turning the apparatus around changes the sign of the torque (31).**

The method of Dicke also uses a torsion balance (Fig. 1.11), but detects the torque, if any, produced by the gravitational force of the sun and the centrifugal force of the earth's motion around the sun. If, for the sake of simplicity, we imagine a torsion balance of the type shown in Fig. 1.10 suspended at the north pole of the earth, then this centrifugal force is entirely horizontal. Assuming equal arms ($l=l'$), the torque about the vertical axis will be

---

* Careful! Values of "g" quoted in handbooks usually *include* the centrifugal effects.
** Mathematically, turning the apparatus around is equivalent to $l \rightarrow -l$ and $l' \rightarrow -l'$ in Eqs. (29) and (30). Incidentally, this shows that any horizontal torque produced by the suspension (i.e., an extra term in Eq. (30)) would only increase the torque (31) by an extra term that *does not* change sign upon rotation and hence would have no effect on the measurement.

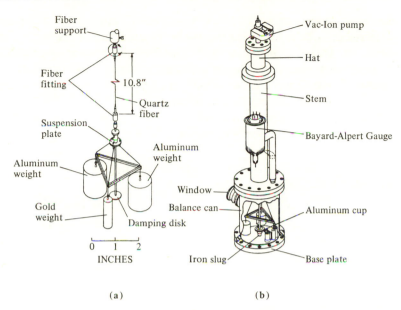

Fig. 1.11 *The apparatus used by Dicke et al. (a) The torsion balance has a triangular beam with a mass suspended at each vertex. (b) The balance is suspended in an evacuated vessel. (From Dicke,* Gravitation and the Universe. *Reprinted with permission of the American Philosophical Society.)*

$$\tau = (m_G g - m_I a)\, l \sin \phi - (m_G' g - m_I' a)\, l \sin \phi \qquad [32]$$

where g is the acceleration of gravity of the *sun* at the place of the experiment, a the centrifugal acceleration in the reference frame of the earth,* and ϕ the angle between the beam of the balance and the sun. In Dicke's experiment, the beam is held stationary with respect to the earth and the torque required to do this is measured. It is obvious from (32) that under these conditions the torque oscillates with a 24-hr period as the sun angle ϕ increases by 2π. Any other effect ("noise") disturbing the experiment with a period different from 24 hr can be filtered out and discarded. Note that Dicke's apparatus actually uses a somewhat more complicated torsion balance with a triangular "beam" holding three masses. This arrangement makes the balance less sensitive to gradients in the gravitational field produced by nearby massive bodies, such as the body of an experimenter (as a further precaution, measurements were carried out by remote control).

The experiments listed in Table 1.3 have tested the ratio m_I/m_G for a wide variety of materials. Thus, Eötvös et al. compared copper,

* To a sufficient approximation, $g = a$.

water, copper sulfate, asbestos, snakewood, etc., etc., with platinum.* The results of the experiment may be interpreted to mean that different kinds of energy contribute to the inertial mass of a system the same amount they contribute to the gravitational mass. Forms of energy that appear in the inertial mass (which, as we know from special relativity, is the same thing as total internal energy) of a sample of ordinary matter are listed in the first column of Table 1.4. The second column shows how these different forms of energy gravitate.

TABLE 1.4 DIFFERENT FORMS OF ENERGY AS SOURCES OF GRAVITY[a]

Form of Energy	$\lvert m_I - m_G \rvert / m_I$
Rest mass, protons	$< 10^{-11}$
Rest mass, electrons	$< 2 \times 10^{-8}$
Electrostatic binding energy and kinetic energy of atomic electrons	$< 10^{-6}$
Electrostatic repulsion of nuclear protons	$< 3 \times 10^{-10}$
Strong binding energy and kinetic energy of nucleons	$< 10^{-8}$

[a] For the purpose of this table, neutrons have been taken as standard with $m_I \equiv m_G$.

Note that this table is based on the assumption that no fortuitous cancellations take place between deviations from $m_I = m_G$. Strictly speaking, we cannot make independent statements about, say, electron and proton rest mass contributions to gravitation, since these particles are present in equal numbers in samples of neutral matter. Free-fall experiments have been performed with individual neutrons and with individual electrons in the gravitational field of the earth, but the precision of these experiments is relatively low. For example, measurement of the downward sag of a horizontal beam of low-velocity neutrons emerging from a nuclear reactor has shown that the downward acceleration of these free neutrons agrees with the standard local value of the earth's gravitational acceleration to within at least 1%.[29]

As concerns antimatter, we have no direct experimental evidence on the value of m_G, but there is some circumstantial evidence in favor of normal gravitational interaction of antimatter. It has been pointed out by Schiff[24] that if positrons had a negative gravitational mass ("fell upward"), then one would expect nuclei to have anomalous ratios of m_I/m_G. The reason for this is that a nucleus is surrounded by a fluctuating cloud of virtual electron-positron pairs (a condition known as vacuum

* However, the extremely precise experiments of Dicke and of Braginsky only tested gold, platinum, and aluminum.

polarization). The number of pairs depends on the charge distribution of the nucleus, and if the gravitational mass of positrons were negative, nuclei surrounded by more pairs would appear to have a lower gravitational mass. This effect would lead to a difference in m_I/m_G between platinum and aluminum amounting to a few parts in 10^{-7}, in clear contradiction to Eötvös's experimental results.

A much more precise statement about gravitational interactions of antimatter can be made by examining the famous case of the decay of the neutral K^0 meson. The decay of this meson is very sensitive to quantum-mechanical phase factors, and even a small change in the (small) phase factor $e^{-im_G \Phi t/\hbar}$ coming from the gravitational energy of the meson in the gravitational potential Φ of the galaxy would lead to drastic changes in the decay.[25] One can conclude that the gravitational masses of the meson K^0 and the antimeson $\bar{K}^0$ differ at most by a few parts in 10^{10}. (The *inertial* masses of K^0 and $\bar{K}^0$ are exactly equal; this is a general theorem for every particle and the corresponding antiparticle.)

According to Table 1.4, the energies associated with the electromagnetic and strong interactions contribute to the gravitational mass in the "right" way. What about the energies associated with the weak interaction (force involved in β-decay) and the gravitational interaction? Does gravitational energy gravitate? It has sometimes been argued that the weak and gravitational interactions contribute very little to the self-energy of the materials that can be and have been tested in the Eötvös experiment, and therefore any anomalous contribution of these self-energies to the gravitational mass would be quite undetectable. This argument is very questionable. In fact, it is not known how much weak and gravitational energy is contained in the rest mass of an electron. Calculation of the self-energy of an electron gives an infinite value. This, of course, only proves that the calculation is wrong, that our understanding of the quantum dynamics is incomplete. But one thing is clear: no proof exists that the gravitational energy of an electron is small. It is conceivable that the gravitational self-energy is very large; it may even be that the very large gravitational energy exactly cancels the very large electromagnetic energy. The observed equality of gravitation and inertia for particle rest masses implies that either the weak and gravitational self-energies are small, *or else* they contribute in the normal way to gravitation. We will assume that the second alternative holds: all forms of energy gravitate equally. Newton's principle of equivalence,

$$m_I = m_G$$

will be postulated as an exact relation that our gravitational theory must satisfy.

Although a direct experimental test of the hypothesis that gravity gravitates cannot be carried out with laboratory masses, it now seems that such a test can be carried out with the mass of the earth. If we treat the earth as a continuous mass distribution, then the gravitational self-energy of the earth comes out as 4×10^{-10} times the rest-mass energy. (This number is actually a lower bound since the internal gravitational self-energy of the particles in the earth has been ignored in the calculation.) If gravitational self-energy does not contribute in the normal way to the gravitational mass, then the earth and the moon will not fall at the same rate in the gravitational field of the sun. This leads to a distortion of the orbit of the moon relative to the earth. Although the effect is small, very precise measurements of the earth-moon distance can now be carried out by the laser ranging technique. A pulse of light emitted by a laser on the earth is sent to the moon and is reflected back to the earth by the corner reflectors installed on the moon by the Apollo 11 mission. Measurement of the travel time of the pulse determines the distance to within a few centimeters. These experiments are expected to place a direct limit of 10^{-2} or 10^{-3} on the value of $|m_I - m_G|/m_I$ associated with gravitational energy (compare Table 1.4).

1.6 TIDAL FORCES

Consider a spaceship in orbit around the earth.* The spaceship is in free fall. The astronauts find themselves in a zero-g environment, they are weightless, they feel no gravitational force. The interior of the spaceship is an inertial reference frame: a test particle at rest relative to the spaceship remains at rest, a test particle in motion remains in motion with uniform velocity. However, for an observer at rest relative to the fixed stars, the spaceship will be in accelerated motion, and such an observer will refute the astronauts' claim that they are in an inertial frame by saying that both the spaceship and the test particle in it are falling at the same rate and that the astronauts are being fooled by appearances. Note that the elimination of the earth's gravitational field by free fall is impossible unless the equivalence of inertial and gravitational mass holds for all bodies; any object with $m_I \neq m_G$ would appear to accelerate spontaneously with respect to the freely falling spaceship.

We now ask: are the gravitational effects of the earth *completely* eliminated by free fall? Is there some local experiment that permits the astronauts to find out that they are falling in a gravitational field rather than at rest in some region far away from any attracting mass? The

---

* We assume that the spaceship is not spinning about its axis. To be precise: we assume that the rate of spin of the spaceship has been adjusted so that the axis of a gyroscope carried by the spaceship remains at rest relative to the spaceship. In the Newtonian approximation, the axis of spin of a gyroscope will of course maintain its direction relative to the fixed stars. But, as we will see later, relativistic effects produce a precession of the axis of spin; these effects are a generalization of the well-known Thomas precession of special relativity.

answer is that the astronauts *can detect the gravitational field by the tidal effects* it produces. If the astronauts place a drop of liquid at the center of their spaceship, they will find that this drop is not exactly spherical, but has two bulges (Fig. 1.12). One bulge points towards the earth, one away. Since, in the absence of external forces, surface tension would make the drop spherical, the deviation from a sphere indicates the existence of a gravitational field. The bulges result from the in-homogeneity of the gravitational field: the end of the drop nearer the earth is pulled too much by gravitation, the other end is not pulled enough.

The force that produces the bulges is called the tidal force. We can calculate it as follows:

Suppose you have a reference point which is moving in free fall; take this point as the origin of a coordinate system with the z-axis parallel to the radial line (Fig. 1.13). A particle at position $(0,0,z)$ experiences a gravitational acceleration

$$-\frac{GM}{(r_0 + z)^2} \qquad [33]$$

where r_0 is the distance from the center of the earth to the origin of our coordinates and M the mass of the earth. Since the origin has acceleration $-GM/r_0^2$, the acceleration of the particle *relative* to the origin is, in the limit of small z,

$$-\frac{GM}{(r_0 + z)^2} + \frac{GM}{r_0^2} \simeq 2z\frac{GM}{r_0^3} \qquad [34]$$

Hence, relative to our origin the particle moves as though subjected to a force

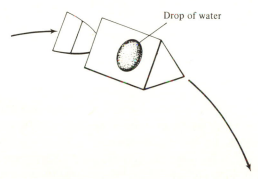

Drop of water

Earth

Fig. 1.12 *A spaceship in orbit.*

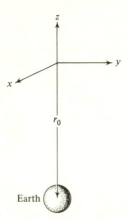

Fig. 1.13

$$f_z = 2z \frac{GMm}{r_0^3} \tag{35}$$

This is a tidal force. Note that it is directly proportional to the distance of the particle from the origin and is repulsive.

For a particle at a position $(0,y,0)$ the tidal force points in the $-y$ direction and is given by

$$f_y = -y \frac{GMm}{r_0^3} \tag{36}$$

This is a tidal restoring force, toward the origin. For a particle at the position $(x,0,0)$ there is a similar restoring force,

$$f_x = -x \frac{GMm}{r_0^3} \tag{37}$$

If the particle has x, y, and z displacements, then all three forces—(35), (36), and (37)—are of course present simultaneously. Note that these expressions for the tidal force only apply if the displacements x, y, z are small compared to r_0.

EXERCISE 11. Derive (36) and (37).

From these results it is obvious that a drop of liquid, consisting of many particles, will be stretched in the radial direction and compressed in the transverse direction.

In general, given an arbitrary gravitational field, the tidal force in a reference frame whose origin is in free fall can be expressed as follows*

* In Eq. (38) we use the *summation convention:* a repeated Latin index (such as l) is to be summed over all its values ($l = 1,2,3$). Thus (38) stands for $f^k = \sum\limits_{l=1}^{3} x^l \, \partial F^k / \partial x^l = -m \sum\limits_{l=1}^{3} x^l \, \partial^2 \Phi / \partial x^k \partial x^l$.

$$f^k = x^l \frac{\partial F^k}{\partial x^l} = -x^l m \frac{\partial^2 \Phi}{\partial x^k \partial x^l} \qquad [38]$$

where F^k is the gravitational force on the particle and Φ the potential and where the derivatives are evaluated at the origin ($x^l = 0$). Eq. (38) is of course only valid near the origin (small x^l).

EXERCISE 12. Derive Eq. (38).

We will call the quantity

$$R^k_{\ 0l0} \equiv -\frac{1}{mc^2} \frac{\partial F^k}{\partial x^l} = \frac{1}{c^2} \frac{\partial^2 \Phi}{\partial x^k \partial x^l} \qquad [39]$$

the *tidal force tensor*. The extra factor mc^2 and the extra indices $_{00}$ have been introduced for later convenience. It turns out that the nine quantities $R^k_{\ 0l0}$ are components of the Riemann curvature tensor $R^\mu_{\ \nu\alpha\beta}$; the latter tensor describes the tidal field in the four-dimensional spacetime of general relativity (see Section 7.5). The tidal force that acts on a particle placed at x^l (with $x^l \ll r_0$) is then

$$f^k = -mc^2 R^k_{\ 0l0} x^l \qquad [40]$$

which produces an acceleration

$$\frac{d^2 x^k}{dt^2} = -c^2 R^k_{\ 0l0} x^l \qquad [41]$$

In the special case considered above, the tensor $R^k_{\ 0l0}$ has components:

$$R^k_{\ 0l0} = \begin{pmatrix} \dfrac{GM}{r_0^3 c^2} & 0 & 0 \\[2mm] 0 & \dfrac{GM}{r_0^3 c^2} & 0 \\[2mm] 0 & 0 & -2\dfrac{GM}{r_0^3 c^2} \end{pmatrix} \qquad [42]$$

Note that the index k gives the row, l the column.

The tidal force given by Eqs. (35), (36), and (37) satisfies the identity

$$\frac{\partial f_x}{\partial x} + \frac{\partial f_y}{\partial y} + \frac{\partial f_x}{\partial z} = -\frac{GMm}{r_0^3} - \frac{GMm}{r_0^3} + \frac{2GMm}{r_0^3} = 0$$

—that is, the tidal force field has zero divergence. This result holds in general for gravitational fields in empty regions of space. We can prove this by using Eq. (38):

$$\frac{\partial}{\partial x^k} f^k = -\frac{\partial}{\partial x^k}\left(x^l m \frac{\partial^2 \Phi}{\partial x^k \partial x^l}\right)$$

$$= -m\delta_k{}^l \frac{\partial^2 \Phi}{\partial x^k \partial x^l} = -m \frac{\partial^2 \Phi}{\partial x^k \partial x^k} = 0$$

[43]

The last equality holds true because the gravitational potential satisfies the Laplace equation $\nabla^2 \Phi = 0$ (see Eq. (10)). Note that the argument shows that the equations $\nabla\cdot\mathbf{f} = 0$ and $\nabla^2 \Phi = 0$ are equivalent. Hence a gravitational field *in vacuo* may be characterized by the vanishing of the divergence of the tidal force.

EXERCISE 13. Show that in the presence of a mass density ρ, the divergence of the tidal force is given by

$$\frac{\partial}{\partial x^k} f^k = -4\pi m G \rho$$

As we will see later, the tidal forces given by general relativity are somewhat more complicated.* They depend on the velocity of the reference point and on the velocity of the particle with respect to that reference point, and furthermore, the gravitational field is not that given by Newton. However, the general relativistic tidal force agrees with the above Newtonian value in the limit of weak gravitational fields ($GM/r_0 c^2$ << 1) and low velocity (v/c << 1).

The "tidal forces" take their name from the tides they generate on the oceans of the earth. These tidal forces are exerted by the moon, and to a lesser extent, by the sun; we will ignore the sun for now. The center of "spaceship earth" is in free fall towards the moon. Consider a particle on the surface of the earth, with coordinates x, y, z (see Fig. 1.14). According to (35), (36), and (37), the tidal force exerted by the *moon* is

$$f_z = 2z \frac{GMm}{r_0{}^3}; f_y = -y \frac{GMm}{r_0{}^3}; f_x = -x \frac{GMm}{r_0{}^3}$$

where M is now the mass of the moon and r_0 the earth-moon distance. The potential that will give this force is

$$-\frac{GMm}{r_0{}^3}(z^2 - \tfrac{1}{2}x^2 - \tfrac{1}{2}y^2) = -\frac{GMm}{r_0{}^3}\left(-\tfrac{1}{2}R^2 + \frac{3}{2}z^2\right)$$

---

* These forces are contained in the equation of geodesic deviation of Section 7.5.

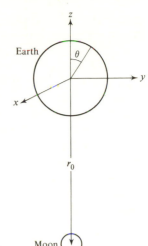

Fig. 1.14

$$= -\frac{GMm}{r_0^3}\left(\frac{3\cos^2\theta - 1}{2}\right)R^2 \qquad [44]$$

where $R = (x^2 + y^2 + z^2)^{1/2}$ is the radius of the earth and θ is the angle shown in Fig. 1.14.

A particle on the surface of the earth also feels the gravitational attraction of the earth. Since we are concerned with particles that remain near the surface, we can write for the potential produced by the earth

$$mgh \qquad [45]$$

where h is the height above the surface.* The net gravitational potential is then

$$V(h, \theta) = mgh - \frac{GMmR^2}{r_0^3}\left(\frac{3\cos^2\theta - 1}{2}\right) \qquad [46]$$

To find the height of the tide, let us assume that the water is in static equilibrium; then the surface must be an equipotential

$$gh - \frac{GMR^2}{r_0^3}\left(\frac{3\cos^2\theta - 1}{2}\right) = \text{constant} \qquad [47]$$

This gives us the height h of the water as a function of θ. High tides occur at $\theta = 0$ and $\theta = \pi$, low tides at $\theta = \pi/2$. The difference between high and low tide (the "tidal range") is given by

* To be precise, we should write $(R + h)$ in Eq. (44) rather than R. But for $h \ll R$, this makes little difference.

$$\Delta h = \frac{GMR^2}{gr_0^3} \left[\frac{3\cos^2\theta - 1}{2} \right] \begin{array}{l} \theta = 0 \\ \theta = \pi/2 \end{array} = \frac{3GMR^2}{2gr_0^3} \qquad [48]$$

The numerical value of the tidal range given by Eq. (48) is 53 cm (the sun produces an effect about half as large). This is a reasonable order of magnitude estimate for the tide in the open sea. But our results are not accurate because the water does *not* reach static equilibrium with the tidal force. As a consequence of the rotation of the earth about its axis, the tidal field appears to move once around the earth every lunar* day and we must seek the response of water to an oscillating force of period 12 hr. The water in ocean basins has certain natural periods of oscillation, in much the same way as water in a bathtub has a well defined period for sloshing back and forth, and we are therefore dealing with the problem of an oscillator subjected to a periodic driving force. The response of the oscillator depends critically on how the frequency of the driving forces compares to the natural frequency. If the natural frequency is much larger than the driving frequency, the oscillations remain in phase with the driving force and quasistatic conditions prevail (the restoring force is almost in static equilibrium with the driving force). However, for the case of some ocean basins there exist natural modes of oscillation with frequencies *smaller* than 1 cycle per 12 hr; this leads to an inverted tide, out of phase with the driving force. Basins that have a natural frequency very close to the driving frequency can develop enormous tides by resonance. For example, tides with a range of 50 feet are found in the Bay of Fundy.

1.7   THE TIDAL FIELD AS A LOCAL MEASURE OF GRAVITATION

Let us now look a bit more closely at the detection of gravitational fields by tides. If astronauts in orbit wish to detect the gravitational field of the earth by measuring the tide produced by the earth on a drop of water, they will find it desirable to use a very large drop of water. Eq. (48) shows that the height of tide increases with the size of the drop. This suggests that if the astronauts have been ordered to confine their experiments to the interior of a sufficiently small spaceship, then they will not be able to detect the tide, nor the gravitational field. But suppose that the astro-

---

* The time between successive passages of the moon through a given meridian is $24^h\ 50^m$. If the axis of rotation of the earth were perpendicular to the orbital plane of the moon, then the two maxima in the tidal potential would simply travel around the equator of the earth and the period of the tidal driving force would be $12^h\ 25^m$. Actually, the axis of the earth makes an angle with the orbit of the moon and therefore, at any given latitude, the two daily maxima are not exactly equal. The force has a strong Fourier component at $12^h\ 25^m$ ("semidiurnal tide") and a weaker component ("diurnal tide") at $24^h\ 50^m$. The sun contributes other components.

nauts use as measure of the tidal field not the height Δh, but rather the ratio $\Delta h/R$; according to Eq. (48) we have

$$\frac{\Delta h}{R} = \frac{3GMR}{2gr_0^3}$$
[49]

where M is now the mass of the earth, R the radius of the "drop," and g the acceleration of gravity that the "drop" of water produces at its own surface. Strictly speaking, Eq. (48) is only applicable to tides in a thin layer of liquid covering the surface of a dense rigid sphere. For a drop made entirely of liquid the change in the gravitational field of the drop produced by the change of shape must be taken into account; this modifies (48) by a numerical factor of no great interest to us. Note that we are assuming that the drop is held together by gravitation, rather than by surface tension; this is not realistic for small drops of water, but we will pretend that some very powerful chemical has been added to the water so as to make its surface tension very small. In terms of the density of the fluid, we can write $g = G(4\pi\rho R^3/3)/R^2$ and

$$\frac{\Delta h}{R} = \frac{9}{8\pi} \frac{M}{\rho r_0^3}$$
[50]

This shows that the *shape* of the tidal ellipsoid is independent of its size. Even in the limit as $R \to 0$, the tidal deformation remains. We can therefore regard the prolateness of the tidal ellipsoid as a *local* measure of the gravitational force.

We can now give a very pretty graphical description of a gravitational tidal field: at each point of space, draw the tidal ellipsoid that gives the shape of a drop of fluid (with zero surface tension) in free fall at that point. The tidal field of a spherical mass is shown in Fig. 1.15. (It is well known that given an arbitrary symmetric tensor $R^k{}_{0l0}$, one can associate with it a quadratic form $R^k{}_{0l0} x^k x^l$. The surface $R^k{}_{0l0} x^k x^l = $ constant can then serve as a graphical representation of the tensor. Our method of using the shape of a drop of fluid as a graphical representation of $R^k{}_{0l0}$ is similar but has the advantage that it yields a *closed* surface, which the usual method does not because the quadratic surface $R^k{}_{0l0} x^k x^l = $ constant is a hyperboloid.)

EXERCISE 14. Suppose that the symmetric tensor $R^k{}_{0l0}$ has been reduced to diagonal form. Show that the components of the tensor cannot all have the same sign. (Hint: Φ cannot have a maximum or a minimum.)

We can also give an alternative graphical representation by appealing to Eq. (43). According to this equation, the tidal force field *in vacuo* has zero divergence; this means that the tidal force can be repre-

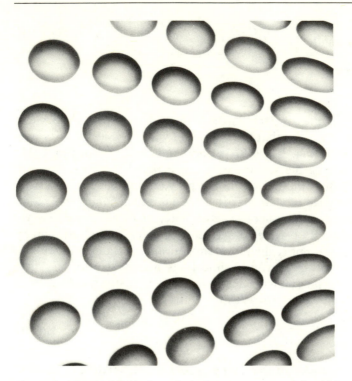

Fig. 1.15 *The tidal field of a spherical mass represented by tidal ellipsoids. The mass is beyond the right edge of the drawing.*

sented by *lines of force* in exactly the same way as an electric field *in vacuo*. Fig. 1.16 shows the pattern of lines of force near points in the space surrounding a spherical mass. Note that a separate pattern must be drawn for the vicinity of each point because our expressions for the tidal field are only valid for small displacements.

Although the graphical representations of Figs. 1.15 and 1.16 are mathematically equivalent, the picture of tidal ellipsoids has the advantage of emphasizing the local character of the tidal effects. The picture of lines of force suggests that at the center of each pattern (see Fig. 1.16) the tidal effects are absent. We must firmly reject this suggestion. Although some tidal effects do vanish near the origin of a freely falling reference frame, some other effects remain finite and it is these that ultimately count.

A more familiar graphical representation of gravitational fields uses the lines of force of the ordinary gravitational force rather than the tidal force (see Fig. 1.17). This is of course very useful to an outside observer who wishes to study the motion of a spaceship, or whatever, through the field. But this representation fails to describe what happens locally. If we want a description of what the astronauts experience in-

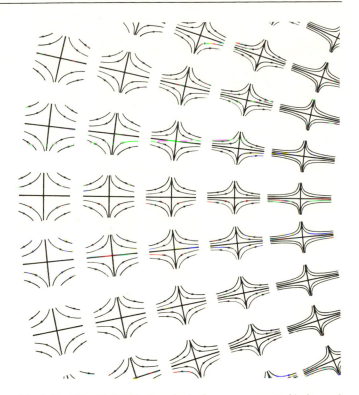

Fig. 1.16   *The tidal field of a spherical mass represented by lines of force. Each of the patterns of lines of force is valid only near its center. Note that a two-dimensional drawing does not correctly indicate the density of field lines of a three-dimensional field. The density along the radial axis is actually twice as large as the density along the transverse axis.*

side their spaceship, then the picture of tidal forces is much more relevant than that of the ordinary force.

   The detection of tidal fields by means of the deformation of drops of water is of course not a practical method because for a small drop the surface tension will prevent the formation of tidal bulges. A somewhat more realistic method is the following: Suppose the astronauts place a rigid body in their spaceship. Fig. 1.18 shows a rigid rod at the origin of the coordinates. The tidal force will exert a torque on the rod and give it an angular acceleration. For the tidal field given by (35)–(37), the torque about the x-axis will be

$$\tau_x = \int y \left(2z \frac{GM}{r_0^3} \right) - z \left(-y \frac{GM}{r_0^3} \right) dm$$

$$= \frac{3GM}{r_0^3} \int yz\, dm = \frac{3GM}{r_0^3} (-I_{yz}) \qquad\qquad [51]$$

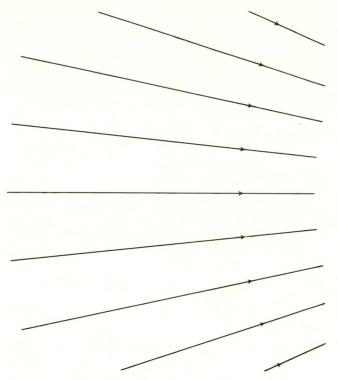

Fig. 1.17 *The inverse square gravitational force represented by
lines of force. The diagram is drawn to the same scale as
the preceding diagrams.*

where $I_{yz} = I^{23}$ is the yz-component of the moment of inertia tensor

$$I^{kl} = \int (r^2 \delta_k{}^l - x^k x^l) dm \qquad [52]$$

Expressions similar to (51) may be found for the torque about the y- and
z-axes.

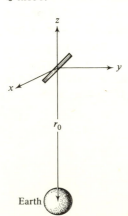

Fig. 1.18 *Rigid rod in free fall, instantaneously at rest.*

EXERCISE 15. Find τ_y, τ_z.

EXERCISE 16. Show that the torque exerted by an arbitrary tidal field $R^k{}_{0s0}$ on a body with an inertia tensor I^{ls} is given by

$$\tau^n = c^2 \epsilon^{nkl} R^k{}_{0s0}\left(-I^{ls} + \frac{1}{3}\,\delta_l{}^s\,I^{rr}\right) \qquad [53]$$

where the quantity ϵ^{nkl} is defined as follows

$$\epsilon^{123} = \epsilon^{231} = \epsilon^{312} = 1$$
$$\epsilon^{321} = \epsilon^{213} = \epsilon^{132} = -1$$

with all other components zero.

    If no other forces act, Eq. (51) implies that the x-component of the spin of the rigid body changes at a rate

$$\frac{dS_x}{dt} = -\frac{3GM}{r_0^3}\,I_{yz} \qquad [54]$$

This rate of change of the spin can serve as measure of the tidal field.
    For a *local* measurement, we must proceed to the limit of small size of the rigid body. The inertia I_{yz} is of the order $I_{yz} \sim ml^2$ (where l is a typical dimension of the body), and the spin is $S \sim ml^2\omega$. In the limit $l \to 0$, both torque and spin vanish, but $d\omega/dt$ remains finite. It is easy to see that $d\omega/dt$ depends only on the *shape* of the rigid body, not on the size. We can therefore use the angular acceleration to detect the tidal force. If the body is originally spinning, then the tidal force will cause a precession of the spin. A well-known example is the equinoctial precession of the spin of the earth which is caused by the lunar and solar tidal forces acting on the equatorial bulge of the earth.

EXERCISE 17. Show that if a body has the moment of inertia tensor of a sphere, then the torque (53) vanishes. Hence instead of measuring the spin precession (54) relative to the fixed stars, we can measure it locally, relative to a spherical gyroscope.

    There exist other methods for local measurement of the tidal field. The Eötvös balance shown in Fig. 1.8 was originally designed to measure gradients in the gravitational field of the earth by means of the torque they produce on the balance. One of the weights of this balance is placed lower than the other to make the balance sensitive to vertical gradients in the horizontal component of gravitational force. With this arrangement, tidal fields as small as $R^k_{0l0} = (1/c^2)\partial^2\Phi/\partial x^k \partial x^l \simeq 10^{-32}/\text{cm}^2$ were detectable.*

The beam of the Eötvös balance was 40 cm long. In order to see how good the balance is at measuring R^k_{0l0} *locally,* let us consider the limit $l \to 0$ (where l is a typical dimension of the balance). The tidal torque on the balance will then tend to zero. The equilibrium position of the beam is determined by the condition that the torque exerted by the tidal forces be equal to the restoring torque exerted by the fiber. If we scale down the size of the balance, we must also decrease the torque constant of the fiber in order to retain the sensitivity of the balance to a given tidal field. This means that the thickness of the fiber must be reduced. (It can be shown that a reduction of the size of the apparatus by a factor of, say, two requires a reduction in fiber thickness by a factor of $2^{3/2}$.) There are, of course, practical limitations which prevent us from reducing the size of our apparatus ad infinitum and obtaining the value of R^k_{0l0} at exactly *one point;* these limitations must, however, not be blamed on the gravitational field, but rather on the properties (ultimately *quantum* properties) of materials.

As a final example of local measurement of tidal fields, we mention a "gravity gradiometer" that was built recently by Forward et al. (1971).[26] The instrument is intended for use aboard an airplane or satellite and will permit very precise mapping of local irregularities in the earth's gravita-

Fig. 1.19   *The gravity gradiometer of Forward et al. (Courtesy of Dr. R. L. Forward, Hughes Research Laboratories.)*

---

* In geodesy it is customary to describe tidal fields by $\partial^2\Phi/\partial x^k \partial x^l$ rather than $(1/c^2)\partial^2\Phi/\partial x^k \partial x^l$. A value of $10^{-32}/\text{cm}^2$ for the latter corresponds to $\sim 10^{-11}/\text{sec}^2$, or 10^{-2} E.U., for the former (the Eötvös Unit (E.U.) is defined as $10^{-9}/\text{sec}^2$).

tional field. The device consists of a Greek cross with four masses at the ends of its arms. The arms are held together at the center by a torsional spring (see Figs. 1.19 and 1.20); when the arms are pressed together and then released they oscillate with a natural frequency of 32 cycles per second. If the cross is placed in a tidal field, it will be deformed; to facilitate the detection of this deformation, one uses the tidal field to drive the cross at resonance. The trick is this: Suppose you place the cross in the tidal field and spin it about an axis perpendicular to the plane of the cross with an angular frequency of 16 cycles per second. In the reference frame of the cross, the tidal driving force will then appear to oscillate at 32 cycles per second. Since the frequency of this oscillation coincides with the natural frequency of the arms, large amplitudes build up by resonance.* Tidal fields of $10^{-30}/cm^2$ have been detected with a prototype of this gradiometer. The device is only 15 cm in size; measurements with it are quite "local." A reduction in the size of the apparatus is possible in principle; the sensitivity does not depend directly on the size since the angular acceleration of the arms is determined by the ratio of torque to inertia and this ratio is independent of the size of the apparatus.

To summarize: we have seen that there exist several methods for measuring the tidal field *locally,* in an arbitrarily small neighborhood of a given point. The limitations on the minimum size of the neighborhood that is needed to perform measurements of a given precision do not arise from any intrinsic properties of the gravitational field; rather, these limitations arise from the quantum nature of matter which prevents us from constructing an apparatus of arbitrarily small size. The tidal field is no less a local quantity than, say, the electric field. Mathematically, this local nature of the tidal field is perfectly obvious: the tidal field is the derivative of the force field, which means that a knowledge of the latter in an arbitrarily small neighborhood of a point is sufficient to determine the former at that point.

Local experiments can distinguish between a reference frame in

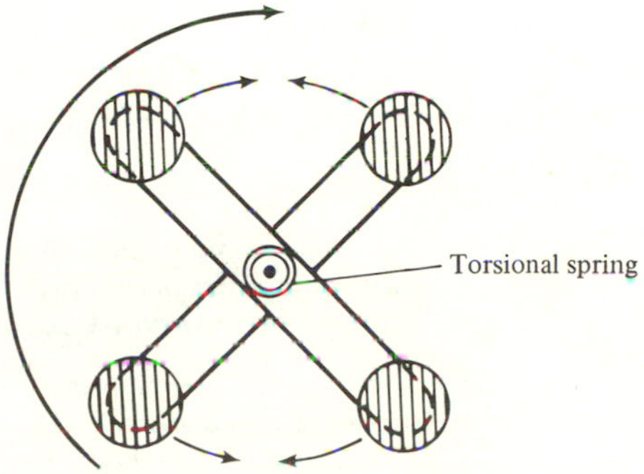

Fig. 1.20 *Schematic diagram of the gradiometer. The crossed arms, with masses at their ends, are pivoted at the center and held at 90° by a spring.*

Torsional spring

---

* This is very similar to the building up of resonant tides on the oceans.

free fall in a gravitational field and a truly inertial reference frame placed far away from all gravitational fields. *Local* experiments can distinguish between a reference frame at rest in a gravitational field and an accelerated reference frame far away from all gravitational fields. Gravitational effects are *not equivalent* to the effects arising from an observer's acceleration.* It is worth repeating this a few times, because in his original paper on the theory of general relativity Einstein wrote:

> [Let K' be a system of reference such that] relative to K' a mass sufficiently distant from other masses has an accelerated motion such that its acceleration and direction of acceleration are independent of its material composition and physical state.
>   Does this permit an observer at rest relative to K' to draw the conclusion that he is on a "really" accelerated system of reference? The answer is in the negative; for the above mentioned behavior of freely moving masses relative to K' may be interpreted equally well in the following way. The system of reference K' is unaccelerated, but the spacetime region being considered is under the sway of a gravitational field, which generates the accelerated motion of the bodies relative to K'.[27]

This statement, which is one formulation of the principle of equivalence of gravitation and acceleration, is only true in a limited sense. Gravitation and acceleration are only equivalent as far as the translational motion of point particles is concerned (this amounts to what we called the Galileo principle of equivalence, sometimes also called the *"weak"* principle of equivalence). If the rotational degrees of freedom of the motion of "masses" are taken into consideration, then the equivalence fails.

Unfortunately, Einstein's statement has often been generalized to sweeping assertions about all laws of physics being the same in a laboratory freely falling in a gravitational field, and in another laboratory far away from any field (this is called the "strong" principle of equivalence). Such generalizations are unwarranted since, as we have seen, even quite simple devices will signal the presence of a true gravitational field by their sensitivity to tidal forces and will permit us to discriminate between a gravitational field and the pseudo-force field of acceleration. The confusion surrounding the principle of equivalence led J. L. Synge to remark:

> . . . I have never been able to understand this principle. . . . Does it mean that the effects of a gravitational field are indistinguishable from the effects of an observer's acceleration? If so, it is false. In Einstein's

---

* It would seem that a perfectly homogeneous gravitational field (zero tidal force) cannot be distinguished from the pseudo-force field of uniform acceleration. This is true, but not very relevant: perfectly homogeneous gravitational fields can only exist under extremely exceptional and unrealistic conditions. It can be shown that uniform fields are only possible in regions (cavities) completely surrounded by a continuous distribution of mass.

theory, either there is a gravitational field or there is none, according as the Riemann tensor* does or does not vanish. This is an absolute property; it has nothing to do with any observer's worldline. . . . The Principle of Equivalence performed the essential office of midwife at the birth of general relativity. . . . I suggest that the midwife be now buried with appropriate honours and the facts of absolute spacetime be faced.[28]

In order to avoid confusion, we will base our further development of gravitational theory on the very precise and unambiguous equality $m_I = m_G$. This equality is necessary and, to a large extent, sufficient for the construction of the relativistic theory.

FURTHER READING

Koestler, *The Sleepwalkers,* is a lively history of the beginning of our understanding of planetary motion as well as a study of the psychology of discovery. The book contains fascinating biographies of Copernicus and Kepler.

North, *Isaac Newton,* is a very short but excellent scientific biography. For a wealth of details from Newton's personal life, see More, *Isaac Newton;* this is written in the leisurely and elegant manner of the nineteenth century and is full of anecdotes and quotations from Newton's letters. Koyré, *Newtonian Studies,* gives a historical analysis of Newton's scientific work. Koyré, *The Astronomical Revolution,* covers the contributions of the predecessors of Newton to the theory of planetary motion and gravitation.

Mach, *Die Mechanik* (of which a not very good translation is available under the title *The Science of Mechanics*), is a classic historical and critical examination of the foundations of Newtonian physics. It contains profound and surprising insights into many fundamental issues.

The exciting stories of the rediscovery of the minor planet Ceres (through calculations of Gauss) and the discovery of Neptune (through calculations of Adams and LeVerrier) are told by Grosser, *The Discovery of Neptune.* Lyttleton, in an interesting article entitled "The Rediscovery of Neptune" (in Beer, ed., *Vistas in Astronomy,* vol. 3), points out that although the calculations of Adams and LeVerrier predicted the position of Neptune to within 1° or 2°, their methods were unnecessarily complicated and gave entirely wrong results for the orbital elements.

A very good textbook on Newton's gravitational theory is Danby, *Fundamentals of Celestial Mechanics,* which covers both the theory of the potential and the solution of the equations of motion.

Macmillan, *The Theory of the Potential,* is a detailed treatise on the Newtonian potential with many interesting and many irrelevant results; *if* the potentials produced by elliptic cylinders, parallelopipeds, etc., are needed, this is the place to find them.

---

* Tidal force tensor.

Mathematically, the theory of the Newtonian potential is equivalent to that of the electrostatic potential; textbooks that treat electrostatics are therefore often helpful, some of the best being Jackson, *Classical Electrodynamics,* and Schwartz, *Principles of Electrodynamics.*

A recent review of the measurement of the gravitational constant is given in Beams's article "Finding a Better Value for *G*" (see ref. 8).

A nice description of the measurement of solar oblateness and of the theoretical background is given in Dicke, *Gravitation and the Universe.* The article "The Oblateness of the Sun and Relativity" by the same author (*Science* **184,** 419 [1974]) covers the same subjects in a more technical way.

The definitions of inertial mass, gravitational mass, and all manner of other "masses" are examined in their historical context in Jammer, *Concepts of Mass.*

The theory of the Eötvös balance, and the effect of gravitational gradients on the measurement of m_I/m_G are discussed in "Beiträge zum Gesetze der Proportionalität von Trägheit und Gravität" by Eötvös, Pekar, and Fekete (see ref. 17).

The article "The Equivalence of Inertial and Passive Gravitational Mass" by Roll, Krotkov, and Dicke (see ref. 22) is strongly recommended. It is one of the few articles on experimental physics in a research journal that is so well written that it is fun to read. The authors describe the extreme precautions that had to be taken to prevent magnetic contamination, electrostatic effects, temperature gradients, etc., from spoiling their precision experiment. An abbreviated description of the experiment may be found in Dicke, *Gravitation and the Universe.*

Elementary, nonmathematical, but nevertheless very interesting descriptions of the tides are given by Clancy, *The Tides,* and by Defant, *Ebb and Flow.* The static equilibrium theory of Section 1.6 represents the most oversimplified approach conceivable. For a discussion of the dynamics of tidal oscillations and the very important effect produced by Coriolis forces, see Defant, *Physical Oceanography* (vol. 2).

REFERENCES

1. I. Newton, *Principia,* Book III, Proposition VII and Corollary.
2. See ref. 1, Book III, Rule I. (Italics added.)
3. Exact references for the first ten entries of this table may be found in the article "Gravitation" by J. H. Poynting in the *Encyclopedia Britannica,* 11th ed.
4. M. V. Crémieu, Comptes Rendus **149,** 700 (1909).
5. P. R. Heyl, J. Res. Natl. Bur. Std. (U.S.) **5,** 1243 (1930).
6. J. Zahradnicek, Phys. Zeits. **34,** 126 (1933).
7. P. R. Heyl and P. Chrzanowski, J. Res. Natl. Bur. Std. (U.S.) **29,** 1 (1942).
8. R. D. Rose, H. M. Parker, R. A. Lowry, A. R. Kuhltau, J. W. Beams, Phys. Rev. Lett. **23,** 655 (1969); J. W. Beams, Physics Today, May 1971.
9. Exact references for the first four entries of this table are given by J. Ashbrook, Sky and Telescope **34,** 229 (1967).

10. R. H. Dicke and H. M. Goldenberg, Phys. Rev. Lett. **18**, 313 (1967); R. H. Dicke, Science **184**, 419 (1974).

11. H. A. Hill and R. T. Stebbins, University of Arizona Preprint (1974); H. A. Hill, P. D. Clayton, D. L. Patz, A. W. Healy, R. T. Stebbins, J. R. Oleson, and C. A. Zanoni, Phys. Rev. Lett. **33**, 1497 (1974).

12. C. Brans and R. H. Dicke, Phys. Rev. **124**, 925 (1961).

13. Galileo Galilei, *Dialogues Concerning Two New Sciences.* Translated from *Le Opere di Galileo Galilei,* Edizione Nazionale, Florence, 1898, pp. 128, 129.

14. See ref. 1, "The System of the World," paragraph 19. Originally published by the University of California Press; reprinted by permission of The Regents of the University of California.

15. F. W. Bessel, Pogg. Ann. **25**, 401 (1832).

16. R. V. Eötvös, Math. Nat. Ber. Ungarn **8**, 65 (1890).

17. R. V. Eötvös, D. Pekar, and E. Fekete, Ann. der Phys. **68**, 11 (1922).

18. L. Southerns, Proc. Roy. Soc. (London) **84**, 325 (1910).

19. P. Zeeman, Proc. K. Akad. Amsterdam **20.4**, 542 (1918).

20. H. H. Potter, Proc. Roy. Soc. (London) **104**, 588 (1923); **113**, 731 (1927).

21. J. Renner, Hung. Acad. Sci. **53**, 542 (1935).

22. P. G. Roll, R. Krotkov, and R. H. Dicke, Ann. Phys. (N.Y.) **26**, 442 (1967).

23. V. B. Braginsky and V. I. Panov, Zh. Eksp. and Teor. Fiz. **61**, 873 (1971); translated in Sov. Phys. JETP **34**, 463 (1972).

24. L. I. Schiff, Proc. Natl. Acad. Sci. (U.S.) **45**, 69 (1959).

25. M. L. Good, Phys. Rev. **121**, 311 (1961).

26. R. L. Forward, *Review of Artificial Satellite Gravity Gradiometer Techniques for Geodesy* (Hughes Research Laboratories, 1973); C. C. Bell, R. L. Forward, and H. P. Williams, *Simulated Terrain Mapping with the Rotating Gravity Gradiometer: Advances in Dynamic Gravimetry* (Instrument Society of America, Pittsburgh, 1970).

27. A. Einstein, Ann. der Phys. **49**, 769 (1916); another translation is given in H. A. Lorentz, A. Einstein, H. Minkowski, and H. Weyl, *The Principle of Relativity* (Dover, New York), p. 114. Quoted with the permission of the Estate of Albert Einstein.

28. J. L. Synge, *Relativity; The General Theory* (North-Holland, Amsterdam, 1971), pp. IX–X. Reprinted with the permission of North-Holland Publishing Company.

29. For a review of these and other free fall experiments, see V. B. Braginskii and V. N. Rudenko, Soviet Physics Uspekhi **13**, 165 (1970).

PROBLEMS

1. Evaluate the quadrupole moment tensor of the earth assuming that the earth is an ellipsoid of revolution of constant density. Compare the magnitudes of the gravitational accelerations produced by the monopole and quadrupole fields of the earth at an altitude of 100 km above the equator.

2. Express the quadrupole moment tensor Q^{kl} in terms of the moment of

inertia tensor I^{kl} and the trace I^{kk}. The moment of inertia of the earth about the polar axis is

$$I^{33} = 0.3306 \, MR^2$$

and that about an equatorial axis is

$$I^{11} = I^{22} = 0.3295 \, MR^2$$

where M is the mass and R the equatorial radius of the earth. Find all components of Q^{kl}. Evaluate the dimensionless parameter $J_2 = -Q^{33}/2MR^2$.

3. Estimate (give numbers) the fractional difference between the equatorial and polar radii of the earth. Assume that the earth is entirely covered with water. (Hint: This difference is due to centrifugal effects. Show that the potential energy of a particle at a height h above the mean radius R is

$$mgh - \tfrac{1}{2}m\omega^2(R + h)^2 \sin^2 \theta$$

where the second term is the centrifugal potential due to the rotation of the earth; neglect h compared to R in this term. This calculation ignores the change in the gravitational field of the earth produced by the change of shape and only gives an order of magnitude estimate.)

4. Repeat the preceding problem for the sun, assuming $\omega = 2\pi$ radians/25 days. If the oblateness of the sun is as reported by Dicke and Goldenberg, what ought to be the value of ω?

5. Two particles of mass M on the x-axis are connected by a spring (see Fig. 1.7). Their relative distance undergoes simple harmonic motion: $s = s_0 + b_0 \sin \omega t$. Find the mass quadrupole tensor if the origin of coordinates is at the center of mass. Show that the time-dependent part of this quadrupole tensor is independent of the choice of origin.

6. A symmetric dumbbell rotates about an axis perpendicular to the line joining the masses (see Fig. 1.7). Treating the masses as particles, find the quadrupole moment tensor (with respect to the center) as a function of time.

7. In spherical coordinates centered on the earth, what are the radial and tangential components of the centrifugal acceleration due to the earth's rotation at Budapest? By how much does the plumb line at Budapest deviate from the radial direction? Suppose that one plumb bob has $m_I = m_G$ and another bob has $m_I - m_G = 10^{-9} \, m_I$. What is the angular difference between the corresponding plumb lines?

8. What is the ratio of gravitational self-energy to rest mass energy for a platinum sphere of radius 10 cm? For the earth? For the sun? For a neutron star? Treat the objects as continuous mass distributions of uniform density (ignore the structure of elementary particles). The radius of the neutron star is ~ 20 km and the mass equal to that of the sun.

9. An accelerometer is a device that is used to measure gravitational and inertial accelerations. Essentially, it consists of a spring balance with a

standard mass attached to it; the balance is calibrated so that when placed at a point it directly reads the acceleration that a free particle would have (in the rest frame of the accelerometer) when released at this point.

In a recent test, a spacecraft in circular orbit around the earth carried an accelerometer which was placed at a distance of 2 m from the center of mass (see Fig. 1.21). The acceleration measured by the accelerometer was 10^{-6} m/sec^2 with the spacecraft oriented as shown in the figure.

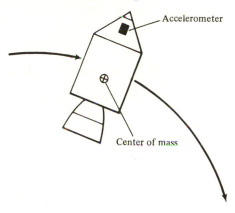

Fig. 1.21 *Spacecraft with accelerometer.*   Earth

What is the radius of the orbit? The spacecraft is not rotating relative to the fixed stars.

Explain how the output from several such accelerometers, placed at different points in the spacecraft, could be used to determine the orientation of the spacecraft relative to the earth.

10. Calculate the height of the tide produced by the sun according to our simple equilibrium model. Taking both sun and moon into account, what is the height of the tide when the moon is full or new (spring tide)? In the first quarter or last quarter (neap tide)?

11. Estimate the fractional tidal deformation produced by the earth on a drop of water of radius ~3 mm orbiting the earth at an altitude of 100 km. [Hint: Assume that the drop has the shape of a spheroid with semi-minor axis b, and semi-major axis $a = 3V/4\pi b^2$ (where the constant V is the volume of the drop). Estimate the work dW_t done by tidal forces if the semi-minor axis changes by db. Find the work dW_s done by surface tension. For equilibrium, $dW_t + dW_s = 0$.]

12. The Bay of Fundy is ~ 250 km long and ~ 100 m deep. Show that the natural frequency of the lowest mode of oscillation of the water is of the order of 12 hr. The velocity of waves in water of depth h is $v = \sqrt{gh}$ (this formula assumes that the depth is small compared to the wavelength, a condition that is always satisfied for tidal waves).

13. In the estuary of the river La Rance (on the coast of Brittany) the tidal

range is as much as 44 ft. The tidal bulge is captured behind a dam and as the tide ebbs the water is gradually allowed to run out through turbine generators. The total volume of the flow is 6×10^9 ft³. Estimate the time average electric power (kW) generated by this tidal power plant. Does the energy come from the moon or from the earth? How is angular momentum conserved?

14. The mean angular velocity of the moon is decreasing at the rate of $\sim 40''/$ (century)² (measured in ephemeris time).* Assuming that this deceleration is entirely due to tidal friction on the earth, find the rate at which the spin angular velocity of the earth must decrease. (Hint: Treat the moon's orbit as circular and use conservation of the orbital angular momentum of the moon plus spin angular moment of the earth; neglect the spin of the moon.) Find the rate at which rotational kinetic energy of the earth is transformed into orbital energy of the moon and find the rate at which energy is dissipated by tidal friction. Give *numerical* answers.

15. Use the result quoted in Table 1.3 to estimate the minimum torque that the Eötvös balance can detect. The balance beam is 40 cm long and the masses are 25 g each. What tidal field would produce such a torque?

16. What is the magnitude of the tidal field of the earth at a distance equal to the earth-moon distance? What is the tidal field of a 1-kg lead sphere at a distance of 100 cm? Compare these numbers with the sensitivity of the Eötvös balance (10^{-32} cm⁻²).

17. A spherical solid moon of mass m and radius R is orbiting a planet of mass M. Show that if the moon approaches the planet closer than

$$r_{\text{crit}} = R \left(\frac{2M}{m} \right)^{1/3}$$

then rocks lying on the surface of the moon will be lifted off by the tidal force. The distance r_{crit} is called *Roche's limit;* a moon approaching a planet closer than this limit will be disrupted by the tidal force.

* There is some disagreement on this number; a value of $\sim 20''/$(century)² is sometimes quoted.

2. THE FORMALISM OF SPECIAL RELATIVITY

**Raffiniert ist der Herrgott,
aber boshaft ist er nicht.
[God is cunning, but not
malicious.]**

Albert Einstein

We will be concerned with fields, that is, functions of space and time. It is therefore quite clear that we must first get to know the structure of spacetime. Unfortunately, because the velocity of light is so large, everyday experience leads us to acquire a certain number of misconceptions about the structure of spacetime. This set of misconceptions goes under the name of Newtonian (or Galilean) spacetime. The true structure of spacetime was discovered by Einstein in a study of electrodynamics (1905). As Einstein wrote in his autobiographical notes:

> After ten years of reflection such a principle [the principle of special relativity] resulted from a paradox upon which I had already hit at the age of sixteen: If I pursue a beam of light with the velocity c (velocity of light in a vacuum), I should observe such a beam as an electromagnetic field constant in time, periodic in space. However, there seems to exist no such thing, neither on the basis of experience, nor according to Maxwell's equations. . . .[1]

This left Einstein with two alternatives: either Maxwell's equations were wrong in all reference frames except one (the ether frame), or else something was wrong with the structure of spacetime suggested by everyday experience. Einstein decided for the second alternative: Newton's spacetime is wrong. He was then faced with the task of finding a structure of spacetime compatible with Maxwell's equations. In fact, it is not hard to find that geometry of spacetime which will make Maxwell's equations true in all inertial reference frames. The needed geometry follows quite directly from the requirement that the velocity of light be the same in all reference frames. The profound modification of our concepts of space and time requires a corresponding modification of mechanics. Newton's laws of mechanics are true in all inertial reference frames if, and only if, spacetime is Newtonian. Einstein and his followers had to develop a new set of mechanical equations, designed so as to hold in all inertial reference frames in the relativistic spacetime.

In this chapter we review the special theory of relativity and develop the mathematical formalism of four-vectors and tensors. Detailed discussions of the physical basis of special relativity and of the experiments which compel us to accept the relativistic spacetime structure may be found in the books listed at the end of the chapter.

2.1   THE STRUCTURE OF SPACETIME

Before we concern ourselves with the geometrical structure of spacetime, we should look at the topological structure. The topological properties of a space are those that are left unchanged by arbitrary "smooth" deformations of the space; topology tells us which points are the same, which distinct, which are in what neighborhood. Experience teaches us that any event (or happening) can be described by stating when and where it happened. We will therefore suppose that the topology of spacetime is that of Euclidean four-dimensional space (R^4). This means that we can describe points in spacetime by using four-dimensional rectangular coordinates, each point corresponding uniquely to a set of four numbers (t,x,y,z). The neighborhood of a given point is then the set of all points such that their coordinates differ only little from those of the given point. We can afford from now on to use the careless terminology "the point (t,z,y,z)" rather than the more precise "the point whose coordinates are (t,x,y,z)."

All of this seems trivial, but is not because it is quite possible that the topology of spacetime is not Euclidean. One possible violation could, and probably does, occur at large distances: our universe may be closed and have the topology of a sphere. A violation of Euclidean topology is also possible at very small distances. When dealing with distances of 10^{-13} cm, or 10^{-23} cm, or even less, it is not possible to set up a rigid coordinate system of clocks and meter sticks. We tend to believe that it cannot be done because of the atomic structure of matter, but maybe there is an extra reason why it cannot be done. Perhaps spacetime at a subnuclear level has a very pathological structure, is multiply connected, full of wormholes and bubbles; perhaps it is not even a *continuum*. We can only test for such diseases indirectly: we assume spacetime is an Euclidean continuum and use coordinates (t,x,y,z). If our theories based on the use of such coordinates are successful, then we feel reasonably confident that we have made the right assumption.

At *really* small distances, of the order of 10^{-33} cm, an Euclidean topological structure is quite unlikely. At such distances the fluctuations of quantum gravitation will be extremely violent and probably produce an ever changing, dynamic topology.[2] The length $\sqrt{\hbar G/c^3} = 1.6 \times 10^{-33}$ cm is called the *Planck length;* it gives the characteristic size of the fluctuations in quantum geometrodynamics.

The topologies of the spacetime of Newton and of special relativity are the same. The difference lies in the definition of distances in spacetime, that is, the metrical structure is different. Before we write down an expression for the distance between two points, we must make a restriction: choose the *t*-, *x*-, *y*-, *z*-axes of spacetime in such a way that we get an inertial reference frame. This means: choose axes in such a way that the orbits of free particles are straight lines, in space *and* in spacetime.* There exists a large class of such inertial frames; any two are in uniform translational motion with respect to one another.

The plot of the position of a particle versus time is called a *worldline*. Fig. 2.1 shows some worldlines of free particles in an inertial reference frame.

Consider now two events separated by a small interval (dt,dx,dy,dz). (The use of infinitesimal intervals will be helpful for our later work.) In order to define the geometry, we write down a quadratic form in dt, dx, dy, dz which gives the distance. Newtonian spacetime has *two distances*, relativistic spacetime *only one:*

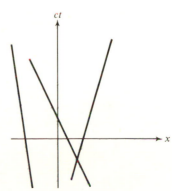

Fig. 2.1 *Worldlines of free particles.*

* Strictly speaking, we *define* straight lines as orbits of free particles and then choose our axes, and units along them, in such a way that the equations of the free particle orbits are those of a straight line. The question of measurement of spacetime will be fully discussed in Chapter 5.

Newtonian spacetime

$dl^2 = dx^2 + dy^2 + dz^2$ (space interval)
dt (time interval)   [1]

Relativistic spacetime

$ds^2 = c^2dt^2 - dx^2 - dy^2 - dz^2$ (spacetime interval)   [2]

In Eq. (2), c is the velocity of light in vacuum.

Thus, Newtonian spacetime contains two separate geometries: a three-dimensional Euclidean geometry for space and a one-dimensional geometry for time. Relativistic spacetime contains only one geometry which combines both space and time. As Minkowski expressed it: "Henceforth space by itself, and time by itself, are doomed to fade away into mere shadows, and only a kind of union of the two will preserve an independent reality."[3]

Note that the spacetime interval of the relativistic geometry is not always positive. Intervals with a positive, negative, or zero value of ds^2 are called, respectively, timelike, spacelike, and lightlike. Fig. 2.2 shows examples of displacements that have time-, space-, and lightlike intervals; in this figure lightlike displacements make an angle of 45° with the spacetime axes. Along the worldline of a massive particle, the interval must be timelike. This follows from Eq. (2) if we take into account that the speed $(dx^2 + dy^2 + dz^2)^{1/2}/dt$ for such a particle is less than c.

Along the worldline of a light signal, the interval is of course lightlike. The set of all worldlines of light signals leaving or arriving at the origin forms a three-dimensional "surface" in the four-dimensional spacetime. This surface has the equation

$t^2 - x^2 - y^2 - z^2 = 0$

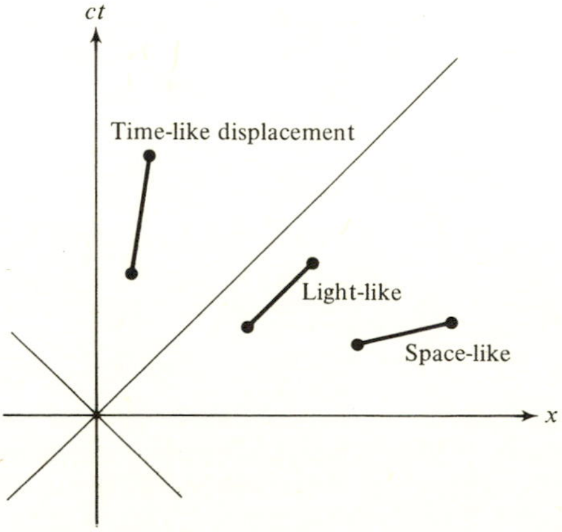

Fig. 2.2 *Time-, light-, and spacelike displacements.*

and is called the *light cone*. Since we cannot make a drawing of a four-dimensional space, let us omit the z-coordinate; then the light cone is simply the ordinary cone $t^2 - x^2 - y^2 = 0$ shown in Fig. 2.3. The upper $(t > 0)$ and lower $(t < 0)$ parts of this cone are called the forward and backward light cones, respectively.* Particles that leave the origin must have worldlines inside the forward light cone; particles that arrive at the origin must have worldlines inside the backward light cone.

The important thing about distance is that it has the same value no matter what coordinate system is used to measure it: the distance must be invariant under coordinate transformations. The requirement that coordinate transformations between two inertial reference frames be such as to leave the distances invariant, uniquely determines the form that these transformations may take. If the origin of the x'-, y'-, z'-coordinates moves with velocity v along the x-axis of the x-, y-, z-coordinates, then the transformations are:

Newtonian spacetime — Galilean transformation

Invariant distances:
$$dl^2 = dl'^2 \text{ (for } dt = 0) \Rightarrow \begin{cases} t' = t \\ x' = x - vt \\ y' = y \\ z' = z \end{cases}$$
$$dt = dt'$$

[3]

Relativistic spacetime — Lorentz transformation

Invariant distance:
$$ds^2 = ds'^2 \Rightarrow \begin{cases} t' = \dfrac{t - vx/c^2}{\sqrt{1 - v^2/c^2}} \\[2mm] x' = \dfrac{x - vt}{\sqrt{1 - v^2/c^2}} \\[2mm] y' = y \\ z' = z \end{cases}$$

[4]

It can be shown that these are the only *linear* transformations (except for trivial shifts of origin or rotations of the axes) that leave the distances (1) and (2) invariant. The transformations must be linear, in order that the worldlines of free particles be straight lines in both reference frames.

EXERCISE 1. Derive the transformations (3) and (4).

Let us now concentrate on relativistic spacetime and the Lorentz transformation. This transformation imposes severe restrictions on acceptable laws of physics. The reason is this: the spacetime structure

* Also called the future and past light cones.

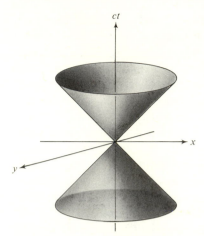

Fig. 2.3 *The light cone.*

has been designed so that as far as the behavior of free particles and light rays is concerned there is no distinction between different inertial reference frames; free particles obey the same laws in all inertial frames. We will now demand that *all* physical systems obey the same laws in all inertial frames. This is a symmetry requirement that we impose on physical laws and, to the best of our knowledge, nature has graciously consented to respect this symmetry. We state the symmetry principle as follows:

Principle of Relativity: All laws of physics must be invariant under Lorentz transformations.

In this context, "invariant" means that the law retains the same mathematical form, and that all numerical constants retain the same values.

The restrictions placed on the laws of physics by this symmetry principle are so stringent that, for example, in electrodynamics we can derive all the Maxwell equations from Coulomb's law alone.[4] The importance of these restrictions is even greater in quantum theory: the principle of relativity and the structure of spacetime tell us that only particles of integer or half-integer spin can exist (spin $\frac{3}{4}\hbar$ is forbidden!), that there must be a connection between spin and statistics, etc.[5] What concerns us most is that the principle of relativity also puts very tight restrictions on possible theories of gravitation. We will exploit these restrictions in Section 3.2.

2.2 TENSORS IN SPACETIME

We begin with some convenient notations. First, we will replace *t, x, y, z* by

$$x^0 \equiv ct, \ x^1 \equiv x, \ x^2 \equiv y, \ x^3 \equiv z \qquad\qquad [5]$$

We can then write Eq. (4) as

$$ds^2 = \sum_{\mu=0}^{3} \sum_{\nu=0}^{3} \eta_{\mu\nu} \, dx^\mu \, dx^\nu \qquad [6]$$

where $\eta_{\mu\nu}$ is a matrix with components

$$\eta_{\mu\nu} = \begin{pmatrix} 1 & 0 & 0 & 0 \\ 0 & -1 & 0 & 0 \\ 0 & 0 & -1 & 0 \\ 0 & 0 & 0 & -1 \end{pmatrix} \qquad [7]$$

This is called the *metric tensor* or the *Minkowski tensor.** We will use Einstein's summation convention, according to which any *repeated Greek vector index* appearing in a term of an equation is to be summed from 0 to 3. Then Eq. (6) can also be written as

$$ds^2 = \eta_{\mu\nu} \, dx^\mu \, dx^\nu \qquad [8]$$

It is also convenient to define

$$x_\mu = \eta_{\mu\nu} \, x^\nu \qquad [9]$$

so that

$$x_0 = ct, \; x_1 = -x, \; x_2 = -y, \; x_3 = -z \qquad [10]$$

Note that the summation convention only affects the repeated index ν in Eq. (9), but not the index μ; the latter simply tells us that the equation has four components ($\mu = 0, 1, 2, 3$). A repeated index such as ν is called a *dummy* index because it disappears when the summation is written out in full,

$$\eta_{\mu\nu} x^\nu \equiv \eta_{\mu 0} x^0 + \eta_{\mu 1} x^1 + \eta_{\mu 2} x^2 + \eta_{\mu 3} x^3 \qquad [11]$$

Hence $\eta_{\mu\nu} x^\nu$, $\eta_{\mu\alpha} x^\alpha$, $\eta_{\mu\beta} x^\beta$, etc., are exactly the same thing. An index such as μ which appears only once in each term of an equation is called a *free* index.

The quantities x^μ and x_μ are called, respectively, the *contravariant* and *covariant* components of the position vector. In terms of x_μ, Eq. (8) becomes

$$ds^2 = dx^\mu \, dx_\mu \qquad [12]$$

---

* The metric tensor is often defined with signs opposite to those used here. However, the signs in Eq. (7) are those used by Einstein in his original work,[6] and there is no good reason for going against his convention.

which resembles a dot product of two vectors.

In general, quantities with Greek superscripts are called contravariant, quantities with subscripts are called covariant. Thus, $\eta_{\mu\nu}$ is the covariant metric tensor. If we invert the relation (9), we obtain

$$x^\nu = \eta^{\nu\mu} x_\mu \qquad [13]$$

where $\eta^{\nu\mu}$ is the matrix inverse to $\eta_{\mu\nu}$. This inverse is defined by

$$\eta^{\nu\mu}\eta_{\mu\alpha} = \eta^\nu{}_\alpha \qquad [14]$$

where $\eta^\nu{}_\alpha$ is the Kronecker delta (we also sometimes write $\delta^\nu{}_\alpha$ for the Kronecker delta). The matrix $\eta^{\nu\mu}$ is the contravariant metric tensor; the tensor $\eta^\nu{}_\mu$ is *mixed*.

EXERCISE 2. Show that $\eta^{\nu\mu}$ has the same numerical value as (7).

In terms of x^μ, the Lorentz transformation (4) can be written

$$x'^0 = \frac{x^0 - vx^1}{\sqrt{1 - v^2}}$$

$$x'^1 = \frac{x^1 - vx^0}{\sqrt{1 - v^2}}$$

$$x'^2 = x^2$$
$$x'^3 = x^3 \qquad [15]$$

In these equations we have adopted the convention $c = 1$, which can always be achieved by a clever choice of units. This convention amounts to the same thing as omission of all factors of c. The restoration of the missing factors of c is easy: insert as many such factors as is necessary to make the equation dimensionally correct in cgs units. For numerical calculations we will always use cgs units; that is, the factors of c are to be restored before inserting numerical values into any equations.

A neater way to write the Lorentz transformation is the following:

$$x'^\mu = a^\mu{}_\nu x^\nu \qquad [16]$$

where

$$a^\mu{}_\nu = \begin{pmatrix} 1/\sqrt{1-v^2} & -v/\sqrt{1-v^2} & 0 & 0 \\ -v/\sqrt{1-v^2} & 1/\sqrt{1-v^2} & 0 & 0 \\ 0 & 0 & 1 & 0 \\ 0 & 0 & 0 & 1 \end{pmatrix} \qquad [17]$$

(The index μ gives the row, ν the column.)

The Lorentz transformation (17) is a particularly simple one in that the velocity is entirely along the x-axis. The matrix $a^\mu_{\ \nu}$ for the most general Lorentz transformation* depends on six parameters: three components of the relative velocity plus three angles which indicate how much the x'-, y'-, z'-axes are rotated with respect to the x-, y-, z-axes. Lorentz transformations with zero rotation angles are called *pure*. Lorentz transformation with zero velocity are nothing but ordinary rotations of the x-, y-, z-axes.

The matrix $a^\mu_{\ \nu}$ for a general Lorentz transformation must satisfy the equation

$$a^\alpha_{\ \mu}\eta_{\alpha\beta}a^\beta_{\ \nu} = \eta_{\mu\nu} \qquad [18]$$

To see this, we substitute $x'^\alpha = a^\alpha_{\ \mu}x^\mu$ into the equation

$$x'^\alpha \eta_{\alpha\beta} x'^\beta = x^\mu \eta_{\mu\nu} x^\nu \qquad [19]$$

that expresses the invariance of the (finite) spacetime interval between the origin and the point x^μ. We get

$$a^\alpha_{\ \mu}x^\mu \eta_{\alpha\beta} a^\beta_{\ \nu}x^\nu = x^\mu \eta_{\mu\nu}x^\nu \qquad [20]$$

or,

$$x^\mu x^\nu (a^\alpha_{\ \mu}\eta_{\alpha\beta}a^\beta_{\ \nu} - \eta_{\mu\nu}) = 0 \qquad [21]$$

In Eq. (21), x^μ and x^ν are arbitrary. Hence the coefficient of $x^\mu x^\nu$ must vanish; this implies Eq. (18).

We will now give the definitions of tensors. The simplest tensor is a *scalar* (tensor of rank zero); this is a one-component object that remains unchanged under Lorentz transformations. Any numerical constant is a scalar. The spacetime interval ds^2 is a scalar since the Lorentz transformation has been designed precisely so as to leave ds^2 invariant. Another closely related scalar is the proper time interval $d\tau$ associated with the motion of a particle. This is defined as equal to ds along the worldline of the particle, i.e.,

$$d\tau = \sqrt{dx^\mu dx_\mu} = \sqrt{dt^2 - dx^2 - dy^2 - dz^2} = dt\sqrt{1 - v^2} \qquad [22]$$

where $v = (dx^2 + dy^2 + dz^2)^{1/2}/dt$ is the particle velocity. In the particle's rest frame, $d\tau = dt$. This gives us the physical interpretation for the proper time: it is the time interval measured by a clock that moves along with the same velocity as the particle.

A *vector* (or tensor of rank one) is a four-component object A^μ

---

* We will deal only with homogeneous transformations, i.e., shifts of the origin of coordinates are excluded (no additive constant in (16)).

which transforms under Lorentz transformations in exactly the same way as x^μ:

$$A'^\mu = a^\mu{}_\nu A^\nu \qquad [23]$$

According to this definition x^μ (and also dx^μ) serves as a prototype for all vectors.

If x^μ is the position vector of a particle, then the quantity

$$u^\mu \equiv \frac{dx^\mu}{d\tau} \qquad [24]$$

is a vector since it consists of the product of a vector dx^μ by a scalar $1/d\tau$. This is called the four-velocity. The components of u^μ are

$$u^\mu = \left(\frac{1}{\sqrt{1-v^2}}, \frac{v_x}{\sqrt{1-v^2}}, \frac{v_y}{\sqrt{1-v^2}}, \frac{v_z}{\sqrt{1-v^2}} \right) \qquad [25]$$

Another important vector is the energy-momentum four-vector, defined as

$$p^\mu \equiv mu^\mu \qquad [26]$$

where m is the particle rest-mass. The components of this vector are

$$p^\mu = \left(\frac{m}{\sqrt{1-v^2}}, \frac{mv_x}{\sqrt{1-v^2}}, \frac{mv_y}{\sqrt{1-v^2}}, \frac{mv_z}{\sqrt{1-v^2}} \right)$$

$$= (E, p_x, p_y, p_z) \qquad [27]$$

These components are the energy (E) and the three components of the momentum (p_x, p_y, p_z) of the particle.

The general definition of a tensor* is as follows: *A tensor of rank r is an object with 4^r components which under a Lorentz transformation transforms according to*

$$A'^{\alpha\beta\cdots\gamma} = a^\alpha{}_\mu a^\beta{}_\nu \ldots a^\gamma{}_\kappa A^{\mu\nu\cdots\kappa} \qquad [28]$$

Note that the object has r indices and there is one Lorentz transformation matrix for each index.

As an example of a second-rank tensor, consider the 16-component object

$$x^\mu x^\nu \qquad [29]$$

* Strictly speaking, what follows is the definition of a *Lorentz* tensor. We will deal with more general tensors in Chapter 6.

From the transformation law for x^μ, it is easy to see that this object does transform as a tensor.

EXERCISE 3. Write down the components of the object (29) as a matrix. Prove that the object transforms as a second-rank tensor.

In general, any object constructed by taking the "product" of two vectors is a tensor. Thus the objects

$$x^\mu p^\nu - x^\nu p^\mu \qquad\qquad [30]$$

and

$$mu^\mu u^\nu \qquad\qquad [31]$$

are tensors. The first is the angular momentum tensor of a particle and the second is related to the energy-momentum tensor (see below).

We have seen that the position vector can be written either in the contravariant or in the covariant form. For an arbitrary contravariant vector A^μ, we define the covariant component A_ν by

$$A_\nu = \eta_{\nu\mu} A^\mu \qquad\qquad [32]$$

This can be inverted to give

$$A^\mu = \eta^{\mu\nu} A_\nu \qquad\qquad [33]$$

EXERCISE 4. Show that (32) implies (33).

The operations performed with the metric tensor on the right sides of Eqs. (32) and (33) are called, respectively, *lowering and raising an index*. The same procedure can be applied to one or more indices of a tensor of arbitrary rank. For example,

$$\begin{aligned}
A_\mu{}^\nu &= \eta_{\mu\alpha} A^{\alpha\nu} \\
A^\mu{}_\nu &= \eta_{\nu\alpha} A^{\mu\alpha} \\
A_{\mu\nu} &= \eta_{\mu\alpha}\eta_{\nu\beta} A^{\alpha\beta} \\
A^{\mu\nu} &= \eta^{\mu\alpha}\eta^{\nu\beta} A_{\alpha\beta}
\end{aligned} \qquad\qquad [34]$$

EXERCISE 5. Write down the components of $u^\mu u^\nu$ as a matrix, expressing everything in terms of the ordinary velocities v_x, v_y, v_z. Repeat for $u_\mu u^\nu$, for $u^\mu u_\nu$, and for $u_\mu u_\nu$.

Our rules for raising and lowering indices can also be applied to the Lorentz transformation matrix a^μ_ν. Thus, (18) can be written

$$a_{\beta\mu} a^\beta_\nu = \eta_{\mu\nu}$$ [35]

If we raise the index μ, we get

$$a_\beta{}^\mu a^\beta_\nu = \eta^\mu_\nu$$ [36]

This shows that $a_\beta{}^\mu$ is the inverse of a^β_ν,

$$(a^{-1})^\mu_\beta = a_\beta{}^\mu$$ [37]

The matrix $a_\beta{}^\mu$ can be used to write the transformation law for covariant tensors in the form

$$A'_{\alpha\beta\ldots\gamma} = a_\alpha{}^\mu a_\beta{}^\nu \ldots a_\gamma{}^\kappa A_{\mu\nu\ldots\kappa}$$ [38]

This equation is obtained from (28) by lowering all the free indices.

EXERCISE 6. Prove this.

In particular,

$$x'_\mu = a_\mu{}^\nu x_\nu$$ [39]

Combining this with Eq. (36) yields

$$x_\nu = a^\mu_\nu x'_\mu$$ [40]

and this in turn gives

$$x^\nu = a_\mu{}^\nu x'^\mu$$ [41]

EXERCISE 7. Derive (40) and (41).

Differentiation of Eq. (16) and (41) gives, respectively,

$$a^\mu_\nu = \frac{\partial x'^\mu}{\partial x^\nu}$$

$$a_\mu{}^\nu = \frac{\partial x^\nu}{\partial x'^\mu}$$ [42]

Given a second-rank tensor $A^{\mu\nu}$, the quantity

$$A^\mu_\mu$$ [43]

is called the *trace* of the tensor. This is a one-component object which transforms as a scalar.

EXERCISE 8. Show that $A^\mu_{\ \mu}$ is left invariant by Lorentz transformations. Also show that

$$A^\mu_{\ \mu} = A_\mu^{\ \mu} \qquad\qquad [44]$$

We will sometimes simply use the notation A for the trace of a tensor $A^{\mu\nu}$,

$$A \equiv A_\mu^{\ \mu} \qquad\qquad [45]$$

A tensor of rank higher than two has more than one trace. For example,

$$A^\mu_{\ \mu}{}^\alpha, \ A^{\mu\nu}_{\ \ \mu}, \ A^{\mu\alpha}_{\ \ \alpha} \qquad\qquad [46]$$

are three different traces of a third-rank tensor $A^{\mu\nu\alpha}$. The operation (46) (and also (43)) is often called *contraction*. Thus $A^\mu_{\ \mu}{}^\alpha$ is the contraction of the first two indices of $A^{\mu\nu\alpha}$, etc. It is obvious that contraction reduces the rank of a tensor by two.

For a tensor of the special type

$$B^\mu C^\nu \qquad\qquad [47]$$

(where B^μ and C^ν are vectors), the contraction, or trace, gives the scalar

$$B^\mu C_\mu \qquad\qquad [48]$$

This is often simply written as $B \cdot C$ and called the *scalar product* of the vectors B^μ and C^ν.

EXERCISE 9. Show that

$$u^\mu u_\mu = 1 \qquad\qquad [49]$$

and that

$$p^\mu p_\mu = m^2 \qquad\qquad [50]$$

Note that in summations, such as those implied by the repeated index in Eqs. (46) and (48), one dummy index is always covariant and one is contravariant. The reason is that if we were to form the sum $B^\mu C^\mu$, then the resulting object would not be a scalar — that is, we would get an answer that depends on the reference frame used in the evaluation; such an object is uninteresting.

EXERCISE 10. Show that $B^\mu C^\mu$ is *not* a scalar. Show that $A^{\mu\mu\alpha}$ is *not* a vector.

A tensor $A^{\mu\nu}$ is said to be *symmetric* in μ,ν if

$$A^{\mu\nu} = A^{\nu\mu} \tag{51}$$

and it is said to be *antisymmetric* in μ,ν if

$$A^{\mu\nu} = - A^{\nu\mu} \tag{52}$$

Given an arbitrary tensor $B^{\mu\nu}$, it is always possible to write it as a linear combination of a symmetric and an antisymmetric part as follows:

$$B^{\mu\nu} = \tfrac{1}{2}(B^{\mu\nu} + B^{\nu\mu}) + \tfrac{1}{2}(B^{\mu\nu} - B^{\nu\mu}) \tag{53}$$

Obviously the first term in parentheses is symmetric, and the second antisymmetric. Eq. (53) can also be written as

$$B^{\mu\nu} = B^{(\mu\nu)} + B^{[\mu\nu]} \tag{54}$$

with the definitions

$$B^{(\mu\nu)} \equiv \tfrac{1}{2}(B^{\mu\nu} + B^{\nu\mu}) \tag{55}$$

$$B^{[\mu\nu]} \equiv \tfrac{1}{2}(B^{\mu\nu} - B^{\nu\mu}) \tag{56}$$

EXERCISE 11. Show that Lorentz transformations change symmetric tensors into symmetric tensors and antisymmetric tensors into antisymmetric tensors.

Finally, given an arbitrary tensor $B^{\mu\nu}$ it is always possible to write it as a combination of a traceless tensor (i.e, a tensor of zero trace) and $\eta^{\mu\nu}$ as follows:

$$B^{\mu\nu} = (B^{\mu\nu} - \tfrac{1}{4}\eta^{\mu\nu}B) + \tfrac{1}{4}\eta^{\mu\nu}B \tag{57}$$

Since $\eta_\mu{}^\mu = \delta_\mu{}^\mu = 4$, the term in parentheses is traceless.

2.3 TENSOR FIELDS

The definitions of the preceding section hold for tensors which are constant and also for tensors which vary in space and time; the latter are called *tensor fields*. For example, a *scalar field* is a function $\phi(P)$ which has different values at different points P of spacetime. At each given point of spacetime, the value of the scalar field is left invariant under Lorentz transformations:

$$\phi'(P) = \phi(P) \qquad\qquad [58]$$

Instead of describing a given point of spacetime by the symbol P we can, of course, use coordinates. If the point P has coordinates x and x', respectively, in the two inertial systems, then Eq. (58) becomes*

$$\phi'(x') = \phi(x) \qquad\qquad [59]$$

If the function $\phi(x)$ is given, then this determines the function $\phi'(x')$ since x can be expressed in terms of x' by Eq. (41):

$$\phi'(x') = \phi(x(x')) \qquad\qquad [60]$$

Note that $\phi'(x')$ is *not* the same function of x' as ϕ is of x. The important thing is that each point of spacetime, ϕ and ϕ' take the same value. This is what is meant by invariance and the notation (58) puts more emphasis on this than (60) does. We will often simply write $\phi = \phi'$, leaving it understood that the functions are to be evaluated at the same point.

In general, a tensor field is a function of spacetime such that the transformation law (28) holds at each given point of spacetime.

It can be shown that all the fields that describe the state of (continuous) physical systems in classical physics must necessarily be tensor fields. This important theorem rests on a group theoretical analysis of the Lorentz transformations, and therefore we must forego the proof.[7] It follows that the *only* building blocks that we can use in the construction of a classical field theory are scalars, vectors, tensors, etc.** We will rely heavily on this fact in the next chapter.

The gradient operator in spacetime is

$$\partial_\mu \equiv \left(\frac{\partial}{\partial x^0}, \frac{\partial}{\partial x^1}, \frac{\partial}{\partial x^2}, \frac{\partial}{\partial x^3} \right) \qquad\qquad [61]$$

The chain rule for differentiation tells us that

$$\partial'_\mu = \frac{\partial}{\partial x'^\mu} = \frac{\partial x^\nu}{\partial x'^\mu} \frac{\partial}{\partial x^\nu} = a_\mu{}^\nu \partial_\nu \qquad\qquad [62]$$

where use has been made of (42). This shows that ∂_μ transforms as a covariant vector. We will also sometimes use

---

* When the spacetime variables appear as arguments of functions, as in Eq. (59), we will use an abbreviated notation without superscripts. Thus, $\phi(x)$ stands for $\phi(x^0,x^1,x^2,x^3)$.

** In quantum theory, besides tensor fields there exist spinor fields. The tensor fields correspond to quanta of spin 0, 1, 2, . . . ; the spinor fields to quanta of spin 1/2, 3/2, ...

$$\partial^{\mu} \equiv \eta^{\mu\nu}\partial_{\nu} = \left(\frac{\partial}{\partial x^0}, \, -\frac{\partial}{\partial x^1}, \, -\frac{\partial}{\partial x^2}, \, -\frac{\partial}{\partial x^3}\right) \qquad [63]$$

This operator has the transformation law of a contravariant vector,

$$\partial'^{\mu} = a^{\mu}_{\nu}\partial^{\nu} \qquad [64]$$

The "Laplacian" operator in spacetime is

$$\partial^{\mu}\partial_{\mu} = \left(\frac{\partial}{\partial x^0}\right)^2 - \left(\frac{\partial}{\partial x^1}\right)^2 - \left(\frac{\partial}{\partial x^2}\right)^2 - \left(\frac{\partial}{\partial x^3}\right)^2 \qquad [65]$$

This operator is invariant under Lorentz transformations:

$$\partial'^{\mu}\partial'_{\mu} = \partial^{\mu}\partial_{\mu} \qquad [66]$$

EXERCISE 12. Prove (64) and (66).

As a consequence of Eq. (64), $\partial^{\mu}\phi$ is a vector whenever ϕ is a scalar. It also follows that $\partial^{\mu}\partial^{\nu}\phi$ is a tensor and $\partial^{\mu}\partial_{\mu}\phi$ a scalar. In general, when ∂^{μ} operates on a tensor of rank r, the result is a tensor of rank $r + 1$.

EXERCISE 13. Prove that $\partial^{\mu}A^{\alpha\beta\cdots\kappa}$ is a tensor of rank $r + 1$ whenever $A^{\alpha\beta\cdots\kappa}$ is a tensor of rank r.

It is convenient to indicate derivatives by commas, in the style of the following examples:

$$\begin{aligned}
\phi_{,\mu} &\equiv \partial_{\mu}\phi \\
\phi^{,\mu} &\equiv \partial^{\mu}\phi \\
\phi^{,\mu,\nu} &\equiv \partial^{\mu}\partial^{\nu}\phi \\
\phi^{,\mu}{}_{,\mu} &\equiv \partial^{\mu}\partial_{\mu}\phi \\
A^{\alpha\beta}{}_{,\mu} &\equiv \partial_{\mu}A^{\alpha\beta} \text{ etc.}
\end{aligned} \qquad [67]$$

This concludes our discussion of the mathematical aspects of special relativity. More on tensor analysis in four-dimensional spacetime will be found in the books listed at the end of the chapter. We will now turn to some important applications of the formalism.

2.4 THE ENERGY-MOMENTUM TENSOR

In our search for a relativistic theory of gravitation the energy density will play an important role. But it makes little sense to consider energy density by itself, because what is energy density in one reference frame will be some combination of energy density, energy flux density, and momentum flux density as seen from another reference frame. Hence all these quantities must be considered together.

It will be best to explain this for a system consisting of a collection of noninteracting particles (a cloud of dust). Suppose that near some point in this cloud the density of particles is n per unit volume and their velocity **v**. In this case the *energy density* can be expressed in terms of the density and the velocity of the particles:*

$$T^{00} = \frac{nm}{\sqrt{1 - v^2}} \tag{68}$$

This is simply the product of the number n of particles per unit volume by the energy per particle. We use the notation T^{00} for energy density for reasons which will soon be obvious.

The *energy flux density* can be defined in the following way: the energy flux in the x-direction is the amount of energy transported in unit time across a unit yz-area, that is

$$\frac{nmv_x}{\sqrt{1 - v^2}} \tag{69}$$

This is the product of particle current nv_x by the energy per particle. In general, the energy flux density in the k-direction is

$$T^{k0} = \frac{nmv^k}{\sqrt{1 - v^2}} \tag{70}$$

As an alternative notation for (70) we will use T^{0k}, so that $T^{k0} \equiv T^{0k}$. Note that T^{k0} can also be regarded as the *density of momentum*.

Finally, let us define the *momentum flux*. The xy-momentum flux is defined as the amount of x-momentum that flows in the y-direction per unit area and unit time. Since the x-momentum per particle is

$$\frac{mv_x}{\sqrt{1 - v^2}}$$

the xy-momentum flux must obviously be

$$\left(\frac{mv_x}{\sqrt{1 - v^2}}\right)nv_y \tag{71}$$

The general expression for the kl-momentum flux is

---

* This expression follows the convention $c = 1$. Restoration of factors of c would give $T^{00} = nmc^2/\sqrt{1 - v^2/c^2}$ in cgs units.

$$T^{kl} = \frac{nmv^k v^l}{\sqrt{1 - v^2}}$$ [72]

It is now easy to show that the 16-component object $T^{\mu\nu}$ given by Eqs. (68), (70), and (72) is a tensor under Lorentz transformations. We can proceed as follows: $T^{\mu\nu}$ can be written as

$$T^{\mu\nu} = n_0 m u^\mu u^\nu$$ [73]

where

$$n_0 \equiv n \sqrt{1 - v^2}$$ [74]

EXERCISE 14. Show that Eq. (73) agrees with Eqs. (68), (70), and (72).

The quantity n_0 is the particle density as measured in a reference frame that moves with the particles. This is a consequence of the well-known volume contraction effect of special relativity: a volume containing a given number of particles and moving along with them is contracted as measured in the laboratory frame; hence the laboratory density is increased.

Because n_0, called the *proper particle density,* is a number measured in the local rest frame, it is an invariant (scalar). Hence $n_0 m u^\mu u^\nu$ is the product of a scalar n_0 by the tensor $m u^\mu u^\nu$ (see Eq. (31)), and therefore $T^{\mu\nu}$ is a tensor. Eq. (73) can also be expressed as

$$T^{\mu\nu} = \rho_0 u^\mu u^\nu$$ [75]

where

$$\rho_0 \equiv n_0 m$$ [76]

is the mass density as measured in the local rest frame of the particles; ρ_0 is called the *proper mass density.* Note that the tensor $T^{\mu\nu}$ is symmetric:

$$T^{\mu\nu} = T^{\nu\mu}$$ [77]

The definition of $T^{\mu\nu}$ may be summarized as follows:

$T^{00} = $ (energy density)
$T^{0k} = T^{k0} = $ (momentum density) $= $ (energy flux density)
$T^{kl} = $ (flux density of k-momentum in l-direction)   [78]

This definition applies in general. For example, if instead of a system of dust particles we have a system consisting of electromagnetic fields, then we can still use (78) to define the corresponding energy-momentum tensor (see next section).

The energy-momentum tensor will usually be a function of space and time. For example, in Eq. (75) both the particle density and the velocity may vary in space and time. The energy-momentum tensor of a complete system* satisfies the (differential) conservation law

$$\partial_\nu T^{\mu\nu} = 0 \qquad\qquad [79]$$

The proof of this is no harder than the well-known proof for the conservation law of charge. Consider the $\mu = 1$ component of this equation:

$$\frac{\partial}{\partial x^0} T^{10} + \frac{\partial}{\partial x^k} T^{1k} = 0 \qquad\qquad [80]$$

To derive this from the conservation of momentum, take a small volume in the system. The x-momentum in the volume is

$$\int T^{10} d^3x \qquad\qquad [81]$$

and the rate of decrease of this x-momentum is

$$-\int \frac{\partial}{\partial x^0} T^{10} d^3x \qquad\qquad [82]$$

This momentum decrease must be the result of x-momentum flowing out of the volume. The total outflow of x-momentum across the boundary of the volume is

$$\int T^{1k} dS_k \qquad\qquad [83]$$

We can apply Gauss' theorem to the three component object T^{1k} ($k = 1, 2, 3$) and convert (83) into

$$\int \frac{\partial}{\partial x^k} T^{1k} d^3x \qquad\qquad [84]$$

The terms (82) and (84) must be equal. If the volume is small enough, the integrands are nearly constant and the integration can be replaced by a multiplication of integrand by volume:

$$\left(-\frac{\partial}{\partial x^0} T^{10}\right) \Delta x \Delta y \Delta z = \left(\frac{\partial}{\partial x^k} T^{1k}\right) \Delta x \Delta y \Delta z \qquad\qquad [85]$$

---

* By a "complete system" is meant a system not subject to external forces (i.e., all forces are already included in the system).

This reduces to Eq. (80) if the volumes are canceled. The proof for the other components of (79) is similar.

According to our definition of $T^{\mu\nu}$, the volume integrals of T^{00} and T^{k0} give, respectively, the total energy and momentum of the system:

$$P_0 = \int T_0^0 d^3x \qquad\qquad\qquad [86]$$

$$P_k = \int T_k^0 d^3x \qquad\qquad\qquad [87]$$

The four-component object P^μ *should* be a four-vector, but it is not at all obvious that the volume integrals appearing on the right side of (86) and (87) will behave correctly. To prove that the four integrals (86) and (87) do transform as a four-vector, we begin with the conservation law

$$\frac{\partial}{\partial x^\mu} T_\nu^\mu = 0 \qquad\qquad\qquad [88]$$

and multiply this by an arbitrary *constant* vector B^ν,

$$\frac{\partial}{\partial x^\mu} (T_\nu^\mu B^\nu) = 0 \qquad\qquad\qquad [89]$$

Integration over a four-dimensional volume gives

$$\int \frac{\partial}{\partial x^\mu} (T_\nu^\mu B^\nu) dtdxdydz = 0 \qquad\qquad\qquad [90]$$

The left side of this equation has the form

$$\int \frac{\partial}{\partial x^\mu} C^\mu dtdxdydz$$

where $C^\mu = T_\nu^\mu B^\nu$. We can apply the four-dimensional analog of the familiar Gauss theorem to convert the volume integral into a "surface" integral,

$$\int \frac{\partial}{\partial x^\mu} C^\mu dtdxdydz = \int C^\mu dS_\mu \qquad\qquad\qquad [91]$$

Since the volume is four-dimensional, its "surface" is actually a three-dimensional hypersurface. The vector dS_μ has a magnitude equal to the corresponding three-dimensional volume element and a direction "perpendicular" to the hypersurface; that is, $dS_\mu \delta x^\mu = 0$ whenever δx^μ is any displacement within the hypersurface.

EXERCISE 15. Prove the four-dimensional Gauss theorem of Eq. (91). (Hint: A volume of arbitrary shape can be regarded as made up of a large number of rectangular boxes. Hence, it suffices to prove Eq. (91) for the case of a volume in the shape of a rectangular box with sides perpendicular to the coordinate axes).

For simplicity we will assume that $T_\nu{}^\mu$ differs from zero only in some finite region of space. If we consider this finite region of space at successive instants of time, then in spacetime $T_\mu{}^\nu$ is nonzero in a four-dimensional tube of infinite length in the time direction and of a finite cross section in the space directions. Fig. 2.4 shows this tube in the x^0,x^1-plane: the two vertical lines form the boundaries of the tube and $T_\mu{}^\nu$ is different from zero only between these boundaries.

Consider now a Lorentz transformation which introduces new coordinates x'^μ according to Eq. (16). The equations

$$x^0 = \text{constant} \tag{92}$$

and

$$x'^0 = \text{constant} \tag{93}$$

define three-dimensional hypersurfaces in spacetime. These surfaces are distinct (if we express (93) in terms of x-coordinates it becomes $x^0 - vx^1 = (\text{constant}) \times \sqrt{1-v^2}$, which differs from (92)). In Fig. 2.4, the hypersurfaces (92) and (93) are represented by the heavy lines; in this figure

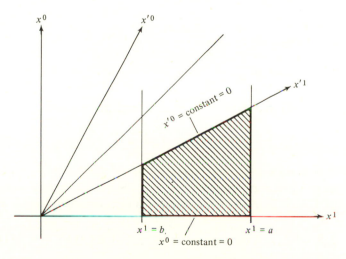

Fig. 2.4 *The energy-momentum tensor is different from zero in the region included between the lines* x¹ = a, x¹ = b.

it has been assumed that constant $= 0$. We now concentrate on that part of the tube included between the surfaces (92) and (93); in Fig. 2.4 this four-dimensional region is shown cross-hatched. If we apply Eqs. (90) and (91) to this region, we obtain

$$0 = \int T_\nu{}^\mu B^\nu dS_\mu$$

$$= \int_I T_\nu{}^\mu B^\nu dS_\mu + \int_{II} T_\nu{}^\mu B^\nu dS_\mu + \int_{III} T_\nu{}^\mu B^\nu dS_\mu \qquad [94]$$

where the surface I is at the top (Eq. (93)), surface II is at the bottom (Eq. 92)), and surface III is the "lateral" surface of the tube (see Fig. 2.4). Since $T_\nu{}^\mu = 0$ on the lateral surface, we have

$$\int_I T_\nu{}^\mu B^\nu dS_\mu = -\int_{II} T_\nu{}^\mu B^\nu dS_\mu \qquad [95]$$

The surface element appearing on the right side is $dS_\mu = (-d^3x, 0, 0, 0)$ and hence the right side reduces to

$$\int T_\nu{}^0 B^\nu d^3x \qquad [96]$$

The surface element appearing on the left side of Eq. (95) is more complicated. However, we can take advantage of the fact that the complete integrand is a scalar ($T_\nu{}^\mu$ is a tensor, B^ν and dS_μ are vectors) so that

$$\int_I T_\nu{}^\mu B^\nu dS_\mu = \int_I T'_\nu{}^\mu B'^\nu dS'_\mu \qquad [97]$$

Since in the new coordinates $dS'_\mu = (d^3x', 0, 0, 0)$, it follows that the left side of Eq. (95) is simply

$$\int T'_\nu{}^0 B'^\nu d^3x' \qquad [98]$$

We have therefore established the equality

$$B'^\nu \int T'_\nu{}^0 d^3x' = B^\nu \int T_\nu{}^0 d^3x \qquad [99]$$

which shows that the quantity $B^\nu \int T_\nu{}^0 d^3x$ is a scalar under Lorentz transformations. Since B^ν is an arbitrary contravariant vector, it must be that $\int T_\nu{}^0 d^3x$ is a covariant vector. This is what we wanted to prove.

2.5 RELATIVISTIC ELECTRODYNAMICS

The principle of relativity demands the invariance of the laws of physics under Lorentz transformations. It is obvious that any equation of the form

$$A^{\mu\nu\cdots\kappa}=0 \qquad\qquad [100]$$

where $A^{\mu\nu\cdots\kappa}$ is a tensor of rank r, will satisfy the principle of relativity. In fact, multiplication of Eq. (100) by $a^{\alpha}{}_{\mu}a^{\beta}{}_{\nu}\ldots a^{\gamma}{}_{\kappa}$ gives

$$A'^{\alpha\beta\cdots\gamma}=0 \qquad\qquad [101]$$

which shows that the law (100) is invariant under the Lorentz transformation that is described by the matrix $a^{\alpha}{}_{\mu}$. Our next objective is to show that the equations of electrodynamics are of the form (100).

Maxwell's equations of electrodynamics are, in cgs (Gaussian) units,

$$\nabla\cdot\mathbf{E}=4\pi\rho \qquad\qquad [102]$$

$$\nabla\times\mathbf{B}-\frac{1}{c}\frac{\partial\mathbf{E}}{\partial t}=\frac{4\pi}{c}\mathbf{j} \qquad\qquad [103]$$

$$\nabla\cdot\mathbf{B}=0 \qquad\qquad [104]$$

$$\nabla\times\mathbf{E}+\frac{1}{c}\frac{\partial\mathbf{B}}{\partial t}=0 \qquad\qquad [105]$$

These equations do not seem to be of the type (100) and their Lorentz invariance is not at all obvious.

In order to discuss the Lorentz transformation properties of Eqs. (102)–(105), we begin by rewriting them in four-dimensional component notation as follows:

$$\partial_{\mu}F^{\mu\nu}=\frac{4\pi}{c}j^{\nu} \qquad\qquad [106]$$

$$\partial^{\alpha}F^{\mu\nu}+\partial^{\mu}F^{\nu\alpha}+\partial^{\nu}F^{\alpha\mu}=0 \qquad\qquad [107]$$

where the four-component object j^{ν} is defined as

$$j^{\nu}\equiv(c\rho,j_x,j_y,j_z) \qquad\qquad [108]$$

and the 16-component object $F^{\mu\nu}$ is defined as

$$F^{\mu\nu}\equiv\begin{pmatrix}0 & -E_x & -E_y & -E_z\\ E_x & 0 & -B_z & B_y\\ E_y & B_z & 0 & -B_x\\ E_z & -B_y & B_x & 0\end{pmatrix} \qquad\qquad [109]$$

With these definitions, it is a straightforward exercise to check that the components of Eq. (106) are equivalent to Eqs. (102), (103) and the components of Eq. (107) are equivalent to Eqs. (104), (105). For example, if $\nu = 0$, Eq. (107) gives

$$\partial_0 F^{00} + \partial_1 F^{10} + \partial_2 F^{20} + \partial_3 F^{30} = \frac{4\pi}{c} j^0$$

According to (108) and (109) this is simply

$$\frac{\partial}{\partial x} E_x + \frac{\partial}{\partial y} E_y + \frac{\partial}{\partial z} E_z = 4\pi\rho$$

which agrees with Eq. (102).

EXERCISE 16. Check the other components of (106) and (107).

The object j^μ is a four-vector. To see this, recall the definition of charge and current density. Suppose that we have a cloud of particles (a fluid), all of the same kind. In the vicinity of any given point, we describe this distribution of particles by the number per unit volume and by the velocity. Both the number density and the velocity are functions of space and time:

$$n = n(x)$$

$$\mathbf{v} = \mathbf{v}(x) \tag{110}$$

The charge density is then defined by

$$\rho(x) = qn(x) \tag{111}$$

and the current density by

$$\mathbf{j} = qn(x) \mathbf{v}(x) \tag{112}$$

In these equations q is the charge of one particle. For example, $q = -e$ for electrons, $q = +e$ for protons, $q = 0$ for neutrons. If several types of particles are present, then we must combine their charge and current densities additively.

Incidentally, as a consequence of the conservation of the number of particles, the charge is conserved:

$$\nabla \cdot \mathbf{j} + \frac{\partial \rho}{\partial t} = 0 \tag{113}$$

EXERCISE 17. Show this.

Conservation of the number of each kind of particle is sufficient, but not necessary, for the validity of (113). Obviously all that is necessary is that in any reaction which destroys or creates particles, the net number of charges should be the same before and after.

In view of Eqs. (111) and (112), the four-component object j^μ can be expressed as

$$j^\mu \equiv (c\rho, j_x, j_y, j_z)$$
$$= (cqn,\ qnv_x,\ qnv_y,\ qnv_z) \qquad [114]$$

This can also be written as

$$j^\mu = n_0 q u^\mu$$

or as

$$j^\mu = \rho_0 u^\mu \qquad [115]$$

where

$$n_0 = n\sqrt{1 - v^2} \qquad [116]$$

and

$$\rho_0 = qn_0 \qquad [117]$$

where n_0 is the proper particle density (see Eq. (74)) and ρ_0 the proper charge density;* these proper quantities are invariants. Since j^μ consists of u^μ multiplied by an invariant, it must have the transformation law of a four-vector. We will call j^μ the electric *current four-vector.*

In terms of the four-vector j^μ, the conservation law (113) can be written

$$\partial_\mu j^\mu = 0 \qquad [118]$$

This equation is of the general form (100) and its Lorentz invariance is obvious.

Next, we want to discover the Lorentz transformation properties of the 16-component object $F^{\mu\nu}$. For this we appeal to the principle of relativity which tells us that under a Lorentz transformation the law (106) must become

* The symbol ρ_0 has also been used for mass density in an earlier section. Which meaning is intended will be clear from the context.

$$\partial'_\alpha F'^{\alpha\beta} = \frac{4\pi}{c} j'^\beta \qquad\qquad [119]$$

Since

$$\partial'_\alpha = a_\alpha{}^\mu \partial_\mu \qquad\qquad\cdot \qquad\qquad [120]$$

and

$$j'^\beta = a^\beta{}_\sigma j^\sigma \qquad\qquad [121]$$

Eq. (119) can also be expressed as

$$a_\alpha{}^\mu \partial_\mu F'^{\alpha\beta} = \frac{4\pi}{c} a^\beta{}_\sigma j^\sigma \qquad\qquad [122]$$

If we multiply this by $a_\beta{}^\nu$, and use Eq. (36), we obtain

$$a_\beta{}^\nu a_\alpha{}^\mu \partial_\mu F'^{\alpha\beta} = \frac{4\pi}{c} j^\nu \qquad\qquad [123]$$

We now compare Eqs. (123) and (106). Agreement between these equations demands*

$$a_\beta{}^\nu a_\alpha{}^\mu F'^{\alpha\beta} = F^{\mu\nu} \qquad\qquad [124]$$

The tensor $F^{\mu\nu}$ is called the *electromagnetic field* tensor.

Given the tensor character of $F^{\mu\nu}$, it is then obvious that both Eqs. (106) and (107) are tensor equations of the general type of Eq. (100). The electrodynamic equations therefore satisfy the principle of relativity.

The equation of motion for a charged particle can be written in terms of the field tensor $F^{\mu\nu}$. The equation of motion is

$$\frac{d\mathbf{P}}{dt} = q\mathbf{E} + \frac{q}{c}\mathbf{v} \times \mathbf{B} \qquad\qquad [125]$$

where $\mathbf{p} = m\mathbf{v}/\sqrt{1 - v^2}$ is the relativistic momentum. A simple calculation shows that (115) is equivalent to

$$\frac{dp^\mu}{d\tau} = \frac{q}{m} p_\nu F^{\mu\nu} \qquad\qquad [126]$$

---

* Mathematically, these equations would agree even if $a_\beta{}^\nu a_\alpha{}^\mu F'^{\alpha\beta}$ and $F^{\mu\nu}$ differ by a term $B^{\mu\nu}$ provided that $\partial_\mu B^{\mu\nu} = 0$. This means that $B^{\mu\nu}$ is a free electromagnetic wave. Physically, such an extra term makes no sense because a Lorentz transformation cannot create an electromagnetic wave where there was none before.

The Lorentz invariance of Eq. (126) is obvious.

EXERCISE 18. Check that the components of (126) give (125).

The energy-momentum tensor of the electromagnetic field is

$$T^{\mu\nu} = -\frac{1}{4\pi}\left[F^{\mu\alpha}F^{\nu}{}_{\alpha} - \frac{1}{4}\eta^{\mu\nu}F^{\alpha\beta}F_{\alpha\beta}\right] \qquad [127]$$

A complete derivation of this expression is given in Appendix 1. The T^{00} component is the familiar energy density of the electromagnetic field,

$$T^{00} = \frac{1}{8\pi}(\mathbf{E}^2 + \mathbf{B}^2) \qquad [128]$$

and the T^{0k} component is the Poynting vector,

$$T^{0k} = \frac{1}{4\pi}(\mathbf{E} \times \mathbf{B})^k \qquad [129]$$

The trace of $T^{\mu\nu}$ is zero:

$$T^{\mu}{}_{\mu} = 0 \qquad [130]$$

EXERCISE 19. Obtain (128) and (129) from (127).

EXERCISE 20. Prove (130).

If it is taken for granted that the energy density in the electromagnetic field is $(\mathbf{E}^2 + \mathbf{B}^2)/8\pi$, then a very simple derivation of the expression (127) can be given. The argument is this: Suppose Eq. (128) for the energy density is true in every reference frame. Since (127) agrees with (128), it follows that $T^{\mu\nu}$ can differ from the expression (127) only by a (symmetric) tensor $B^{\mu\nu}$ whose B^{00} component vanishes identically in *every reference frame*. It is easy to see that in this case all components of $B^{\mu\nu}$ must vanish.

EXERCISE 21. To see that $B^{11} = 0$, take a Lorentz transformation of the type (17). Then

$$B'^{00} = a^0{}_0 a^0{}_0 B^{00} + a^0{}_1 a^0{}_1 B^{11} + 2a^0{}_1 a^0{}_0 B^{01}$$

and if $B'^{00} = B^{00} = 0$, it follows that

$$0 = a^0{}_1 B^{11} + 2a^0{}_0 B^{01}$$

This cannot be true for all possible choices of $a^0{}_1$ and $a^0{}_0$ unless $B^{11} = B^{01} = 0$. Given that B^{00}, B^{11}, and B^{01} vanish in every reference frame, construct similar arguments to prove that all the remaining components also vanish.

Finally, we present the relativistic formulation of the electromagnetic potentials. The usual connection between fields and potentials is

$$\mathbf{E} = -\nabla\phi - \frac{1}{c}\frac{\partial\mathbf{A}}{\partial t} \tag{131}$$

$$\mathbf{B} = \nabla \times \mathbf{A} \tag{132}$$

where ϕ is the "scalar" potential and $\mathbf{A}$ the "vector" potential (these are scalars and vectors only in the three-dimensional sense!). If we define a four-component object

$$A^\mu = (\phi, A_x, A_y, A_z) \tag{133}$$

then Eqs. (131), (132) can simply be written

$$F^{\mu\nu} = \partial^\mu A^\nu - \partial^\nu A^\mu \tag{134}$$

EXERCISE 22. Check that (134) is equivalent to (131) and (132).

Since the left side of Eq. (134) is a tensor, it follows that A^μ must be a four-vector, usually called the four-vector potential.

If one inserts the expression (134) into the Maxwell equations (106) and (107), one finds that the second of these is satisfied identically while the first gives

$$\partial_\mu\partial^\mu A^\nu - \partial^\nu\partial_\mu A^\mu = \frac{4\pi}{c} j^\nu \tag{135}$$

This is the wave equation for the potential.

We will not attempt the solution of the relativistic Maxwell equations, because the application of these equations to electromagnetic problems is beyond the scope of our discussion. We are only interested in electrodynamics as an example of a field theory. In the next chapter we will see how the understanding of the electromagnetic field leads to an understanding of the gravitational field.

2.6 LOCAL FIELDS VERSUS ACTION-AT-A-DISTANCE

Although Newton's formulation of universal gravitation suggests action-at-a-distance, Newton himself had considerable misgivings about this interpretation. In his words:

> It is inconceivable that inanimate brute Matter should without the Mediation of something else which is not material, operate upon, and affect other Matter without mutual Contact. . . . That Gravity should be innate, inherent and essential to Matter so that one Body may act upon another at a Distance thro' a *Vacuum* without the Mediation of any thing else, by and through which their Action and Force may be conveyed from one to another, is to me so great an Absurdity that I believe no Man who has in philosophical Matters a competent Faculty of thinking can ever fall into it. . . .[8]

Nevertheless, the followers of Newton, and in particular Laplace and his contemporaries, regarded Newton's gravitational forces as action-at-a-distance. Because of the tremendous success of the theory, it became very fashionable throughout the eighteenth and most of the nineteenth centuries to regard all forces as action-at-a-distance. This trend was reversed by the work of Faraday and Maxwell in electromagnetism. Faraday's picture of lines of force had irresistible appeal, and with Maxwell's brilliant mathematical formulation of the field equations of electrodynamics, the interpretation of forces by means of fields and action-by-contact gained prevalence. Attempts were then made at describing gravitation as a field with a finite velocity of propagation. Einstein's theory may be regarded as the most successful such attempt.

Nowadays, all the fundamental interactions are regarded as due to local fields. The interaction of two particles that are widely separated is regarded as an action-by-contact of one particle on the field followed by a second action-by-contact of the field on the second particle. Why do we prefer fields to action-at-a-distance? The answer is simple: we need fields in order to uphold the law of conservation of energy and momentum.

In nonrelativistic physics, the conservation of momentum is a trivial consequence of Newton's law of equality of action and reaction. In relativistic physics, the conservation of momentum is much harder to achieve. For example, consider two particles that interact but do not come into contact—that is, the interaction takes place while the particles are separated by some distance. The momenta of the particles are $mu_{(1)}^{\mu}$ and $mu_{(2)}^{\mu}$, and as a result of the interaction these momenta change in some way. The total momentum of the particles is

$$P^{\mu} = mu_{(1)}^{\mu} + mu_{(2)}^{\mu} \qquad\qquad [136]$$

We will prove that this total momentum is *not* constant during the inter-action.

The proof is by contradiction. Suppose that the total momentum *is* conserved. We may suppose further that the total momentum transforms as a four-vector, that is,*

$$P'^{\mu} = a^{\mu}{}_{\nu} P^{\nu} \qquad\qquad\qquad [137]$$

By conservation, each side of this equation is a constant and it makes no difference at which time t' or t we evaluate P'^{μ} or P^{μ}. Let us evaluate P'^{μ} at the time $t' = 0$ and P^{μ} at the time $t = 0$. If the worldlines and co ordinates are as shown in Fig. 2.5, then we obtain

$$mu'_{(1)}{}^{\mu}|_{P} + mu'_{(2)}{}^{\mu}|_{Q'} = a^{\mu}{}_{\nu} \left(mu_{(1)}{}^{\nu}|_{P} + mu_{(2)}{}^{\nu}|_{Q} \right)$$

But at the point P, the Lorentz transformation law for the four-velocity is

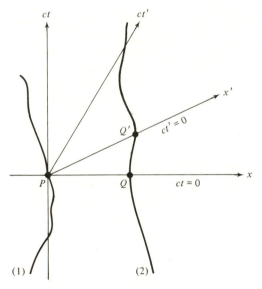

Fig. 2.5 *Worldlines of two inter-acting particles. Particle 1 is at the origin* P *at time* t = t' = 0. *Particle 2 is at the point* Q *at time* t = 0, *and at the point* Q' *at time* t' = 0.

---

* Since P^{μ} is the sum of two four-vectors (see Eq. (136)), at first sight it seems obvious that it should be a four-vector. However, there is a catch: in Eq. (136) it is implicitly assumed that the four-velocities are evaluated at the same time. Since Lorentz transformations do not preserve simultaneity, the meaning of "same time" is different in the two reference frames and this difference could complicate the transformation law of the sum (136). However, for the special case of *free* particles, (137) is obviously correct since in this case the four-velocities are constant and it is irrelevant whether the velocities are evaluated simultaneously or not. It then follows that (137) must also be correct for any particles that were free (did not interact) at some time in their past: the momenta P'^{μ} and P^{μ} are *constant* by hypothesis and if they had the transformation law (137) at one time, they must keep it forever.

$$u'_{(1)}{}^{\mu}|_P = a^{\mu}{}_{\nu} u_{(1)}{}^{\nu}|_P$$

and at the point Q'

$$u'_{(2)}{}^{\mu}|_{Q'} = a^{\mu}{}_{\nu} u_{(2)}{}^{\nu}|_{Q'}$$

Hence

$$ma^{\mu}{}_{\nu} u_{(1)}{}^{\nu}|_P + ma^{\mu}{}_{\nu} u_{(2)}{}^{\nu}|_{Q'} = a^{\mu}{}_{\nu}(mu_{(1)}{}^{\nu}|_P + mu_{(2)}{}^{\nu}|_Q)$$

This reduces to

$$ma^{\mu}{}_{\nu} u_{(2)}{}^{\nu}|_{Q'} = ma^{\mu}{}_{\nu} u_{(2)}{}^{\nu}|_Q$$

Since $a^{\mu}{}_{\nu}$ is arbitrary, it follows that

$$mu_{(2)}{}^{\mu}|_{Q'} = mu_{(2)}{}^{\mu}|_Q$$

which means that the four-momentum of the particle remains constant
as the particle moves from Q to Q'. We have therefore shown that the
total momentum is conserved if and only if the *momentum of each
particle remains constant,* that is, if and only if the particles *do not
interact.* This no-interaction theorem is due to Van Dam and Wigner.[9]

This theorem tells us that in order to preserve the law of conservation of energy-momentum for interacting particles, we must introduce another entity which can serve as a storehouse of energy-
momentum. This entity is the field.

To gain some understanding of how the field achieves the balance
of momentum, we begin with the observation that an interaction which
propagates instantaneously, with infinite velocity, is ruled out by the
requirements of special relativity. Interactions must be retarded; the
effects of a change in the source of, say, gravitation must propagate outwards with finite velocity. The only velocity left invariant by Lorentz
transformations is the velocity of light (see Problem 1), and hence if
gravitational effects propagate with a definite velocity, it will have to be
the velocity of light. Of course, this strongly suggests that what is propa-
gated is a field, but it is also possible to describe the situation by means of
"retarded action-at-a-distance." By the latter is meant that the action of
a source of gravitation (or whatever) on some particle is always assumed
to be delayed by a time r/c, where r is the distance between source and
particle.

Let us now take a look at momentum conservation in the earth-
sun system. In this case Newton's law of action and reaction states that
the force exerted by the sun on the earth is equal and opposite to the
force exerted by the earth on the sun; under these conditions the mo-

mentum of the earth-sun system would be conserved. However, it is easy to see that the equality of action and reaction is in contradiction with the requirement that gravitational signals propagate at the velocity of light. Consider the following thought-experiment: at some time the earth is suddenly given an acceleration (by e.g., receiving a series of kicks from the impact of a stream of meteors). This changes the position of the earth in the field of the sun and therefore changes the force of the sun on the earth. The force of the earth on the sun will *not change simultaneously*. It takes light signals $r/c \cong 500$ sec to travel the earth-sun distance r, and hence any signal will be delayed by at least as much. The gravitational signal that "adjusts" the gravitational pull of the earth moves from earth to sun at the speed of light and until it reaches its destination, action and reaction will certainly be out of balance. This conclusion is of course in agreement with the no-interaction theorem.

Our thought-experiment is less farfetched than it seems. The earth is in continuous acceleration as it orbits the sun, and action and reaction are in fact out of balance all the time. We can give a rough estimate of the difference between action and reaction. The force of the sun on the earth is $GM_\odot M/r^2$, where $M_\odot$ and M are the masses of sun and earth, respectively. The force of the earth on the sun differs from this by a term depending on the acceleration of the earth. In the time r/c, the radial displacement of the earth caused by the acceleration is

$$\Delta r \cong \frac{1}{2} \frac{GM_\odot}{r^2} \left(\frac{r}{c}\right)^2 = \frac{1}{2} \frac{GM_\odot}{c^2} \qquad [138]$$

Hence, *without* the acceleration the earth would have moved off tangentially and increased its distance by this amount. Fig. 2.6 shows the present position of the earth and the retarded position, that is, the position at a time r/c before the present position. The figure also shows the "extrapolated" position, i.e., the position that the earth would have reached if it had moved at constant velocity along a straight line beginning

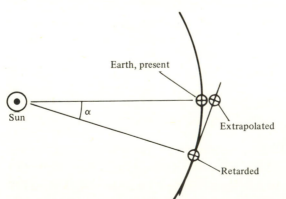

Fig. 2.6 *Present, retarded, and extrapolated positions of the earth. The angle α is $\sim 10^{-4}$ radians.*

at the retarded position. This extrapolated position is nearly on the same radius as the present position, but the radial distance is larger by the amount given by Eq. (138).

As a rough estimate for the force felt at the sun, we take that force which the earth would exert if stationary at the extrapolated position. This estimate for the force gives

$$\frac{GM_\oplus M}{(r + \Delta r)^2} \cong \frac{GM_\oplus M}{r^2} - \frac{1}{2}\frac{GM_\odot}{c^2}\left(\frac{2GM_\oplus M}{r^3}\right) \qquad [139]$$

Of course, Eq. (139) is only an educated guess. Essentially, we are assuming that the force depends mainly on the velocity of the earth at the retarded time, and that to this the acceleration only contributes some insignificant corrections (which have to do with the emission of gravitational radiation). If only the velocity is important, then the force exerted by the earth is the same as that for a mass that moves with uniform velocity from the retarded position and reaches the extrapolated position at the present time. For a mass moving with uniform velocity, the center of force will coincide with the position of the mass, as becomes immediately obvious if one asks for the force as seen from the rest frame of that mass. (However, the transformation to the rest frame of that mass introduces relativistic corrections which are typically of the same order of magnitude as the extra term in (139).)

The difference between action and reaction is then of the order

$$\left(\frac{GM_\oplus M}{rc^2}\right)\left(\frac{GM_\odot}{r^2}\right) \qquad [140]$$

The inequality of action and reaction implies a violation of momentum conservation. The question is now: where does the momentum go? There is only one place it can go . . . it must be stored in the *gravitational field*. What is conserved is not the momentum of the earth-sun system, but rather the momentum of the system consisting of earth, sun, and gravitational fields. This interpretation is in accord with our estimate (140). To see this, note that $GM_\oplus M/rc^2$ may be interpreted as the "mass" associated with the interaction energy (potential energy) between the sun and the earth. This energy is stored in the gravitational fields,* and as the earth orbits the sun, so will this extra "mass." To obtain the rate of momentum transfer to the fields, we must take the product of this "mass" by a characteristic acceleration. A reasonable estimate for the acceleration is $GM_\odot/r^2$, and the product has then the form (140).

Note that the momentum transfer to the "convective" fields considered above averages to zero over a period of the motion. However,

_____
* According to Eq. (1.12), this energy is in part pure field energy and in part interaction energy of field and mass.

under some circumstances it is also possible for a particle system to lose momentum (and energy) continuously. In this case we say that the momentum and energy are carried away by "radiation" fields.

Fields provide us with a very elegant way to uphold the law of momentum conservation. In action-at-a-distance theories, momentum conservation looks quite ugly. At first sight it seems impossible to satisfy momentum conservation at all: when fields are absent there is no place to store the energy and the momentum lost by particles. However, it is possible to get around this by a trick: one changes the definition of momentum of the particles in such a way that it *is* conserved. This means that one adds an extra term (interaction momentum) to the usual expression for momentum. The extra term has a horrible appearance: it is an integral over the entire history of each particle. In fact this integral is precisely what we would regard as field momentum in a field theory (the integration over history arises from writing the field in terms of the past motion of all the particles that contributed to the field).

The contrast between field theory and action-at-a-distance theory is brought out sharply by an example that goes somewhat beyond classical physics: the annihilation of particle-antiparticle pairs. If an electron annihilates with a positron, their energy and momentum must continue to exist. According to field theory, the energy-momentum exists in the electromagnetic radiation (photons) emitted in the annihilation. According to action-at-a-distance theory, the energy-momentum exists as an interaction energy-momentum associated with particles; but since the particles have annihilated, we are forced to associate this momentum with the ghosts of particles that once were. This can be done, even though it looks rather artificial; in fact, it can be shown that fields and action-at-a-distance are two mathematically equivalent formulations of an interaction.[10] But fields provide us with the most natural picture of the conservation of energy and momentum and we therefore prefer to describe electromagnetism, and gravitation, in terms of local fields.

FURTHER READING

As Bertrand Russell said, "Einstein's theory of relativity is probably the greatest synthetic achievement of the human intellect up to the present time." Einstein's recollections of how he arrived at his discoveries are contained in his "Autobiographical Notes" (in Schilpp, ed., *Albert Einstein: Philosopher-Scientist*). A brilliantly written biography of Einstein as a scientist is Hoffmann, *Albert Einstein, Creator and Rebel*.

Einstein's article, "Zur Elektrodynamik bewegter Körper" (Ann. d. Physik, **17**, 891, 1905), *must* be read. A translation is available in the paperback Lorentz, Einstein, Minkowski, and Weyl, *The Principle of Relativity*. This is the original article on relativity and it is amazing for its completeness. It contains just about everything: from the definition of simultaneity and the Lorentz transformations to the transformation of electric and magnetic fields and the relativistic equations for the motion and energy of a particle. It is not unfair to say that the only essential features of special relativity that remained to be added by the followers of Einstein were the geometrical interpretation of spacetime and the theory of the representations of the Lorentz group.

A delightful discussion of spacetime geometry, with a minimum of mathematics, is given in Synge, *Talking about Relativity*.

Qualitative estimates of the effects of quantum fluctuations on the geometry and topology of spacetime will be found in Wheeler, *Einsteins Vision*.

The best introduction to special relativity is Taylor and Wheeler, *Spacetime Physics*. This book strongly emphasizes that the theory of relativity is a theory of the geometry of spacetime; it contains a wealth of intriguing puzzles and paradoxes and their solutions. Another very good introduction is French, *Special Relativity*. These books lay the physical foundation for the mathematical formalism that we have developed in the present chapter.

More advanced and complete treatments of special relativity may be found in:

Anderson, *Principles of Relativity Physics* (includes a careful discussion of the spacetime structures underlying Newtonian and relativistic physics).

Bergmann, *Introduction to the Theory of Relativity* (contains a thorough presentation of general tensor calculus).

Einstein, *The Meaning of Relativity* (a very concise presentation of the mathematical formalism).

Fock, *The Theory of Space, Time, and Gravitation* (develops special relativity in general coordinates by means of tensor notation, a good preliminary exercise leading into general invariance).

Lightman, Press, Price, and Teukolsky, *Problem Book in Relativity and Gravitation* (nearly 500 interesting problems, and their solutions, covering many puzzling questions in special relativity, general relativity, gravitation, astrophysics, and cosmology).

Misner, Thorne, and Wheeler, *Gravitation* (emphasizes spacetime geometry and gives a beautiful introduction to the language of differential forms which is an alternative to the language of tensors).

Møller, *The Theory of Relativity* (very complete, includes a general discussion of relativistic energy, momentum, angular momentum, and center of mass for arbitrary closed systems).

Pauli, *Theory of Relativity* (a classic written when Pauli was a student of only 21; it was originally written for the *Mathematical Encyclopedia* as a complete review of all the advances in relativity up to 1921 and although some parts are by now outdated, the mathematical material remains valid and informative).

Robertson and Noonan, *Relativity and Cosmology* (comparable to Møller's book).

Synge, *Relativity: The Special Theory* (discusses systems of particles and collisions, and contains a simple definition of the energy-momentum tensor).

Tolman, *Relativity, Thermodynamics, and Cosmology* (a very readable exposition; however, tensor analysis is disposed of in four pages and that is rather too short).

Weinberg, *Gravitation and Cosmology* (a short, neat review of special relativity is given in Chapter 2; it includes some remarks on representations of the Lorentz group).

The relativistic formulation of Maxwell's equations is discussed in these books and also in many books on electrodynamics such as, e.g., Jackson, *Classical Electrodynamics,* and Schwartz, *Principles of Electrodynamics.* Unfortunately, many of the above books follow the abominable custom of introducing a factor of $\sqrt{-1}$ into the time-component of four-vectors. This simplifies the notation somewhat, because it eliminates the metric tensor $\eta_{\mu\nu}$ from all equations; but since the metric is precisely the most important feature of special relativity it makes little sense to pretend that it does not exist.

The historical role of the field concept in gravitation is discussed in the very readable book *The Measure of the Universe,* by North. Electrodynamics was the first field theory (and remains the best understood one); the contributions of Maxwell, Faraday, and their predecessors to this theory are presented with a wealth of detail in Whittaker, *A History of the Theories of Aether and Electricity.* Incidentally, Whittaker expresses the rather curious opinion that the theory of special relativity was invented by Poincaré and Lorentz. The fact is that even the best efforts of Lorentz (see his article "Electromagnetic Phenomena in a System Moving with Any Velocity Less Than That of Light" in ref. 3) amounted to no more than an, admittedly very ingenious, generalization of the Lorentz-Fitzgerald contraction hypothesis and were in no sense a theory of relativity.

REFERENCES

1. A. Einstein *in* P. A. Schilpp, ed., *Albert Einstein: Philosopher-Scientist* (Tudor Publishing Company, New York, 1951), p. 53. Reprinted with the permission of the Estate of Albert Einstein and Amiel Book Company.
2. J. A. Wheeler, *Einsteins Vision* (Springer-Verlag, Berlin, Heidelberg, 1968).
3. H. A. Lorentz, A. Einstein, H. Minkowski, and H. Weyl, *The Principle of Relativity* (Dover, New York), p. 75.
4. M. Schwartz, *Principles of Electrodynamics* (McGraw-Hill, New York, 1972).
5. R. F. Streater and A. S. Wightman, *PCT, Spin, Statistics and All That* (W. A. Benjamin, New York, 1964).
6. A. Einstein, Ann. Physik **49,** 769 (1916); translated in H. A. Lorentz, A. Einstein, H. Minkowski, and H. Weyl, *The Principle of Relativity* (Dover, New York), p. 119.
7. E. Wigner, Ann. Math., **40,** 39 (1939).
8. I. B. Cohen, ed., *Isaac Newton's Papers and Letters on Natural Philosophy* (Harvard University Press, Cambridge, 1958), Third Letter to Bentley.
9. H. Van Dam and E. P. Wigner, Phys. Rev. **142,** 838 (1966).
10. E. C. G. Sudarshan, Fields and Quanta **2,** 175 (1972).

PROBLEMS

1. Show that the Lorentz transformation (4) implies the following transformation laws for the (ordinary) velocity

$$u_x = \frac{u_x' + v}{1 + vu_x'/c^2} \qquad u_y = \frac{\sqrt{1 - v^2/c^2}\, u_y'}{1 + vu_x'/c^2} \qquad u_z = \frac{\sqrt{1 - v^2/c^2}\, u_z'}{1 + vu_x'/c^2}$$

where $u_x' = dx'/dt'$, etc.

Prove that the only velocity whose magnitude is left invariant by all Lorentz transformations is the velocity of light.

2. At $12^h 0^m 0^s$ eastern standard time a boiler explodes in the basement of the Museum of Modern Art in New York City. At $12^h 0^m 0.0003^s$ a similar boiler explodes in the basement of a soup factory in Camden, N.J., at a distance of 100 miles from the first explosion. Find a Lorentz frame in which the first explosion occurs *after* the second.

3. Suppose that there exist particles ("tachyons") which move with a velocity *larger* than that of light. Show that you can then send signals into your own past. (Hint: Suppose you are at rest at the origin. Send a tachyon signal, with $v > c$ in your reference frame, towards a spaceship moving away from you. When this spaceship receives the signal, it immediately sends a similar signal, with $v > c$ in the spaceship's frame, back to you. Show that the signal arrives at your position *before* the time at which you sent the original signal.)

4. For an ordinary rotation about the x-axis,

$$a^\mu{}_\nu = \begin{pmatrix} 1 & 0 & 0 & 0 \\ 0 & \cos\theta & \sin\theta & 0 \\ 0 & -\sin\theta & \cos\theta & 0 \\ 0 & 0 & 0 & 1 \end{pmatrix}$$

Show that this matrix satisfies Eq. (18).

5. Express both the energy and momentum of a photon in an electromagnetic wave in terms of the frequency of the wave. Show that as a consequence $p_\mu p^\mu = 0$ for the photon.

6. The *invariant mass M* of a system of particles is defined by

$$M^2 = (p^{(1)}{}_\mu + p^{(2)}{}_\mu + \ldots)(p^{(1)\mu} + p^{(2)\mu} + \ldots)$$

where $p^{(1)\mu}$, $p^{(2)\mu}$, etc., are the energy-momentum vectors of particles (1), (2), . . . , etc.

Show that M equals the energy of the system in that reference frame in which the (spatial) momentum is zero (center of momentum, or C.M., frame).

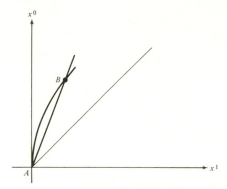

Fig. 2.7 *Worldlines of two clocks.*

7. Fig. 2.7 shows the worldlines of two clocks that move from A to B. The first clock moves along the straight worldline

$$x = v_0 t$$

where v_0 is a constant. The second clock moves along the curved worldline

$$x = \tfrac{1}{2} a_0 t^2$$

where a_0 is a constant. For each clock find the elapsed proper time between A and B. Which clock shows the longer proper time?

8. Two spacetime points, A and B, have a timelike separation. Consider the straight worldline connecting A and B, and consider curved worldlines. Show that the proper time between A and B,

$$\Delta \tau = \int_A^B d\tau = \int_A^B \sqrt{dx^\mu dx_\mu}$$

is largest when evaluated along the straight worldline. Hence, the worldline of a free particle can be characterized as the spacetime curve of *maximum* proper time. (Hint: Use a reference frame in which the particle moving along the straight worldline is at rest.)

9. Assume that $A^\mu(x)$ is a vector field with components

$$A^\mu(x) = (0, \sin(5x^1 - 3x^0), \sin(5x^1 - 3x^0), 0)$$

Find the vector field $A'^\mu(x')$ if x and x' are related by the Lorentz transformation (16). (Your answer should be a function of x'.)

10. Suppose that $\Phi(x)$ is a scalar field. Show that the integral

$$\int \Phi(x) \, d^3x \, dt$$

evaluated over a given (four-dimensional) region of spacetime, is a scalar. (Hint: Use volume contraction and time dilation to transform $d^3x \, dt$; alternatively, use the Jacobian determinant.)

11. Suppose that the current density j^μ differs from zero only in a region of finite extent in the x,y,z-directions (see Fig. 2.4). Use arguments similar to those given in Eqs. (88)–(99) to show that the *total* charge

$$Q = \int j^0 d^3x$$

is a Lorentz scalar.

12. A vector field $C^\mu(x)$ satisfies the differential equation

$$\partial_\mu C^\mu = 0$$

Assuming that C^μ is different from zero only in a region of finite extent in the x,y,z-directions (see Fig. 2.4), show that

$$\frac{d}{dt}\left(\int C^0 d^3x\right) = 0$$

13. Write down all components of $F_\mu{}^\nu$, $F^\mu{}_\nu$, and $F_{\mu\nu}$ in terms of **E** and **B**. Write down all components of the electromagnetic energy-momentum tensor in terms of **E** and **B**.

14. Show that $F^{\mu\nu} F_{\mu\nu} = 2(\mathbf{B}^2 - \mathbf{E}^2)$ where $F^{\mu\nu}$ is the electromagnetic field tensor. Hence, show that $\mathbf{B}^2 - \mathbf{E}^2$ is a scalar under Lorentz transformations.

15. Show that the equation of motion (126) has a first integral

$$p_\mu p^\mu = \text{constant}$$

16. A plane electromagnetic wave has fields

$$\mathbf{E} = E_0 \hat{\mathbf{x}} \cos(\omega t - kz)$$
$$\mathbf{B} = E_0 \hat{\mathbf{y}} \cos(\omega t - kz)$$

Find all components of the energy-momentum tensor for this wave. Check that the energy flux equals the flux of z-momentum. Is this what you expect for a wave made of photons?

17. Show that if the energy of a system of particles in collision is conserved as seen from *all* possible Lorentz frames, then the momentum must necessarily also be conserved. (Hint: Consider the four-vector ΔP^μ that represents the loss of energy-momentum. You are given that $\Delta P^0 = 0$ in *all* reference frames.)

18. Consider a perfect fluid, e.g. a gas, whose state is described by density and pressure (the fluid can support no shear). Show that in the rest frame of the fluid, in which the macroscopic velocity of the fluid is zero, the energy-momentum tensor is

$$T^{00} = \rho_0, \qquad T^{0k} = 0, \qquad T^{kl} = p\,\delta_k{}^l$$

where ρ_0 is the (macroscopic) mass density measured in this rest frame and p is the pressure.

Show that a Lorentz transformation to a reference frame in which the fluid has a velocity **V** gives

$$T'^{\mu\nu} = (\rho_0 + p)V^\mu V^\nu - p\eta^{\mu\nu}$$

*19. Show that the set of all Lorentz transformations forms a group and show that scalars, vectors, and tensors form representations of this group.

* This problem requires some knowledge of group theory.

3. THE LINEAR FIELD APPROXIMATION

Twinkle, twinkle little star
How I wonder where you are.
"1.75 seconds of arc from where I seem to be
For $ds^2 \simeq (1 - 2GM/r)dt^2 -$
$$(1 + 2GM/r)dr^2 - r^2d\theta^2 - r^2\sin^2\theta d\phi^2."$$

Source unknown

We will now seek out the relativistic field equations for gravitation. We begin with what is simple by assuming a linear approximation and neglecting the effects of the gravitational field on itself. Of course, if Newton's principle of equivalence ($m_I = m_G$) is to hold as an *exact* statement, gravitational energy must gravitate and the exact field equations must be nonlinear. Although it is true that the most spectacular results of gravitational theory depend in a crucial way on the nonlinearity of the field equations, almost all of the results that have been the subject of experimental investigation can be described by the linear approximation. For example, the deflection of light, the retardation of light, the redshift, and gravitational radiation emerge in the linear approximation. Furthermore, this approximation applies to all phenomena that lie in the region of overlap between Newton's and Einstein's theories.

Most discussions of gravitational theory begin by *postulating* that spacetime is curved and carry on from there; the linear approximation arises in the end from the full nonlinear equations. The great disadvantage of this approach is that it never makes clear just why anybody would entertain the preposterous notion that our beautiful flat spacetime should be curved, bent, and deformed. The reason why we believe in curved space can best be understood by beginning with flat spacetime and seeking out field equations that describe the gravitational field (in linear approximation) in this flat spacetime. As we will see, the behavior of particles in a gravitational field, and the behavior of "clocks" and "meter sticks" built out of these particles, are then such that measurements of spacetime indicate that the real geometry is curved. Thus, the flat spacetime with which we begin, turns out to be an unobservable, fictitious geometry. But, nevertheless, flat spacetime is a useful starting point in the case of weak fields.

3.1   THE EXAMPLE OF ELECTROMAGNETISM

Since electromagnetism is the linear theory par excellence, it will be instructive to play a game. Let us pretend that we do know that there is a long-range interaction between charges, but we do not know the field equations (Maxwell's equations) that govern this interaction. How could we go about finding the equations?

We will, of course, demand that the equations that describe the interaction between charges are Lorentz invariant. This requirement of Lorentz invariance tells us that an interaction that involves only the charge density is impossible. The charge density ρ and the current density $\mathbf{j}$ taken together form a four-vector (see Eq. (2.114)) and any Lorentz invariant interaction which involves one component of this four-vector must also involve the other components. For example, suppose we have a reference frame in which there are charge densities but no currents. In a new moving reference frame there will then be both a charge density $j'^0 = a^0{}_0 j^0$ and a current density $j'^k = a^k{}_0 j^0$. Thus, although an observer in the old reference frame attributes the interaction entirely to charges, an observer in the new frame will attribute it to both charges and currents. Any law that pretends to be valid in all reference frames must therefore include both the interaction of charge densities and that of current densities. Hence, if the charge density j^0 generates some kind of force field, so will the current density $\mathbf{j}$. To put it briefly: the four-vector j^μ is the source of the electromagnetic field.

The field equations are supposed to establish the link between the field and its source j^μ. As a first step in our search for the field equations, we must decide what kind of field will be used to describe the interaction. According to Section 2.3, our choices are between a scalar, a vector, and a tensor of higher rank. Since the source of the field is the four-vector j^μ, the most natural choice is a *vector field*, and this is what we will try.

This choice of a vector field can be made plausible by an analogy with Lagrangian mechanics. The equation of motion for a generalized coordinate q_k under the influence of a (generalized) driving force Q_k is typically of the form

$$\frac{d^2q_k}{dt^2} + \cdots \propto Q_k$$

where the extra terms indicated by dots represent the coupling between q_k and other generalized coordinates. In our case, the field plays the role of the generalized coordinate and the current density plays the role of the driving force. The equation of motion should then have the form

$$\frac{\partial^2 A}{\partial t^2} + \cdots \propto j^\mu \tag{1}$$

Here the field variable has simply been written as A, without any index. This quantity A could be a scalar, a vector, or a tensor of higher rank. One obvious defect of our tentative equation of motion is that while a Lorentz transformation changes A into A' and j^μ into j'^μ, it does *not* change $\partial^2/\partial t^2$ into $\partial^2/\partial t'^2$; hence the equation is not Lorentz invariant and fails to satisfy the principle of special relativity. To achieve invariance, we recall that $\partial^2/\partial t^2$ is only one part of the operator $\partial^\nu \partial_\nu$. If we take the complete operator $\partial^\nu \partial_\nu$, then a Lorentz transformation changes it into $\partial'^\nu \partial'_\nu$, as desired. We therefore arrive at an equation of motion

$$\partial^\nu \partial_\nu A + \cdots \propto j^\mu \tag{2}$$

Since the right side of this equation is a four-vector, it follows that A must also be a four-vector. We designate the vector field by $A^\mu(x)$.*

We now look for the most general linear differential equation for A^μ, with j^μ acting as source. This means that the equation should have the form of Eq. (2): on the left side there should appear a linear differential operator acting on the field, and on the right side should appear the current. For example,

$$\partial_\mu \partial^\mu A^\nu + b\partial^\nu \partial_\mu A^\mu + aA^\nu = 4\pi j^\nu \tag{3}$$

where a and b are constants, is an acceptable equation. In fact, Eq. (3) is the *only* acceptable equation if we insist that the differential order should be no more than two. To see this, note that Lorentz invariance demands that the left side be a four-vector. But the only four-vectors

---

* Although the analogy with the dynamical equations of mechanics suggests that the field equation should be of second differential order (and involve a vector field), it is also possible to write a first-order field equation involving a second-rank antisymmetric tensor field. It turns out that these alternatives are equivalent (see Eqs. (16) and (17)).

linear in A^μ that can be constructed out of A^μ and no more than two of the differential operators ∂^μ are

$$\partial_\mu \partial^\mu A^\nu, \quad \partial^\nu \partial_\mu A^\mu, \quad A^\nu \tag{4}$$

Of course, we would generalize Eq. (3) slightly by inserting an extra arbitrary constant in front of the first term on the left side, but this amounts to no more than a redefinition of A^ν and is therefore not needed. Likewise, the extra constant that could be inserted in front of j^ν on the right side can always be absorbed in the definition of the current. In fact, the current already contains an adjustable constant q which represents the charge on a particle (see the definition given by Eqs. (2.111) and (2.112)). This is a coupling constant; it characterizes the strength of the coupling between matter and field. The value of this constant q must be determined from experiment.* On the other hand, the values of the constants a and b are completely determined by the following arguments.

As concerns the constant a, we can immediately say that it must not be negative. This can best be seen in the special case of vacuum solutions of Eq. (3); if $j^\nu = 0$, we have

$$\partial_\mu \partial^\mu A^\nu + b \partial^\nu \partial_\mu A^\mu + a A^\nu = 0 \tag{5}$$

This is a wave equation describing the propagation of the vector field.

EXERCISE 1. Show that

$$A^\mu = (0, \cos(kz - \omega t), 0, 0) \tag{6}$$

with

$$\omega = \sqrt{k^2 + a} \tag{7}$$

is a possible solution of Eq. (5).

The group velocity of the wave (6) is given by

$$v_g = \frac{d\omega}{dk} = \frac{k}{\sqrt{k^2 + a}} \tag{8}$$

If a is negative, the velocity exceeds the velocity of light.**

---

* This can only be done after we discover the field equations. If we do not know Maxwell's equations and the exact form of the law of force between charges, then we cannot give an operational procedure for measuring the constant q that appears in (2.111) and (2.112). But we *can* write down the expressions (2.111) and (2.112), leaving q as a temporarily unknown parameter.

** A nonzero value of a corresponds to a nonzero mass of the photon; a negative value of a corresponds to an imaginary mass and propagation with speeds larger than c (tachyons).

To decide what value of a we should adopt, let us look at another special solution of Eq. (3): take a time-independent, spherically symmetric charge distribution. If we consider the $\nu = 0$ component of Eq. (3) and omit all the terms containing time derivatives, we obtain a differential equation for A^0,

$$-\nabla^2 A^0 + aA^0 = 4\pi\rho \tag{9}$$

In the region exterior to the charge distribution this has the solution

$$A^0 \sim \frac{e^{\pm\sqrt{a}\,r}}{r} \tag{10}$$

EXERCISE 2. Show this.

Only the solution with the negative sign is relevant, and this solution shows that the field decays exponentially with distance outside of the charge distribution. By the rules of our game, we do not know Maxwell's equations, but we do know that the interaction is *long range;* we therefore will take $a = 0$. (See Section 1.1 for experimental limits on a.) Our field equation then reduces to

$$\partial_\mu\partial^\mu A^\nu + b\partial^\nu\partial_\mu A^\mu = 4\pi j^\nu \tag{11}$$

The only unknown constant remaining in our field equation is b. To determine the value of this constant, apply ∂_ν to both sides of (11):

$$\partial_\nu(\partial_\mu\partial^\mu A^\nu + b\partial^\nu\partial_\mu A^\mu) = 4\pi\partial_\nu j^\nu \tag{12}$$

If we exchange the (dummy) μ and ν labels in the second term on the left side, we obtain

$$(1 + b)\partial_\mu\partial^\mu(\partial_\nu A^\nu) = 4\pi\partial_\nu j^\nu \tag{13}$$

The right side of this equation vanishes because the current satisfies the conservation law $\partial_\nu j^\nu = 0$ (Eq. (2.118)). Hence the consistency of our field equation with current conservation demands that the left side of Eq. (13) vanish. In order to meet this demand, we will take $b = -1$ so that the left side vanishes *identically.**

The final form of our field equation is then

$$\partial_\mu\partial^\mu A^\nu - \partial^\nu\partial_\mu A^\mu = 4\pi j^\nu \tag{14}$$

---

* This is the most elegant way to make the field equations compatible with current conservation. There are other ways. For example, we can leave b arbitrary and impose the condition $\partial_\nu A^\nu = 0$; this is a gauge condition (see Eq. (20)).

This last equation has the same form as the Maxwell equation for the potential that we wrote down in Section 2.5 (see Eq. (2.135)). Obviously, our vector field A^μ is to be identified with the four-vector potential. To obtain the Maxwell equations for the fields, we *define* the tensor $F^{\mu\nu}$ as

$$F^{\mu\nu} \equiv \partial^\mu A^\nu - \partial^\nu A^\mu \qquad [15]$$

and then Eq. (14) becomes

$$\partial_\mu F^{\mu\nu} = 4\pi j^\nu \qquad [16]$$

Furthermore, the tensor (15) satisfies the identity

$$\partial^\alpha F^{\mu\nu} + \partial^\mu F^{\nu\alpha} + \partial^\nu F^{\alpha\mu} = 0 \qquad [17]$$

These equations are, of course, exactly the Maxwell equations (2.106) and (2.107).

Although in some sense the preceding arguments may be regarded as a derivation of Maxwell's equations, the limitations of this approach should be kept in mind. Clearly, we have had to make quite a few assumptions to reach Eq. (14). The objective of our game with electrodynamics was to obtain a prescription for finding the field equations in the hope that an analogous prescription will lead us to the relativistic field equations for gravitation. The *number* of ingredients in this prescription need not concern us, as long as each ingredient in the electromagnetic case has an analogous ingredient in the gravitational case. As we will see in the next section, the preceding arguments can be reproduced step-by-step in the gravitational case and lead to a reasonable result.

We will now look at some properties and consequences of the electromagnetic field equation (14). One of the remarkable properties of this equation is the following: *Whenever A^μ satisfies Eq. (14), then so does $A^\mu + \partial^\mu\Lambda$, where Λ is any arbitrary scalar function.* This is so because in the expression

$$\partial_\mu\partial^\mu(A^\nu + \partial^\nu\Lambda) - \partial^\nu\partial_\mu(A^\mu + \partial^\mu\Lambda) \qquad [18]$$

the terms involving Λ cancel identically and therefore the differential equation (14) is left unchanged.

The transformation

$$A^\mu \to A^\mu + \partial^\mu\Lambda \qquad [19]$$

is called a *gauge transformation*. We have shown that the field equation is invariant under this gauge transformation. In general, whenever an object is left unchanged (invariant) by some operation, one says that the operation is a symmetry of the object. Thus, gauge invariance is a *symmetry* of the field equation.

The gauge function Λ is completely arbitrary; this means that our equations do not determine the field uniquely. There is no harm in this because an ambiguity of the form $\partial^\mu\Lambda$ in the field leads to no observable consequences. The gauge transformations of A^μ leave the field tensor $F^{\mu\nu}$ unchanged; this implies that the observable quantities **E** and **B** are gauge independent.

EXERCISE 3. Show that the gauge transformation (19) produces no change in $F^{\mu\nu}$.

However, it is helpful to make a particular choice of the gauge function and eliminate some of the ambiguity in A^μ. We can do this by imposing the *Lorentz gauge condition* on the fields

$$\partial_\mu A^\mu = 0 \qquad [20]$$

To see how this works, suppose that A^μ does *not* satisfy (20). Then we can always find a new field $\bar{A}^\mu$ which does. We simply take

$$\bar{A}^\mu = A^\mu + \partial^\mu\Lambda \qquad [21]$$

and choose Λ in such a way that $\partial_\mu \bar{A}^\mu = 0$:

$$0 = \partial_\mu \bar{A}^\mu = \partial_\mu A^\mu + \partial_\mu\partial^\mu\Lambda$$

that is,

$$\partial_\mu\partial^\mu\Lambda = -\partial_\mu A^\mu \qquad [22]$$

This is a differential equation which can be solved for Λ. Eq. (22) always has a solution. In fact, it has *several solutions* since to any solution Λ, one can always add a solution of the homogeneous equation (the Lorentz condition does not remove the ambiguity in the field entirely).

We will from now on assume that the Lorentz condition holds. Note that with this condition the field equation reduces to

$$\partial_\mu\partial^\mu A^\nu = 4\pi j^\nu \qquad [23]$$

This form of the field equation is often the most convenient when one seeks solutions for a given current.

In vacuo, where $j^\nu = 0$, Eq. (23) is a simple wave equation with a propagation velocity c.

EXERCISE 4. Show that Eq. (23) (with $j^\nu = 0$) and Eq. (20) have the following linearly independent solutions

$$A^\mu = \epsilon_{(n)}{}^\mu \cos(\omega z - \omega t) \qquad [24]$$

where $\epsilon_{(n)}{}^\mu$ is one of the following four-vectors:

$$\epsilon_{(1)}{}^\mu = (0,1,0,0) \qquad [25]$$

$$\epsilon_{(2)}{}^\mu = (0,0,1,0) \qquad [26]$$

$$\epsilon_{(3)}{}^\mu = (1,0,0,1) \qquad [27]$$

(The subscript n has been enclosed in parentheses to indicate that it is not a tensor index; it is only a label that distinguishes between (25)–(27).) Show that any other solution of the type $\cos(\omega z - \omega t)$ can be written as a linear combination of the three solutions given above. Find the **E** and **B** fields for each solution and show that the $\epsilon_{(3)}{}^\mu$ solution gives $\mathbf{E} = \mathbf{B} = 0$; hence only the first two solutions have any physical significance.

    To complete our electromagnetic theory, we need to find the equation of motion of a charged particle in an electromagnetic field. This equation of motion can be derived from the conservation law of energy-momentum. If we have a charged particle interacting with an electromagnetic field, neither the energy-momentum of the particle nor that of the field are separately conserved; only the *total* energy momentum is conserved:

$$\partial_\nu (T^{\mu\nu}_{(m)} + T^{\mu\nu}_{(em)}) = 0 \qquad [28]$$

where $T^{\mu\nu}_{(m)}$ and $T^{\mu\nu}_{(em)}$ are, respectively, the energy-momentum tensors for particles and for the field.

    The expression for $T^{\mu\nu}_{(m)}$ is given by Eq. (2.75). Note that in the present context we want to use this expression to describe a single particle. Hence, we must assume that ρ_0 has a sharp peak (strictly speaking, a delta function) at the position of the particle and vanishes everywhere else.

    The expression for $T^{\mu\nu}_{(em)}$ is given by Eq. (2.127). Unfortunately, according to the rules of our game we must pretend that we do not know Eq. (2.127). How can we derive this expression for $T^{\mu\nu}_{(em)}$?

    For a field with a given field equation there exists a general prescription for constructing the energy-momentum tensor ("canonical energy-momentum tensor"). An important feature of this prescription is that it does not make use of the equation of motion of a particle. Since we ultimately want to derive this equation of motion from the energy-momentum tensor, this feature of the canonical prescription is crucial for the consistency of our approach.* In the construction of the canonical energy-momentum tensor we must use the tools of Lagrangian field theory; because of this, the derivation of the expression (2.127) will be left to Appendix 1.

Note that without any derivation we can make the expression (2.127) plausible by asking what properties we expect of the energy-momentum tensor. The first and most obvious property is that for a free field, with $j^\nu = 0$, the differential conservation law $\partial_\nu T^{\mu\nu} = 0$ must hold; that this is indeed so follows from Eq. (29) and Exercise 5. Furthermore, we expect that the tensor is quadratic in the derivatives of the fields. This property is a generalization of what we know to be the case for a system of particles: the (kinetic) energy is quadratic in the time derivatives of the generalized coordinates. Thus we expect time derivatives, and hence (by Lorentz invariance) also space derivatives, of the fields to appear quadratically in the energy-momentum tensor. Finally, the tensor should be symmetric and its T^{00} component (energy density) should be positive. Since (2.127) satisfies all these requirements, we expect it to be the energy-momentum tensor, and nature would have to be very malicious to ruin our expectation.

A straightforward calculation using Eqs. (16) and (17) shows that

$$\partial_\nu T^{\mu\nu}_{(em)} = -F^\mu{}_\nu j^\nu \qquad [29]$$

EXERCISE 5. Derive Eq. (29).

Hence, Eq. (28) becomes

$$\partial_\nu T^{\mu\nu}_{(m)} - F^\mu{}_\nu j^\nu = 0 \qquad [30]$$

This equation tells us how much energy and momentum the electromagnetic field transfers to the matter on which it acts. Thus, the equation determines the rate at which the momentum of a particle changes and therefore determines the equation of motion of a particle acted upon by an electromagnetic field.

To derive this equation of motion, we begin by integrating the expression (30) over the volume of the particle. (In the present context, "particle" simply means an arbitrary system of small size.) The first term gives

$$\int \partial_\nu T_{(m)\mu}{}^\nu d^3x = \int \partial_0 T_{(m)\mu}{}^0 d^3x + \int \partial_k T_{(m)\mu}{}^k d^3x \qquad [31]$$

---

* Other methods for constructing the energy-momentum tensor rely on knowledge of the equation of motion.

The second integral on the left side of Eq. (31) can be converted into a surface integral of the form

$$\int T_{(m)\mu}{}^{k} dS_k$$

over a surface surrounding the particle; since $T_{(m)\mu}{}^{k}$ is zero everywhere outside of the particle, the surface integral is zero. The first integral in (31) is simply the rate of change of the momentum of the particle:

$$\frac{d}{dt} \int T_{(m)\mu}{}^{0} d^3x = \frac{d}{dt} p_\mu \qquad [32]$$

To evaluate the volume integral of the second term in (30), we make the assumption that $F^\mu{}_\nu$ is roughly constant over the volume of the particle; this is a good approximation for a sufficiently small particle. We then have

$$-F^\mu{}_\nu \int j^\nu d^3x \qquad [33]$$

EXERCISE 6. Show that integration over the volume of the particle gives

$$\int j^\nu d^3x = \int \rho_0 u^\nu d^3x = q u^\nu \sqrt{1 - v^2} \qquad [34]$$

Assume that u^ν is constant over the volume of the particle. (Hint: take the volume contraction into account.)

With the results (32) and (34), we obtain the following for the volume integral of (30):

$$\frac{d}{dt} p_\mu - q \sqrt{1 - v^2} F_{\mu\nu} u^\nu = 0 \qquad [35]$$

which is the same as

$$\frac{d}{d\tau} p_\mu = q F_{\mu\nu} u^\nu \qquad [36]$$

This equation of motion agrees with Eq. (2.126), and completes our program of deriving electrodynamics from nothing (or nearly nothing).*
    It may seem strange that the equation of motion of a particle should emerge as a consequence of the electromagnetic field equations. But this

---

* The equation of motion we have derived is that of a point-particle without electric or magnetic multipole moments. If we drop the assumptions that $F^\mu{}_\nu$ and u^ν are constant over the volume of the particle, then equations for particles with multipoles can be derived.

becomes much less surprising if one realizes that the equation of motion of a particle is no more than a statement about the exchange of momentum between particle and field. Hence the equation of motion cannot be independent of the conservation law for the energy-momentum tensor of the fields. Note that our equations completely determine the force between charged particles. For example, if we have two positive charges at rest, then we can solve Maxwell's equations (Eqs. (16) and (17)) to obtain the **E** field produced by one of these charges, and then calculate the force (Eq. (36)) that this field exerts on the other charge. This force is of course repulsive. The fact that a vector field, such as A_μ, necessarily leads to a repulsive force between like charges, shows that the gravitational field cannot be a vector field. For this, and other reasons, we will use a second-rank tensor field as the carrier of the gravitational interaction.

3.2 THE LINEAR FIELD EQUATIONS FOR GRAVITATION

In order to discover the field equations of gravitation, we proceed in a manner entirely analogous to what we have done for electromagnetism.* The analogy is summarized in Table 3.1. The first thing we must do is decide what the *source* of gravitation is going to be; that is, we ask what quantity plays a role analogous to that of the electromagnetic j^μ. The answer is given by Newton's principle of equivalence according to which the (total) inertial mass of a system gravitates, i.e., the energy gravitates. The source of gravitation must then be the *energy density*. However, it is impossible to construct a Lorentz-invariant theory of gravitation in which the energy density is the *only* source of gravitation. We recall that the energy density T^{00} is only one component of the second-rank tensor $T^{\mu\nu}$. What is pure T^{00} in one reference frame becomes $T'^{\mu\nu} = a^\mu{}_0 a^\nu{}_0 T^{00}$ in another reference frame and, in general, this is not pure T'^{00}. Thus, what is an energy density in one reference frame will be some combination of energy density, energy flux density, and momentum flux density as seen from another reference frame. If the laws are to have the same form in all Lorentz frames, then *all* these quantities must be sources of gravitation, i.e., the tensor $T^{\mu\nu}$ must be the source of gravitation.

Before we proceed, a remark on a possible alternative. We could try to use the invariant trace of the energy-momentum tensor as the source of gravitation. Since

$$T_\mu{}^\mu = T_0{}^0 + T_1{}^1 + T_2{}^2 + T_3{}^3 \qquad [37]$$

we see that $T_\mu{}^\mu$ contains T^{00} and therefore it would seem that the expression (37) satisfies the requirement that the energy density be among the

* This, of course, means that for a start we assume that our spacetime is the flat spacetime of special relativity.

TABLE 3.1 ANALOGY BETWEEN THE ELECTROMAGNETIC AND GRAVITATIONAL FIELD THEORIES

	Electromagnetism	Gravitation (Linear Approximation)
Source of field	j^ν	$T^{\mu\nu}$
Conservation law	$\partial_\nu j^\nu = 0$	$\partial_\nu T^{\mu\nu} = 0$
Field	A^ν	$h^{\mu\nu}$
Field equation	$\partial_\mu\partial^\mu A^\nu - \partial^\nu\partial_\mu A^\mu = 4\pi j^\nu$	$\partial_\lambda\partial^\lambda h^{\mu\nu} - 2\partial_\lambda\partial^{(\nu}h^{\mu)\lambda} + \partial^\mu\partial^\nu h$ $-\eta^{\mu\nu}\partial_\lambda\partial^\lambda h + \eta^{\mu\nu}\partial_\lambda\partial_\sigma h^{\lambda\sigma} = -\kappa T^{\mu\nu}$
Gauge transformation	$A^\mu \to A^\mu + \partial^\mu\Lambda$	$h^{\mu\nu} \to h^{\mu\nu} + \partial^{(\nu}\Lambda^{\mu)}$
Preferred gauge condition	$\partial_\mu A^\mu = 0$	$\partial_\mu(h^{\mu\nu} - \tfrac{1}{2}\eta^{\mu\nu}h) = 0$
Field equation in preferred gauge	$\partial_\mu\partial^\mu A^\nu = 4\pi j^\nu$	$\partial_\lambda\partial^\lambda(h^{\mu\nu} - \tfrac{1}{2}\eta^{\mu\nu}h) = -\kappa T^{\mu\nu}$
Energy-momentum exchange between field and particle	$\partial_\nu T_{(m)\mu}{}^\nu = F_{\mu\nu}j^\nu$	$\partial_\nu T_{(m)\mu}{}^\nu = \dfrac{\kappa}{2} m h_{\alpha\beta,\mu} T_{(m)}{}^{\alpha\beta}$
Equation of motion of particle	$\dfrac{d}{d\tau}p_\mu = qF_{\mu\nu}u^\nu$	$\dfrac{d}{d\tau}P_\mu = \dfrac{\kappa}{2} m h_{\alpha\beta,\mu}u^\alpha u^\beta$
Energy-momentum of particle	$p_\mu = mu_\mu$	$P_\mu = mu_\mu + m\kappa h_{\mu\alpha}u^\alpha$
Proper time interval	$d\tau^2 = \eta_{\alpha\beta}dx^\alpha dx^\beta$	$d\tau^2 = (\eta_{\alpha\beta} + \kappa h_{\alpha\beta})dx^\alpha dx^\beta$

sources of gravitation. The trouble with (37) is that for electromagnetic fields, $T_\mu{}^\mu \equiv 0$ (see Eq. (2.130)). Hence electromagnetic energy would not gravitate, in contradiction with the principle of equivalence.

We must now decide what type of field will be used as carrier of the interaction. Since the source of gravitation is a second-rank, symmetric tensor, arguments analogous to those given in the electromagnetic case tell us that the obvious and natural choice is a second-rank, symmetric tensor field. We will use the notation $h^{\mu\nu}(x)$ for the gravitational field tensor.*

The most general field equation which is linear in $h^{\mu\nu}$, is of second differential order, and contains $T^{\mu\nu}$ as source must have the form

$$\partial_\lambda\partial^\lambda h^{\mu\nu} - a'(\partial_\lambda\partial^\nu h^{\mu\lambda} + \partial_\lambda\partial^\mu h^{\nu\lambda}) + a\partial^\mu\partial^\nu h_\sigma{}^\sigma$$
$$+ b\eta^{\mu\nu}\partial_\lambda\partial^\lambda h_\sigma{}^\sigma + b'\eta^{\mu\nu}\partial_\lambda\partial_\sigma h^{\lambda\sigma} = -\kappa T^{\mu\nu} \quad [38]$$

where a, a', b, b', and κ are constants. We arrive at this form by writing down all possible symmetric tensors that can be constructed out of $h^{\mu\nu}$

---

* As an alternative we might try a third-rank tensor. This gives results equivalent to those obtained by using a second-rank tensor (it corresponds to the possibility of writing Einstein's equations in terms of $\Gamma^\mu_{\alpha\beta}$ rather than the usual $g_{\alpha\beta}$; see Chapter 7).

and no more than two gradient operators ∂^α. The only such tensors, linear in $h^{\mu\nu}$, are the following:

$$\partial_\lambda \partial^\lambda h^{\mu\nu}, \ \partial_\lambda \partial^\nu h^{\mu\lambda} + \partial_\lambda \partial^\mu h^{\nu\lambda}, \ \partial^\mu \partial^\nu h_\sigma{}^\sigma,$$
$$\eta^{\mu\nu} \partial_\lambda \partial^\lambda h_\sigma{}^\sigma, \ \eta^{\mu\nu} \partial_\lambda \partial_\sigma h^{\lambda\sigma}, \ h^{\mu\nu}, \ \eta^{\mu\nu} h_\sigma{}^\sigma$$

The left side of Eq. (38) is a general linear combination of these tensors, except that we have left out $h^{\mu\nu}$ and $\eta^{\mu\nu}h$. The presence of such terms in Eq. (38) would produce an exponential decrease with distance in the static gravitational field (see Eq. (10)); we will assume that the interaction is long range and that therefore such terms are undesirable.*

Our next task will be to determine the values of the constants a, a', b, b'; we will see that our theory determines all these values unambiguously. The constant κ characterizes the strength of the coupling between matter and gravitation, and its value must be determined by experiment. This constant plays a role analogous to that of the electric charge; the only difference is that the electric charge q is included in the definition of j^μ, while the constant κ is written separate from $T^{\mu\nu}$. This difference in notation is sanctified by tradition.

It is convenient to introduce the notation

$$\partial^{(\nu} h^{\mu)\lambda} \equiv \tfrac{1}{2}(\partial^\nu h^{\mu\lambda} + \partial^\mu h^{\nu\lambda}) \tag{39}$$

This is a special case of the general definition given by Eq. (2.55). Furthermore, we will use the notation h for the trace of $h^{\mu\nu}$:

$$h \equiv h_\sigma{}^\sigma$$

With this, (38) becomes

$$\partial_\lambda \partial^\lambda h^{\mu\nu} - 2a' \partial_\lambda \partial^{(\nu} h^{\mu)\lambda} + a \partial^\mu \partial^\nu h + b \eta^{\mu\nu} \partial_\lambda \partial^\lambda h$$
$$+ b' \, \eta^{\mu\nu} \partial_\lambda \partial_\sigma h^{\lambda\sigma} = - \kappa T^{\mu\nu} \tag{40}$$

If we operate with ∂_μ on both sides of this equation we obtain

$$\partial_\lambda \partial^\lambda \partial_\mu h^{\mu\nu} - a' \partial_\lambda \partial_\mu \partial^\nu h^{\mu\lambda} - a' \partial_\lambda \partial_\mu \partial^\mu h^{\nu\lambda}$$
$$+ a \partial_\mu \partial^\mu \partial^\nu h + b \partial^\nu \partial_\lambda \partial^\lambda h + b' \partial^\nu \partial_\lambda \partial_\sigma h^{\lambda\sigma} = - \kappa \partial_\mu T^{\mu\nu} \tag{41}$$

The right side of this equation vanishes because of the conservation law for the energy-momentum tensor (Eq. (2.79)). In analogy with the electromagnetic case, we then demand that the left side vanish identically.

---

* Incidentally, we could attempt to generalize (38) by using $\kappa T^{\mu\nu} + \kappa' \eta^{\mu\nu} T$ on the right side. But this amounts to no more than a redefinition of the constants a, a', b, b'.

The constants should therefore satisfy the equations

$$-a' + b' = 0$$
$$1 - a' = 0$$
$$a + b = 0 \qquad [42]$$

Hence $a' = 1$, $b' = 1$, and $b = -a$. Our field equation reduces to

$$\partial_\lambda \partial^\lambda h^{\mu\nu} - 2\partial_\lambda \partial^{(\nu} h^{\mu)\lambda} + a\partial^\mu \partial^\nu h - a\eta^{\mu\nu} \partial_\lambda \partial^\lambda h$$
$$+ \eta^{\mu\nu} \partial_\lambda \partial_\sigma h^{\lambda\sigma} = - \kappa T^{\mu\nu} \qquad [43]$$

The value of the constant a is of no physical relevance. Different values of a merely correspond to different ways of describing the same gravitational theory by means of different, but related, tensors. Thus, if instead of using the tensor $h^{\mu\nu}$ to describe the gravitational field we use a new tensor $\bar{h}^{\mu\nu}$ related to the former by

$$h^{\mu\nu} = \bar{h}^{\mu\nu} - C\eta^{\mu\nu}\bar{h} \qquad [44]$$

(where C is a constant), then the field equation for $\bar{h}^{\mu\nu}$ has exactly the same form as (43), but a is replaced by $\bar{a}$:

$$\bar{a} = a(1 - 4C) + 2C \qquad [45]$$

EXERCISE 7. Substitute (44) into (43) and verify (45).

In particular, if we choose

$$C = (1 - a)/(2 - 4a) \qquad [46]$$

then $\bar{a} = 1$ and we obtain the field equation

$$\partial_\lambda \partial^\lambda \bar{h}^{\mu\nu} - 2\partial_\lambda \partial^{(\nu} \bar{h}^{\mu)\lambda} + \partial^\mu \partial^\nu \bar{h} - \eta^{\mu\nu} \partial_\lambda \partial^\lambda \bar{h}$$
$$+ \eta^{\mu\nu} \partial_\lambda \partial_\sigma \bar{h}^{\lambda\sigma} = - \kappa T^{\mu\nu} \qquad [47]$$

Note that $a = \frac{1}{2}$ creates trouble in Eq. (46). However, this troublesome value of a must in any case be excluded, because for $a = \frac{1}{2}$ our field equation (43) contains insufficient information for determining the field.

EXERCISE 8. Show that if $a = \frac{1}{2}$, then

$$h^{\mu\nu} + \eta^{\mu\nu} f$$

is a solution of Eq. (43) whenever $h^{\mu\nu}$ is a solution. The function f is arbitrary.

Since the bars in Eq. (47) serve no further purpose, we omit them and write our final equation as

$$\partial_\lambda \partial^\lambda h^{\mu\nu} - 2\partial_\lambda \partial^{(\nu} h^{\mu)\lambda} + \partial^\mu \partial^\nu h - \eta^{\mu\nu} \partial_\lambda \partial^\lambda h$$
$$+ \eta^{\mu\nu} \partial_\lambda \partial_\sigma h^{\lambda\sigma} = - \kappa T^{\mu\nu} \quad [48]$$

This is the relativistic field equation for gravitation in the linear approximation. This equation plays a role analogous to that of the Maxwell equations: Eq. (48) determines the gravitational field generated by a given matter distribution.

The assumption that the gravitational field is entirely described by a single tensor field entered in a crucial way into our construction of the field equation. This is the simplest of all possible assumptions and, since all the available experimental data is in agreement with the tensor theory of gravitation, it would be contrary to Newton's First Rule (see Section 1.1) to assume anything else. If, nevertheless, one were to hypothesize that besides the tensor field there exist one or several *extra* gravitational scalar, vector, or tensor fields, then one could construct no end of endlessly complicated theories. Among these alternatives the least complicated is the scalar-tensor theory of Jordan[21] and Brans and Dicke[22] which contains an extra scalar field; qualitatively, this scalar field has the effect of making the gravitational constant dependent on position. The experimental evidence (see Section 3.8) speaks against the scalar-tensor theory; we will therefore not bother with it.

In the electromagnetic case we were able to obtain a simplified field equation (Eq. (23)) by taking advantage of gauge invariance. We can also do this in the gravitational case. It is easy to show that Eq. (48) is invariant under the gauge transformation

$$h^{\mu\nu} \rightarrow h^{\mu\nu} + \partial^{(\nu} \Lambda^{\mu)} \tag{49}$$

where $\Lambda^\mu(x)$ is an arbitrary vector field.

EXERCISE 9. Prove this.

Exactly as in the electromagnetic case, gauge invariance implies that the field equations do not determine the field uniquely, but the ambiguity in the field is not serious and leads to no observable consequences.* We find it convenient to (partially) eliminate the ambiguity by imposing the following gauge condition on the fields:

$$\partial_\mu (h^{\mu\nu} - \tfrac{1}{2} \eta^{\mu\nu} h) = 0 \tag{50}$$

---

* Further discussion of the gauge transformation will be given in Section 7.1.

This is known as the *Hilbert condition*. If we are given a field $h^{\mu\nu}$ which *does not* satisfy (50), then, by a gauge transformation, we can always obtain a new field $\bar{h}^{\mu\nu}$

$$\bar{h}^{\mu\nu} = h^{\mu\nu} + \partial^{(\mu}\Lambda^{\nu)} \qquad [51]$$

such that the latter *does* satisfy (50).

EXERCISE 10. Show that if Λ^{ν} is a solution of the differential equation

$$\tfrac{1}{2}\partial_{\mu}\partial^{\mu}\Lambda^{\nu} = -\partial_{\mu}h^{\mu\nu} + \tfrac{1}{2}\partial^{\nu}h \qquad [52]$$

then $\bar{h}^{\mu\nu}$ will satisfy (50).

We now assume that the gauge condition holds; then the field equation simplifies to

$$\partial_{\lambda}\partial^{\lambda}(h^{\mu\nu} - \tfrac{1}{2}\eta^{\mu\nu}h) = -\kappa T^{\mu\nu} \qquad [53]$$

EXERCISE 11. Derive this from (48) and (50).

It is convenient to define a new field variable

$$\phi^{\mu\nu} = h^{\mu\nu} - \tfrac{1}{2}\eta^{\mu\nu}h \qquad [54]$$

In terms of $\phi^{\mu\nu}$, the field equation becomes

$$\partial_{\lambda}\partial^{\lambda}\phi^{\mu\nu} = -\kappa T^{\mu\nu} \qquad [55]$$

with the gauge condition

$$\partial_{\mu}\phi^{\mu\nu} = 0 \qquad [56]$$

EXERCISE 12. Show that (54) implies

$$h^{\mu\nu} = \phi^{\mu\nu} - \tfrac{1}{2}\eta^{\mu\nu}\phi \qquad [57]$$

Eq. (55) expresses the field equation (48) in a particularly simple and convenient form. The remainder of this chapter and the next chapter are largely concerned with the investigation of different solutions of Eq. (55). Thus, in Section 3.7 we will find the field surrounding a static, spherically symmetric mass, and in Section 3.10 the field surrounding a rotating mass. In Chapter 4, we will look at the wave solutions of our field equation and investigate the emission of gravitational radiation.

However, it does us little good to calculate the field $h^{\mu\nu}$ (or $\phi^{\mu\nu}$) generated by some matter distribution if we do not know how this field acts on a particle that is placed in it. It is imperative that we try to discover the equation of motion of a particle in a given gravitational field.

3.3   THE INTERACTION OF GRAVITATION AND MATTER

In writing the field equation (48) we have assumed that the quantity $T^{\mu\nu}$ is the energy-momentum tensor of *matter*. In order to obtain a linear field equation we have left out the effect of the gravitational field upon itself. Because of this omission, our linear field equation has several (related) defects: (1) According to (48) matter acts on the gravitational field (changes the fields), but there is no mutual action of gravitational fields *on* matter; that is, the gravitational field can acquire energy-momentum from matter, but nevertheless the energy-momentum of matter is conserved ($\partial_\nu T^{\mu\nu} = 0$). This is an inconsistency. (2) Gravitational energy does not act as source of gravitation, in contradiction to the principle of equivalence. Thus, although Eq. (48) may be a fair approximation in the case of weak gravitational fields, it cannot be an exact equation.

The obvious way to correct for our sin of omission is to include the energy-momentum tensor of the gravitational field in $T^{\mu\nu}$. This means that we take for the quantity $T^{\mu\nu}$ the *total energy-momentum tensor* of matter plus gravitation:

$$T^{\mu\nu} = T^{\mu\nu}_{(m)} + t^{\mu\nu} \qquad [58]$$

Here $T^{\mu\nu}_{(m)}$ and $t^{\mu\nu}$ are, respectively, the energy-momentum tensors of matter and gravitation. We assume that the interaction energy of matter and gravitation is always *included in* $T^{\mu\nu}_{(m)}$; this is a reasonable convention since the interaction energy-density will only differ from zero at those places where there is matter.*

Our field equation now becomes

$$\partial_\lambda \partial^\lambda \phi^{\mu\nu} = - \kappa(T^{\mu\nu}_{(m)} + t^{\mu\nu}) \qquad [59]$$

Let us operate with ∂_ν on this equation; the left side is zero because of the gauge condition and therefore

$$0 = \partial_\nu(T^{\mu\nu}_{(m)} + t^{\mu\nu}) \qquad [60]$$

This simply expresses the conservation of the total energy-momentum and shows that the modified field equation is free from the inconsistencies

---

* For example, consider the energy density $(\nabla\Phi)^2/8\pi G + \rho\Phi$ given by the Newtonian approximation (see Eq. (1.12)). According to the above convention, the first term would be identified with t^{00}, but the second term would be regarded (together with the density of kinetic and rest mass energy) as part of $T^{00}_{(m)}$.

mentioned at the beginning of this section. Unfortunately, Eq. (60) does not get us very far because we do not know what $t^{\mu\nu}$ is. We have already mentioned that there exists a general procedure for constructing the energy-momentum tensor of a field, if the field equation is known. The trouble with (59) is that the field equation is *not known* explicitly unless we already have some expression for $t^{\mu\nu}$ in terms of ϕ and its derivatives. In Chapter 7 we will find ways of solving, or rather bypassing, this diffi-culty. For now we will adopt a technique of successive approximations.

We begin by writing down the field equation for a *free* gravitational field,

$$\partial_\lambda \partial^\lambda \phi^{\mu\nu} = 0 \qquad [61]$$

Given this equation, we can make use of the canonical prescription to obtain the energy-momentum tensor that goes with the field described by (61). It turns out that the energy-momentum tensor that corresponds to such a free gravitational field is (using the comma notation for deriva-tives)*

$$t_{(1)}^{\mu\nu} = \tfrac{1}{4}[2\phi^{\alpha\beta,\mu}\,\phi_{\alpha\beta}{}^{,\nu} - \phi^{,\mu}\phi^{,\nu} - \eta^{\mu\nu}(\phi^{\alpha\beta,\sigma}\phi_{\alpha\beta,\sigma} - \tfrac{1}{2}\phi_{,\sigma}\phi^{,\sigma})] \qquad [62]$$

where the subscript $_{(1)}$ serves to indicate that $t_{(1)}^{\mu\nu}$ is only a first approxima-tion to the exact energy-momentum tensor $t^{\mu\nu}$. We will leave the deriva-tion of this result to Appendix 1, and only note that (62) is a quite plausi-ble expression since it has all the desirable properties that we listed in the electromagnetic case.

EXERCISE 13. Show that, as a consequence of Eq. (61),

$$\partial_\nu t_{(1)}^{\mu\nu} = 0 \qquad [63]$$

for a *free* gravitational field.

If we substitute (62) into (59) we obtain a new field equation which, although not exact, includes at least *some* of the nonlinearities of the gravitational field:

$$\partial_\lambda \partial^\lambda \phi^{\mu\nu} = -\kappa(T_{(m)}^{\mu\nu} + t_{(1)}^{\mu\nu}) \qquad [64]$$

The corresponding equation for the conservation of the total energy-momentum is

$$\partial_\nu(T_{(m)}^{\mu\nu} + t_{(1)}^{\mu\nu}) = 0 \qquad [65]$$

---

* This assumes the gauge condition $\partial_\mu \phi^{\mu\nu} = 0$.

We will use this to derive the equation of motion. The second term on the right side can be calculated explicitly by means of (62) and (64):

$$\partial_\nu t^{\mu\nu}_{(1)} = \tfrac{1}{4}[2\phi^{\alpha\beta,\mu}\phi_{\alpha\beta}{}^{,\nu}{}_{,\nu} - \phi^{,\mu}\phi^{,\nu}{}_{,\nu}]$$
$$= -\frac{\kappa}{4}[2\phi^{\alpha\beta,\mu}T_{(m)\alpha\beta} - \phi^{,\mu}T_{(m)}] + \ldots \qquad [66]$$

In Eq. (66) the dots stand for terms cubic in ϕ which we omit in this approximation. We now have

$$\partial_\nu T^{\mu\nu}_{(m)} - \frac{\kappa}{2} T_{(m)\alpha\beta}(\phi^{\alpha\beta,\mu} - \tfrac{1}{2}\eta^{\alpha\beta}\phi^{,\mu}) = 0 \qquad [67]$$

which can also be written as

$$\partial_\nu T_{(m)\mu}{}^\nu - \frac{\kappa}{2} h_{\alpha\beta,\mu} T_{(m)}{}^{\alpha\beta} = 0 \qquad [68]$$

This equation is the gravitational analog (see Table 3.1) of the electromagnetic equation (30).

We will next try to derive the equation of motion by the same method as used in the electromagnetic case. One thing that must be kept in mind is that the quantities $T_{(m)\mu}{}^\nu$ in Eqs. (30) and (68) are *not* the same; in the gravitational case $T_{(m)\mu}{}^\nu$ includes some gravitational interaction energy, and this will make our calculations a bit more difficult. Strictly speaking, we should also have worried about the possibility of interaction energy in the electromagnetic case, but it can be proved that there is no such interaction energy. How one could set up such a proof will become clear from what follows (see also Problem 5).

We begin the derivation by integrating the expression (68) over the volume of the particle. The first term gives

$$\frac{d}{dt}\int T_{(m)\mu}{}^0 d^3x \qquad [69]$$

In the electromagnetic case we knew* that $T_{(m)\mu}{}^\nu = \rho_0 u_\mu u^\nu$, and then the volume integral appearing in (69) is simply $p_\mu = mu_\mu$. In the present gravitational case, $T_{(m)\mu}{}^\nu$ contains some interaction part with an as yet unknown dependence on the gravitational field variables; we therefore cannot evaluate the volume integral explicitly. For convenience, let us designate the integral by P_μ and write

$$\frac{d}{dt}\int T_{(m)\mu}{}^0 d^3x = \frac{d}{dt} P_\mu \qquad [70]$$

---

* Or, to be precise, we assumed.

To evaluate the volume integral of the second term in Eq. (68), we make the assumption that $h_{\alpha\beta,\mu}$ is roughly constant over the volume of the particle. We are then left with

$$-\frac{\kappa}{2} h_{\alpha\beta,\mu} \int T_{(m)}{}^{\alpha\beta} d^3x \qquad [71]$$

In order to evaluate this, we note that $T_{(m)}{}^{\alpha\beta}$ has the form

$$T_{(m)}{}^{\alpha\beta} = \rho_0 u^\alpha u^\beta + \dots \qquad [72]$$

where the dots stand for terms that represent the interaction energy between matter and gravitation. These terms will typically have the form $\sim h u^\alpha u^\beta$ and therefore they contribute to (71) only a small correction of order h^2; we will ignore these interaction terms in the evaluation of (71).*

EXERCISE 14. Show that integration over the volume of the particle gives

$$\int \rho_0 u^\alpha u^\beta d^3x = m u^\alpha u^\beta \sqrt{1 - v^2} \qquad [73]$$

Therefore (71) becomes

$$-\frac{\kappa}{2} \sqrt{1 - v^2} \, m h_{\alpha\beta,\mu} u^\alpha u^\beta \qquad [74]$$

With the results (70) and (74) we obtain the following for the volume integral of (68):

$$\frac{d}{dt} P_\mu - \frac{\kappa}{2} \sqrt{1 - v^2} \, m h_{\alpha\beta,\mu} u^\alpha u^\beta = 0 \qquad [75]$$

that is

$$\frac{d}{d\tau} P_\mu - \frac{\kappa}{2} m h_{\alpha\beta,\mu} u^\alpha u^\beta = 0 \qquad [76]$$

This equation has the form we expect for an equation of motion. There is only one difficulty: we do not know how P_μ is related to the particle four-velocity. We might expect that P_μ takes the form $P_\mu = m u_\mu + \dots$ where the dots stand for an extra term involving both the four-velocity and the gravitational field. Our next concern will be to find a precise expression for this extra term.

---

* But we *cannot* ignore these terms in the calculation of (69) because there they contribute corrections of order h, rather than h^2.

3.4 THE VARIATIONAL PRINCIPLE AND THE EQUATION OF MOTION

The next step in our derivation of the equation of motion hinges on the observation that (76) has the form of an Euler-Lagrange equation. To see what this means, let us recall Hamilton's variational principle according to which the motion of a particle is such as to make the action between two spacetime points P_1, P_2 stationary. Thus, the action evaluated along the actual worldline of the particle is the same as the action evaluated along any nearby worldline with the same endpoints P_1, P_2 (see Fig. 3.1).

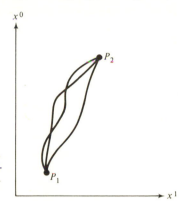

Fig. 3.1 *Worldlines between two spacetime points. The lowest line represents the actual worldline of the particle; the other lines are obtained by variation.*

To formulate the condition of stationary action, we need to introduce a parameter that labels the points along the worldlines. In the relativistic case it is convenient to use the proper time of the *actual* worldline as parameter; we will use the usual notation τ for this parameter.* The action can then be written as an integral over this proper time,

$$I = \int_{\tau_1}^{\tau_2} L\,d\tau \qquad\qquad [77]$$

where the Lagrangian L is a function of the coordinates x^μ and the corresponding "velocities" $u^\mu = dx^\mu/d\tau$.

   If we vary the actual worldline at the point $x_\mu(\tau)$ by an amount $\delta x^\mu(\tau)$, then the velocity will vary by $\delta u^\mu = (d/d\tau)\,\delta x^\mu$. The change in the action will therefore be

---

* It is to be emphasized that although τ is the proper time measured along the actual worldline, it is not the proper time along any of the other nearby worldlines that are considered in the variational principle. Along these other worldlines, τ is to be regarded as a purely mathematical parameter. The reason why we cannot parametrize the other worldlines by their own proper time is that the parameter used in the formulation of the variational equations must range over exactly the same set of values along all worldlines between P_1 and P_2.

$$\delta I = \int \left(\frac{\partial L}{\partial x^{\mu}} \, \delta x^{\mu} + \frac{\partial L}{\partial u^{\mu}} \, \delta u_{\mu} \right) d\tau$$

$$= \int \left(\frac{\partial L}{\partial x^{\mu}} \, \delta x^{\mu} + \frac{\partial L}{\partial u^{\mu}} \frac{d}{d\tau} \, \delta x^{\mu} \right) d\tau$$

The second term of the integrand can be integrated by parts. Since $\delta x^{\mu} = 0$ at the endpoints of the integration interval, we obtain

$$\delta I = \int \left(\frac{\partial L}{\partial x^{\mu}} \, \delta x^{\mu} - \frac{d}{d\tau} \frac{\partial L}{\partial u^{\mu}} \, \delta x^{\mu} \right) d\tau$$

If we are to have $\delta I = 0$ for arbitrary variations δx^{μ}, we need

$$\frac{d}{d\tau} \frac{\partial L}{\partial u^{\mu}} - \frac{\partial L}{\partial x^{\mu}} = 0 \qquad\qquad [78]$$

This is the *Euler-Lagrange equation.* We can also write this equation as

$$\frac{d}{d\tau} \, \pi_{\mu} - \frac{\partial L}{\partial x^{\mu}} = 0 \qquad\qquad [79]$$

where

$$\pi_{\mu} \equiv \frac{\partial L}{\partial u^{\mu}} \qquad\qquad [80]$$

is the momentum canonically conjugate to x^{μ}.

By way of illustration, let us consider the case of a free particle. The equation of motion of a free particle is

$$\frac{d}{d\tau} \, (mu^{\mu}) = 0 \qquad\qquad [81]$$

This equation can be obtained from the Lagrangian

$$L_0 = \frac{1}{2} \, m \, \frac{dx^{\mu}}{d\tau} \frac{dx_{\mu}}{d\tau} \qquad\qquad [82]$$

$$= \frac{1}{2} \, m \, u^{\mu} u_{\mu}$$

EXERCISE 15. Show that (82) gives the equation of motion of a free particle and show that the corresponding canonical momentum is mu_{μ}.

Note that the condition $(dx^{\mu}/d\tau)(dx_{\mu}/d\tau) = 1$ must not be inserted into (82). This condition is only valid along the actual worldline of the particle where the parameter τ coincides with the proper time. The other nearby worldlines that must be considered in the variational principle have their own proper time distinct from τ and hence along them $(dx^\mu/d\tau)(dx^\mu/d\tau) \neq 1$. It is obvious that if we carelessly insert $(dx^\mu/d\tau)(dx^\mu/d\tau) = 1$ into (82), we obtain $L_0 = \frac{1}{2} m$ and then no equation of motion results.

The equation of motion of a free particle can also be obtained from a somewhat different variational principle. We can take

$$I = \int_{P_1}^{P_2} \sqrt{dx^\mu dx_\mu} \qquad [83]$$

as our action. If this integral is evaluated along a given worldline, it gives the proper time that elapses between P_1 and P_2 along that worldline. Again, let us introduce the proper time τ of the actual worldline as a parameter. We can then write the action as

$$I = \int_{\tau_1}^{\tau_2} \sqrt{\frac{dx^\mu}{d\tau} \frac{dx_\mu}{d\tau}}\, d\tau$$

and the Lagrangian as

$$L_0 = \sqrt{u^\mu u_\mu} \qquad [84]$$

This gives an equation of motion

$$\frac{d}{d\tau}\left(\frac{u_\mu}{\sqrt{u^\nu u_\nu}}\right) = 0$$

Since this equation refers to the actual worldline, it is valid to insert the condition $u^\nu u_\nu = 1$ into it. We then recover the equation of motion (81) of a free particle. Thus, the motion of a free particle is such as to make (83) stationary; that is, along the worldline of a free particle the proper time is stationary. It is easy to show that the proper time is actually at a maximum.

EXERCISE 16. Show this. (Hint: See Problem 2.8.)

This concludes our review of the variational principle; we now return to our discussion of the equation of motion of a particle in a gravitational field. A glance at Eq. (76) shows that it has the form of an Euler-Lagrange equation. Comparison shows that Eqs. (76) and (78) will be identical if

$$\frac{\partial L}{\partial x^\mu} = \frac{\kappa}{2}\, m h_{\alpha\beta,\mu}\, u^\alpha u^\beta \qquad [85]$$

and

$$\frac{\partial L}{\partial u^\mu} = P_\mu \tag{86}$$

We can regard (85) as a differential equation for L. If we integrate this equation with respect to x^μ, we obtain

$$L(x,u) = \frac{\kappa}{2} m h_{\alpha\beta} u^\alpha u^\beta + f(u) \tag{87}$$

where $f(u)$ is an arbitrary function of the velocity u^μ, but not a function of the coordinate x^μ. The function f is determined by the condition that in the limit $h_{\alpha\beta} = 0$, the Lagrangian must reduce to that of a *free* particle. The latter is given by Eq. (82),

$$L_0 = \tfrac{1}{2} m \eta_{\alpha\beta} u^\alpha u^\beta \tag{88}$$

and therefore

$$L = \tfrac{1}{2} m (\eta_{\alpha\beta} + \kappa h_{\alpha\beta}) u^\alpha u^\beta \tag{89}$$

According to Eq. (86), the quantity P_μ is then

$$P_\mu = \frac{\partial L}{\partial u^\mu} = m u_\mu + m \kappa h_{\mu\alpha} u^\alpha \tag{90}$$

With this expression for P_μ we can now return to Eq. (76) and obtain a differential equation for the four-velocity (the masses cancel in this equation):

$$\frac{d}{d\tau} (u_\mu + \kappa h_{\mu\alpha} u^\alpha) - \frac{\kappa}{2} h_{\alpha\beta,\mu} u^\alpha u^\beta = 0 \tag{91}$$

This is the equation of motion for a particle in a (weak) gravitational field. Note that according to (91) the motion of a particle in a gravitational field is independent of the mass of the particle, in agreement with the Galileo principle.

We can rewrite Eq. (91) in several ways. For example, we can use

$$\frac{d}{d\tau} h_{\mu\alpha} u^\alpha = h_{\mu\alpha} \frac{d}{d\tau} u^\alpha + h_{\mu\alpha,\beta} u^\alpha u^\beta \tag{92}$$

Since the acceleration of the particle is of order h, the first term on the right side of Eq. (92) is of order h^2 and can be neglected. Therefore (91) is approximately equivalent to

$$\frac{du_\mu}{d\tau} + (\kappa h_{\mu\alpha,\beta} u^\alpha u^\beta - \frac{\kappa}{2} h_{\alpha\beta,\mu} u^\alpha u^\beta) = 0 \qquad [93]$$

The quantity P_0 was originally defined as the volume integral of the energy density associated with the particle (see Eq. (70)). Hence P_0 is the (kinetic plus potential) energy of the particle in the gravitational field:

$$P_0 = mu_0 + m\kappa h_{0\alpha} u^\alpha \qquad [94]$$

According to Eq. (91), this energy is a constant of the motion provided the gravitational field is time independent ($h_{\alpha\beta,0} = 0$). In a time-dependent gravitational field the particle energy is not constant.

EXERCISE 17. Show that if

$$T_{(m)\mu}{}^\nu = \rho_0 u_\mu u^\nu + \kappa \rho_0 h_{\mu\alpha} u^\alpha u^\nu \qquad [95]$$

then the integral $\int T_{(m)\mu}{}^0 d^3x$ agrees with the expression given by Eq. (90). The second term on the right side of Eq. (95) is what we designated by dots in Eq. (72).

It is a general theorem of Lagrangian mechanics that

$$u^\mu \frac{\partial L}{\partial u^\mu} - L \qquad [96]$$

is a constant of the motion regardless of any time dependence of the potential. In our case, this integral of the motion takes the form

$$\tfrac{1}{2} m(u^\mu u_\mu + \kappa h_{\mu\nu} u^\mu u^\nu) = \text{constant} \qquad [97]$$

EXERCISE 18. Show that the left side of Eq. (97) agrees with (96) and use the Lagrangian equation of motion to show that the derivative $d/d\tau$ applied to (96) or (97) gives zero.

The value of the constant in Eq. (97) is $\tfrac{1}{2}m$. This becomes obvious if we suppose that the particle was initially outside of the gravitational field (where $h_{\mu\nu} = 0$ and $u^\mu u_\mu = 1$) or, equivalently, that the gravitational field was initially zero and reached its present value by a gradual increase in time. Hence Eq. (97) can also be written

$$1 = u^\mu u_\mu + \kappa h_{\mu\nu} u^\mu u^\nu$$

or, with $u^\mu = dx^\mu/d\tau$,

$$d\tau^2 = (\eta_{\mu\nu} + \kappa h_{\mu\nu}) dx^\mu dx^\nu \qquad [98]$$

This equation has some remarkable consequences that will be discussed in the next section. For now, we only point out that in the evaluation of Eq. (94) we must use the expression (98) for $d\tau$, rather than the familiar $d\tau^2 = dx_\mu dx^\mu$.* From Eq. (98) it follows that

$$u_0 = \frac{dt}{d\tau} = \frac{dt}{\left[\left(1 - v^2 + \kappa h_{\mu\nu}\frac{dx^\mu}{dt}\frac{dx^\nu}{dt}\right)dt^2\right]^{1/2}}$$

$$\simeq \frac{1}{\sqrt{1-v^2}} - \frac{1}{2}\frac{1}{(1-v^2)^{3/2}}\kappa h_{\mu\nu}\frac{dx^\mu}{dt}\frac{dx^\nu}{dt}$$

so that

$$P_0 \simeq \frac{m}{\sqrt{1-v^2}} - \frac{1}{2}\frac{m}{(1-v^2)^{3/2}}\kappa h_{\mu\nu}\frac{dx^\mu}{dt}\frac{dx^\nu}{dt} + m\kappa h_{0\alpha}u^\alpha \qquad [99]$$

The first term on the right side has the form of the usual kinetic energy; the other terms represent an interaction energy (potential energy).

3.5 THE NONRELATIVISTIC LIMIT AND NEWTON'S THEORY

Our linear gravitational theory will be a good approximation if the gravitational field is weak. We can express this condition more precisely if we compare the linear and nonlinear terms in Eq. (64). The nonlinear terms are contained in $\kappa t^{\mu\nu}_{(1)}$ on the right side; apart from derivatives and tensor indices, these terms are typically of the form $\kappa\phi^{\mu\nu}\phi_{\mu\nu}$ (see Eq. (62)). The linear terms are on the left side of Eq. (64); apart from derivatives they are of the form $\phi^{\mu\nu}$. These two kinds of terms, therefore, typically differ by a factor $\kappa\phi^{\mu\nu}$. The nonlinearities are then insignificant if

$$\kappa\phi^{\mu\nu} << 1$$

As we will see in Section 3.7, the field surrounding a static mass is roughly $\phi^{\mu\nu} \sim GM/rc^2$ (see Eq. (125)), and hence the above condition becomes

$$\frac{GM}{rc^2} << 1$$

---

* Note that in order to derive Eq. (76) from (75) we used $d\tau^2 = dx_\mu dx^\mu = (1-v^2)dt^2$. We now recognize that this is only an approximation; it was a satisfactory approximation in that context.

This is the same as the condition for the validity of the Newtonian theory (see Section 1.1), that is, both the Newtonian and the linear relativistic theories hinge on the assumption that the field is weak. However, Newtonian theory also requires the assumption that the velocities are low ($v \ll c$); this assumption is not made in the relativistic theory. Thus the relativistic theory is an improvement because we can treat time-dependent fields, and we can treat the motion of relativistic particles in (weak) gravitational fields. For example, we can calculate the motion of light (photons) in the field of the sun; this cannot be done in Newtonian theory.

We will now prove that in the special case of static fields and low velocities, Newton's gravitational theory emerges, as it should, from the relativistic theory. In the nonrelativistic limit $v^k \equiv dx^k/dt \ll 1$ and we can neglect terms of order v^2 and vh. In the first term of Eq. (93), we can then approximate $du_k/d\tau = dv_k/dt$, and in the term in parentheses we can approximate $u^0 = 1$, $u^k = 0$. This leads to

$$\frac{d}{dt} v^k = -\kappa(h_{0,0}^k - \tfrac{1}{2} h_{00}{}^{,k})$$

or, if we assume that the field is time independent,

$$\frac{d}{dt} v^k = \tfrac{1}{2}\kappa \, \partial^k h_{00}$$

We can compare this with the equation of motion given by Newton's theory of gravitation (see Eq. (1.20))

$$\frac{d}{dt} v^k = \partial^k \Phi$$

Hence we see that Newton's equation of motion is a consequence of our theory in the nonrelativistic limit. The Newtonian potential is proportional to h_{00}:

$$\Phi = \tfrac{1}{2}\kappa h_{00} \qquad\qquad [100]$$

To complete the proof, we must show that the field equation of Newton's theory,

$$\nabla^2\Phi = 4\pi G\rho \qquad\qquad [101]$$

is a consequence of our field equations. In the nonrelativistic limit, the dominant component of the energy-momentum tensor is T^{00} since this is the component that contains the rest mass density. We therefore have

$$T^{00} \simeq \rho$$

$$T^{k\mu} \simeq 0 \qquad [102]$$

and the field equation (55) gives, for a time-independent field,

$$\nabla^2 \phi^{00} \simeq \kappa\rho \qquad [103]$$

$$\nabla^2 \phi^{k\mu} \simeq 0 \qquad [104]$$

Combining (103) and (104), we see that $\phi \equiv \phi_\mu{}^\mu$ satisfies the differential equation

$$\nabla^2 \phi = \kappa\rho \qquad [105]$$

and hence

$$\nabla^2 h_{00} = \nabla^2(\phi_{00} - \tfrac{1}{2}\phi) = \tfrac{1}{2}\kappa\rho \qquad [106]$$

Eq. (106) has exactly the same form as Eq. (101) and shows that the latter is a consequence of the former.*

Since, in the nonrelativistic limit, the equations of our theory are the same as those of Newtonian theory, the signs of the gravitational force acting between two particles are the same in both theories. Interaction by means of a tensor field automatically gives an attractive force, just as interaction by a vector field automatically gives a repulsive force (between like charges).

To obtain the numerical value of the constant κ, substitute (100) into (106):

$$\frac{2}{\kappa}\nabla^2\Phi = \frac{1}{2}\kappa\rho \qquad [107]$$

Comparison of (107) and (101) gives**

$$\kappa = \sqrt{16\pi G} \qquad [108]$$

In cgs units this becomes $\kappa = \sqrt{16\pi G/c^4} = 2.04 \times 10^{-24}$ sec/(g cm)$^{1/2}$.

---

* When seeking out the values of a, a', b, b' in Section 3.2 we could have saved ourselves some labor by demanding that our equations approach those of Newtonian theory in the nonrelativistic limit. However, it is very satisfying that this correspondence should emerge spontaneously.

** We could equally well take $\kappa = -\sqrt{16\pi G}$. The force between two particles is attractive, no matter what the sign of κ.

3.6 THE GEOMETRIC INTERPRETATION

According to Eq. (98), the proper time interval for a particle moving in a gravitational field is given by

$$d\tau^2 = (\eta_{\mu\nu} + \kappa h_{\mu\nu})dx^\mu dx^\nu \qquad [109]$$

This differs from the expression

$$d\tau^2 = \eta_{\mu\nu}dx^\mu dx^\nu \qquad [110]$$

that is valid in the (flat) relativistic spacetime. In fact, (109) is the spacetime interval for a *curved spacetime*.

In Chapter 6 we will engage in a detailed discussion of curved spacetime. For now we only briefly state that in a curved spacetime, the constant Minkowski tensor $\eta_{\mu\nu}$ is replaced by a function of space and time and the spacetime interval becomes

$$ds^2 = g_{\mu\nu}dx^\mu dx^\nu \qquad [111]$$

with $g_{\mu\nu} = g_{\mu\nu}(x)$. The expression (111) determines the spacetime distances. For example, a coordinate displacement dx^1 along the x-axis has a length $\sqrt{-g_{11}}dx^1$, that is, the measured distance differs from dx^1 by a factor $\sqrt{-g_{11}}$. Likewise a coordinate time (t-time) displacement dx^0 has a duration $\sqrt{g_{00}}dx^0$ when measured by the proper time of a clock. Because $g_{\mu\nu}$ determines the spacetime distances, $g_{\mu\nu}$ is called the *metric tensor* of the curved spacetime. The exact connection between $g_{\mu\nu}$ and the curvature of the spacetime will be given in Chapter 6.*

Obviously, the interval (109) is that of a curved spacetime with a metric tensor

$$g_{\mu\nu} = \eta_{\mu\nu} + \kappa h_{\mu\nu} \qquad [112]$$

Let us now take another careful look at our equation of motion. In terms of $g_{\mu\nu}$, Eq. (91) can be written as

$$\frac{d}{d\tau}\left(g_{\mu\nu}\frac{dx^\nu}{d\tau}\right) - \tfrac{1}{2}g_{\alpha\beta,\mu}\frac{dx^\alpha}{d\tau}\frac{dx^\beta}{d\tau} = 0 \qquad [113]$$

---

* Under exceptional circumstances the interval (111) may merely represent flat spacetime in curvilinear coordinates. To check this possibility it is necessary to calculate the curvature (see Chapter 6).

This equation has a very interesting interpretation: it can be regarded as the equation of motion of a *free* particle in the *curved* spacetime; this means we can replace the effects of the gravitational field by purely geometrical effects associated with motion in a curved geometry. To see how this comes about, we note that in the curved spacetime we can define a *geodesic* curve (or "straightest worldline") by the condition that the invariant length between two points remain stationary with respect to arbitrary (but small) variations of the curve; this can be expressed by the variational principle:

$$\delta \int ds = 0 \qquad [114]$$

This definition is motivated by what is known about a straight worldline in flat spacetime: it is the curve of maximum length connecting two points (see Exercise 16).

To obtain an explicit differential equation for the geodesic, we note that the proper time measured along a worldline connecting P_1 and P_2 is

$$\int_{P_1}^{P_2} ds = \int_{P_1}^{P_2} \sqrt{g_{\mu\nu} dx^\mu dx^\nu} = \int_{\tau_1}^{\tau_2} \sqrt{g_{\mu\nu} \frac{dx^\mu}{d\tau} \frac{dx^\nu}{d\tau}} \, d\tau$$

where, as in Eq. (84), the parameter τ is the proper time along the geodesic. This shows that the Lagrangian function is

$$L = \sqrt{g_{\mu\nu} \frac{dx^\mu}{d\tau} \frac{dx^\nu}{d\tau}} \qquad [115]$$

and the Euler-Lagrange equation is

$$0 = \frac{d}{d\tau} \frac{\partial L}{\partial(dx^\mu/d\tau)} - \frac{\partial L}{\partial x^\mu}$$

$$= \frac{d}{d\tau}\left(\frac{1}{L} g_{\mu\nu} \frac{dx^\nu}{d\tau}\right) - \frac{1}{2L} g_{\alpha\beta,\mu} \frac{dx^\alpha}{d\tau} \frac{dx^\beta}{d\tau} \qquad [116]$$

Since this equation refers to the geodesic, we can insert $g_{\mu\nu}(dx^\mu/d\tau)(dx^\mu/d\tau) = 1$ and obtain the *geodesic equation:*[*]

$$\frac{d}{d\tau}\left(g_{\mu\nu} \frac{dx^\nu}{d\tau}\right) - \tfrac{1}{2} g_{\alpha\beta,\mu} \frac{dx^\alpha}{d\tau} \frac{dx^\beta}{d\tau} = 0 \qquad [117]$$

---

[*] Because the numerical value $g_{\mu\nu}(dx^\mu/d\tau)(dx^\nu/d\tau) = 1$, there are many different Lagrangians which give the same differential equation. For example, $L = \tfrac{1}{2} g_{\mu\nu}(dx^\mu/d\tau)(dx^\nu/d\tau)$ will do just as well (this should be compared with Eq. (89)).

This differential equation has the same form as Eq. (113). Our equation of motion for a particle in a gravitational field can therefore be interpreted to mean that the particle moves along the "straightest" possible worldline in a curved spacetime.

In the logical development of "general relativity," Eq. (109) plays a role similar to that of Maxwell's equations in the development of special relativity. Maxwell's equations tell us that the velocity of light is the same in all inertial reference frames; this leads us to discard the Newtonian spacetime geometry and to replace it by the (flat) relativistic spacetime geometry. Eq. (109) tells us that the metric tensor is a function of spacetime; this leads us to discard the flat spacetime geometry and to replace it by a curved spacetime geometry.

This geometrical interpretation of the gravitational field hinges on the interpretation of the parameter τ defined by Eq. (109). In the equation of motion this parameter certainly appears to play the role of proper time. This is why we expect that it actually is the proper time measured by a clock carried along by the particle. Our expectation is based on what we know to be the case in Newtonian physics: the quantity t that appears in the Newtonian equation of motion does indeed agree with the time measured by a clock. However, relativistic physics is more complicated and since the physical interpretation of the quantity τ has such drastic consequences, we had better be very careful. In a purely mathematical sense, the variable τ enters the equation of motion (93) merely as a parameter and Eq. (109) relates this parameter to spacetime displacements. It might, therefore, be questioned whether we necessarily have to regard τ as the proper time of the particle. Could it be that the variable τ given by Eq. (109) is simply some mathematical parameter having nothing to do with proper time? If so, then there would be no need to regard spacetime as curved.

In order to confirm the physical interpretation of Eq. (109), we must check that a clock carried through a gravitational field shows a proper time $\sqrt{g_{\mu\nu}dx^\mu dx^\nu}$ rather than $\sqrt{\eta_{\mu\nu}dx^\mu dx^\nu}$. For this, we must study the behavior of clocks in a gravitational field. This means that we must begin with a detailed model of the clock and write down the equation of motion of each piece of the clock. For example, if we wish to investigate the rate of an atomic clock, say a hydrogen atom, then we need the equations of motion of charged particles and of the electromagnetic field in the presence of gravitation. Only then can we figure out whether measurements with such an atomic standard of time (and length) will give us a flat or a curved spacetime.

Although in practice atomic clocks are very convenient, in theory they are unnecessarily complicated. It turns out that it is possible to design a clock which is much simpler than an atomic clock. This clock, known as the geometrodynamic clock, relies for its operation exclusively on the worldlines of free particles. The analysis of this clock, to be given in Chapter 5, shows that the proper time of the geometrodynamic clock necessarily coincides with the parameter τ of Eq. (109). Thus, all measurements based on this clock indicate that the metric of spacetime is $g_{\mu\nu}$ rather than $\eta_{\mu\nu}$, and therefore we conclude that the real observable geometry is curved. The flat spacetime geometry which we took as the starting point of our calculations in this chapter is unobservable (except, of course, if gravitational fields are absent).

For our further discussion of the linear approximation, we will take for granted the curved spacetime interpretation with the metric tensor (112). For some parts of our discussion this is not really essential; we can temporarily ignore the question of the geometric interpretation by concentrating on those experiments that involve only measurements with clocks and meter rods placed outside of the gravitational field. The deflection of light by the gravitational field of the sun (see Section 3.8) is a typical experiment of this kind; the deflection angle is measured in a region that is far away from the deflecting body.

3.7   THE FIELD OF A SPHERICAL MASS

For a start, we want to find the gravitational field surrounding a spherically symmetric mass distribution at rest. This means that for the region exterior to the mass we seek the spherically symmetric, static solution of the field equations

$$\partial_\lambda \partial^\lambda \phi^{\mu\nu} = 0 \tag{118}$$

$$\partial_\mu \phi^{\mu\nu} = 0 \tag{119}$$

This special solution is of interest since it describes the gravitational field produced by the sun or by the earth provided we ignore the rotational motion and any oblateness of these bodies.

Under the assumption that the field is static (time independent), Eq. (118) reduces to

$$\nabla^2 \phi^{\mu\nu} = 0 \tag{120}$$

The spherically symmetric solution of this equation is

$$\phi^{\mu\nu} = \frac{C^{\mu\nu}}{r} \tag{121}$$

where $C^{\mu\nu}$ is a constant matrix. An additive constant could be included in (121), but we will assume that $\phi^{\mu\nu} \to 0$ as $r \to \infty$ and therefore there is no such constant. Substitution of (121) into (119) yields

$$C^{k\nu}\partial_k\left(\frac{1}{r}\right) = 0$$

that is,

$$C^{k\nu}\left(\frac{-x^k}{r^3}\right) = 0 \qquad [122]$$

This can hold for all values of x^k only if $C^{k\nu} = 0$. Hence the only non-zero component of $C^{\mu\nu}$ is C^{00}. It then follows that

$$h_{00} = \phi_{00} - \tfrac{1}{2}\eta_{00}\,\phi = \frac{C_{00}}{r} - \tfrac{1}{2}\frac{C_{00}}{r} = \frac{C_{00}}{2r} \qquad [123]$$

This determines the value of C_{00} because we already know that h_{00} is proportional to the Newtonian potential (see Eq. (100)), that is,

$$h_{00} = \frac{2}{\kappa}\left(-\frac{GM}{r}\right) \qquad [124]$$

Hence $C_{00} = -4GM/\kappa$ and

$$\phi^{\mu\nu} = \begin{pmatrix} -4GM/\kappa r & 0 & 0 & 0 \\ 0 & 0 & 0 & 0 \\ 0 & 0 & 0 & 0 \\ 0 & 0 & 0 & 0 \end{pmatrix} \qquad [125]$$

The corresponding expression for $h_{\mu\nu}$ is

$$h_{\mu\nu} = \begin{pmatrix} -2GM/\kappa r & 0 & 0 & 0 \\ 0 & -2GM/\kappa r & 0 & 0 \\ 0 & 0 & -2GM/\kappa r & 0 \\ 0 & 0 & 0 & -2GM/\kappa r \end{pmatrix} \qquad [126]$$

We could use (126) and our relativistic equation of motion to study the motion of planets in the gravitational field of the sun. However, we will not bother with this because the relativistic deviations from the Newtonian orbits are extremely small. The most noticeable deviation between Newtonian and relativistic celestial mechanics is the perihelion precession. The only other deviation that we might hope to detect is a difference in the predicted instantaneous velocities along the orbit. Ordinary telescopic observation cannot detect these differences. It has recently become possible to measure the instantaneous velocity of the (inner) planets to one part in $\sim 10^6$ by means of the Doppler shift suffered by radar signals sent to these planets and reflected by them back to earth.[1] But even these radar echo experiments are not quite precise enough to detect the velocity deviations between Newtonian and relativistic gravitation. At present, the only available test of the relativistic equation of motion of massive particles comes from the perihelion precession. Although we could calculate the perihelion precession from our linear theory, it turns out that this precession receives comparable contributions from the linear field $h^{\mu\nu}$ and from second-order corrections to $h^{\mu\nu}$ (note that (126) is of order M; by "second-order" is meant M^2). Such second-order corrections come from the nonlinearities of the exact field equations. Let us try to understand how these nonlinearities affect the motion of planets. Consider the equation of motion (76),

$$\frac{d}{d\tau} P_\mu = -\frac{\kappa}{2} m \, h_{\alpha\beta,\mu} u^\alpha u^\beta$$

The right side of this equation may be regarded as a force; since $u^0 \sim 1$ and $u^k \sim v^k$, it contains terms of the type $\kappa m \, h_{00,\mu}$ and $\kappa m \, h_{kl,\mu} v^k v^l$. If we use the linear approximation (126), then the term $\kappa m \, h_{00,\mu}$ is simply the Newtonian force; this yields no perihelion precession. The term $\kappa m \, h_{kl,\mu} v^k v^l$ represents a deviation from the Newtonian force and produces a perihelion precession. Since the velocity of a planet in orbit is roughly $v \sim \sqrt{GM/r}$, this term is of order M^2. We now notice that nonlinear corrections to h_{00} will change $\kappa m \, h_{00,\mu}$ by a term of order M^2. Thus, this nonlinear correction will be roughly of the same importance as $\kappa m \, h_{kl,\mu} v^k v^l$. Hence, for the (approximate) evaluation of the perihelion precession we need the second-order terms in h_{00}, although we only need the first-order terms in h_{kl}. We actually could calculate the required second-order corrections by solving the nonlinear field equation (64), and we could then find the correct perihelion precession (see Problems 3.8 and 8.8). But we prefer to leave the whole problem aside until after we have written down and solved the exact Einstein equations.

However, there is an interesting observable effect which we can calculate from (126): the deflection of light by the gravitational field of the sun.

3.8 DEFLECTION OF LIGHT

To find out how light propagates in a gravitational field, let us write down the relativistic equation of motion (Eq. (76)) in a slightly modified form:

$$dP_\mu - \frac{\kappa}{2} h^\alpha{}_{\beta,\mu} P_\alpha dx^\beta = 0 \qquad\qquad [127]$$

In this equation we have made the *approximation* $P_\alpha \simeq mu_\alpha$ in the second term on the left side; this is consistent with our usual neglect of any term of order κ^2. Eq. (127) can be used to describe the effect of the gravitational field on a light wave if we recognize that the wave vector $k^\mu = (\omega, k_x, k_y, k_z)$ is for a light wave what the momentum vector P^μ is for a particle. This can perhaps be most easily understood in terms of photons: except for a factor of $\hbar$, k^μ is the momentum of one photon in the light wave. The equation that gives the change of the wave vector is then

$$dk_\mu - \frac{\kappa}{2} h^\alpha{}_{\beta,\mu} k_\alpha dx^\beta = 0 \qquad [128]$$

Note that the correspondence between particles and waves which we have used to obtain Eq. (128) relates the *canonical* momentum P_μ to the wave vector k_μ. This correspondence can be rigorously justified for wave propagation in the limiting case of geometrical optics (very short wavelength).

Consider now a ray of light passing by the sun with impact parameter b (see Fig. 3.2). Taking the z-axis along the direction of incidence, we have for the displacement along the ray and for the wave vector, respectively,

$$dx^\beta = (dt,0,0,dz) + \cdots = (dz,0,0,dz) + \cdots \qquad [129]$$

$$k_\alpha = (\omega,0,0,-\omega) + \cdots \qquad [130]$$

Here the dots stand for corrections of order κ. Inserting (129) and (130) into (128) and neglecting terms of order κ^2, we obtain

$$dk_\mu = \frac{\kappa}{2}\omega (h^0{}_{0,\mu} - h^3{}_{0,\mu} + h^0{}_{3,\mu} - h^3{}_{3,\mu})\, dz \qquad [131]$$

The gravitational field of the sun is given by Eq. (126) and hence

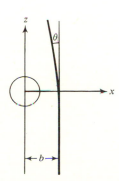

Fig. 3.2 *Path of a ray of light in the field of the sun. The deflection is exaggerated.*

$$dk_x = -dk_1 = -\frac{\kappa}{2}\omega(h^0{}_{0,1} - h^3{}_{3,1})\,dz$$

$$= -\frac{\kappa}{2}\omega\frac{\partial}{\partial x}(h^0{}_0 - h^3{}_3)\,dz$$

$$= -\frac{\kappa}{2}\omega\frac{\partial}{\partial x}\left(-\frac{4GM_\odot}{\kappa r}\right)dz \qquad [132]$$

so that the total change in k_x is

$$\Delta k_x = -\Delta k_1 = -\frac{\omega}{2}\int_{-\infty}^{\infty}\frac{\partial}{\partial x}\frac{-4GM_\odot}{r}\bigg|_{x=b}\,dz$$

$$= -2GM_\odot\omega\int_{-\infty}^{\infty}\frac{b}{(z^2 + b^2)^{3/2}}dz$$

$$= -\frac{4GM_\odot}{b}\omega \qquad [133]$$

The deflection angle is therefore

$$\theta \simeq \frac{\Delta k_x}{k_z} = -\frac{4GM_\odot}{b} \qquad [134]$$

or, expressed in cgs units,

$$\theta = -\frac{4GM_\odot}{bc^2} \qquad [135]$$

EXERCISE 19. Show that, within the same approximations,

$$\Delta k^0 = 0 \qquad [136]$$

$$\Delta k^3 = 0 \qquad [137]$$

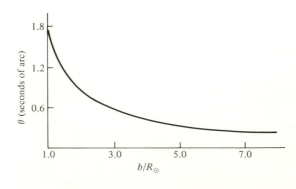

Fig. 3.3 *Deflection angle as a function of impact parameter.*

Fig. 3.3 shows a plot of deflection versus impact parameter. For the case of a light ray grazing the sun, we must set $b = R_\odot = 6.96 \times 10^{10}$ cm, and $M_\odot = 1.99 \times 10^{33}$ gm, which gives

$$\theta = -1.75''\qquad\qquad\qquad [138]$$

Such a deflection of the light reaching the earth from a star results in an (apparent) shift in the position of the star as seen from the earth (see Fig. 3.4). Observations of this effect are difficult to carry out. The stars near the sun are only visible during a total eclipse of the sun, and even then the brightness of the solar corona restricts observations to impact parameters $b \gtrsim 2R_\odot$. The experimental procedure involves taking a photograph of the star field surrounding the eclipsed sun and comparing this photograph with another one taken at night several months before (or after) when the sun is not in this star field. The technical difficulties are that the eclipse photograph must be taken in the very short time interval of totality and, furthermore, disturbances to the instrument must be prevented (or corrected for) during the waiting period between the eclipse photograph and the night photograph. In a temporary field-station set up in the path of the eclipse in some remote part of the world all of this is hard to achieve.

In response to Einstein's prediction of light deflection, Eddington and Dyson organized two expeditions to observe the eclipse of May 29, 1919, on the islands of Sobral (off Brazil) and Principe (in the Gulf of Guinea). Fig. 3.5 shows the instruments installed at Sobral. The experimental results obtained by these expeditions were at the time regarded as a sensational success for Einstein's theory of gravitation.

Since then, observations have been carried out during many solar eclipses. Table 3.2 (based on the review by von Klüber[2]) is a summary of these observations. The numbers given in this table are the deflections extrapolated to impact parameter $b = R_\odot$; they should be compared with Eq. (138).

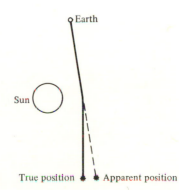

Fig. 3.4 *Apparent shift of position of a star.*   True position ↟ ↟ Apparent position

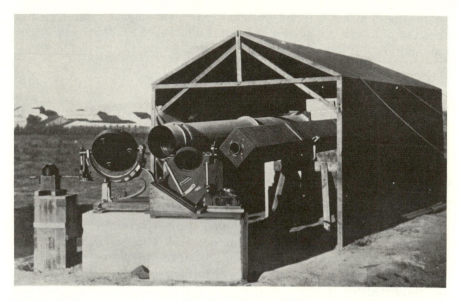

Fig. 3.5 *Eclipse instruments at Sobral. The two separate telescopes are mounted horizontally and mirrors* (center left) *are used to throw the sun's image into them. (From Eddington,* Space, Time, and Gravitation. *Reprinted with permission of Cambridge University Press.*)

Obviously, the data on the light deflection is not very good; all we can say is that a deflection of roughly the right magnitude is present.

More precise results have recently been obtained by the use of radio waves rather than optical light. In this case, it is not necessary to wait for an eclipse; rather one must wait for the sun's limb to ap-

TABLE 3.2 EXPERIMENTAL RESULTS ON THE DEFLECTION OF LIGHT[2]

Observatory	Eclipse	Site	θ (sec)
Greenwich	May 29, 1919	Sobral	1.98 ± 0.16
Greenwich	" "	Principe	1.61 ± 0.40
Adelaide-Greenwich	Sept. 21. 1922	Australia	1.77 ± 0.40
Victoria	" "	Australia	1.42 to 2.16
Lick	" "	Australia	1.72 ± 0.15
Lick	" "	Australia	1.82 ± 0.20
Potsdam	May 9, 1929	Sumatra	2.24 ± 0.10
Sternberg	June 19, 1936	U.S.S.R.	2.73 ± 0.31
Sendai	" "	Japan	1.28 to 2.13
Yerkes	May 20, 1947	Brazil	2.01 ± 0.27
Yerkes	Feb. 25, 1952	Sudan	1.70 ± 0.10
U. of Texas[3]	June 30, 1973	Mauritania	1.58 ± 0.16[a]

[a] According to a preliminary data analysis.

TABLE 3.3 EXPERIMENTAL RESULTS ON THE DEFLECTION OF RADIO WAVES

Radio Telescope	Wavelength (cm)	Baseline (km)	θ (sec)
Owens Valley[4]	3.1	1.07	1.77 ± 0.20
Goldstone[5]	12.5	21.56	1.82 $\begin{cases} +0.26 \\ -0.17 \end{cases}$
National RAO[6]	11.1 and 3.7	~ 2	1.64 ± 0.10
Mullard RAO[7]	11.6 and 6.0	~ 1	1.87 ± 0.30
Cambridge[8]	6.0	4.57	1.82 ± 0.14
Westerbork[9]	6.0	1.44	1.68 ± 0.09
Haystack and National RAO[10]	3.7	845	1.73 ± 0.05
National RAO[11]	11.1 and 3.7	35.6	1.76 ± 0.02
Westerbork[12]	21.2 and 6.0	~ 1	1.82 ± 0.06

proach a radio source. The quasistellar source 3C279 is occulted by the sun on the eighth of October and radio interferometers, consisting of two or more antennas, have been used to measure the apparent shift in the the position of this source. The shift is measured relative to the quasi-stellar source 3C273 which is about 10° farther away and therefore suffers a much smaller shift. The experimental results of Table 3.3 have been expressed in terms of the corresponding extrapolated deflections for a radio signal that just grazes the sun.

These measurements can be, and hopefully will be, much improved. In principle, the resolving power of a radio interferometer is limited by the diffraction angle

$$\Delta\theta \simeq (\text{wavelength})/(\text{baseline}) = \lambda/d \qquad [139]$$

Very long baselines can be attained by use of two widely separated antennas. For $\lambda = 3$ cm and $d = 3000$ km, (139) gives $\Delta\theta \simeq 0.002''$. The results in Table 3.3 are obviously nowhere near this diffraction limit.

Incidentally, the results of the radio wave deflection experiments give evidence against the scalar-tensor theory of gravitation (see Section 3.2). In this alternative theory of gravitation, the predicted deflection is about 1.66″, which is excluded by the data of Table 3.3.

3.9   RETARDATION OF LIGHT

Another observable effect that we can obtain from the linear approximation is the time delay suffered by a radar signal sent from the earth to a planet and reflected back to earth. The gravitational field of the sun affects this time delay because it reduces the speed of propagation of light signals.

The speed of propagation can be deduced from Eq. (131). Consider

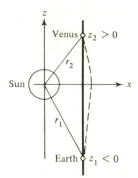

Fig. 3.6 *Path of light signal between earth (z_1) and a planet (z_2). The true path is indicated by the dotted line. The straight line is an approximation.*

a light signal that moves along a straight line connecting two planets (see Fig. 3.6; we take the z-axis parallel to this line). Of course, because of the deflection of the path of light, the true path will not be exactly straight. But this has next to no effect on the travel time because the difference in length between the straight and curved paths shown in Fig. 3.6 is only of order θ^2 (where θ is the deflection angle); this difference is therefore a second-order correction which we can ignore. Along the straight line, the change in k_z given by (131) and (126) is

$$dk_z = -dk_3 = -\frac{\kappa}{2}\,\omega(h^0{}_{0,3} - h^3{}_{3,3})dz$$

$$= -\frac{\kappa}{2}\,\omega\,\frac{\partial}{\partial z}\left[-\frac{4GM_\odot}{\kappa r}\right]dz \qquad [140]$$

where $r = (z^2 + b^2)^{1/2}$. Integration of this yields

$$k_z(z) - k_z(\infty) = 2GM_\odot\omega\int_\infty^z \frac{\partial}{\partial z}\left(\frac{1}{r}\right)dz$$

$$= \frac{2GM_\odot}{r}\,\omega$$

that is,

$$k_z(z) = \omega\left(1 + \frac{2GM_\odot}{r}\right) \qquad [141]$$

The component k_0 of the wave vector does not change.

EXERCISE 20. Show this.

Therefore, the frequency of the light wave remains constant at $k_0 = \omega$ and the velocity of the wave is

$$v = \frac{\omega}{k_z(z)} = \frac{1}{1 + \frac{2GM_\odot}{r}} \tag{142}$$

Note that this velocity is frequency independent; hence there is no dispersion and the group velocity (signal velocity) coincides with the phase velocity, that is, $\partial\omega/\partial k_z(z) = \omega/k_z(z)$. The result (142) shows that the gravitational field of the sun decreases the velocity of propagation of light. In fact, the deflection of light, which we calculated earlier, can be regarded as a refraction of the wave front: the part of the wave front nearer the sun is slowed down more than the part farther away, and hence the direction of advance of the wave changes.

Fig. 3.6 shows the earth (at $z_1 < 0$) and the target planet (at $z_2 > 0$) in the gravitational field of the sun. The travel time for a light or radio signal between z_1 and z_2 is

$$\Delta t = \int_{z_1}^{z_2} \frac{dz}{v} = \int \left(1 + \frac{2GM_\odot}{r}\right) dz$$

$$= \int \left(1 + \frac{2GM_\odot}{\sqrt{z^2 + b^2}}\right) dz$$

$$= z_2 - z_1 + 2GM_\odot \ln \frac{z_2 + \sqrt{z_2^2 + b^2}}{z_1 + \sqrt{z_1^2 + b^2}} \tag{143}$$

The second term on the left side of (143) represents the extra time delay produced by the gravitational field. The extra delay is large when the earth and planet are on (nearly) opposite sides of the sun ("superior conjunction"). Then b/z is small and the extra time delay becomes approximately

$$2GM_\odot \ln \left(\frac{4|z_1|z_2}{b^2}\right) \tag{144}$$

EXERCISE 21. Derive Eq. (144) from (143).

Clearly, b cannot be smaller than the radius of the sun (grazing signal). If we set $b = R_\odot = 6.96 \times 10^{10}$ cm, $|z_1| =$ (radius of earth orbit) $\simeq 14.9 \times 10^{12}$ cm, and $z_2 =$ (radius of Mercury orbit) $\simeq 5.8 \times 10^{12}$ cm, we obtain for the maximum possible extra time delay $\sim 110 \times 10^{-6}$ sec; the extra delay for the signal to go to Mercury and return is twice as large.

Before we can compare this with experiment, we must make a correction because the clocks on the earth used to measure the time delay give the proper time rather than the t-time which appears in Eq. (143). If we assume that the spacetime distances are given by Eq. (111), then the connection between proper time and *t*-time for a clock at rest in the gravitational field of the sun is

$$\Delta\tau = \sqrt{g_{00}}\,\Delta t = \sqrt{1 + \kappa h_{00}}\,\Delta t \simeq \left(1 - \frac{GM_\odot}{r}\right)\Delta t$$

To convert the expression (143) to proper time at the earth, we must therefore multiply it by $1 - GM_\odot/r$, where r is the radius of the earth's orbit.* This correction changes the maximum possible extra time delay by about 10% for the case of Mercury. Further (nongravitational) corrections must be made for the change in the velocity of propagation of the radio signal caused by the solar corona and interplanetary plasma, and for distortion of the return signal caused by topographic features on the reflecting surface of the target planet.

The delays of radar echoes from Mercury, Venus, and Mars have been measured by Shapiro et al.[13,14] using the Haystack and Arecibo radio telescopes. In order to separate the *extra* time delay due to gravitation from the total measured delay, it is necessary to know the position of the reflecting surface very precisely. This means that both the position and the size of the reflecting planet must be known with an accuracy that is beyond what can be achieved by ordinary observational astronomy. This difficulty was resolved by performing a large number of time delay measurements over a period of several years (beginning in 1966). The orbital elements of the planets, their sizes, topographical features, masses, and finally the gravitational delay were then obtained from this data. The curve in Fig. 3.7 (from ref. 14) is a plot of the predicted extra time delay produced by gravitational effects as a function of time. The data points obviously fit this curve very well. To be precise, the theoretical and experimental values agree to within a factor 1.02 ± 0.02.

Time delays have also been measured by Anderson et al.[15] for signals transmitted to the Mariner 6 and 7 spacecrafts in orbit around the sun; signals received by the spacecraft were retransmitted to the earth by a transponder on the spacecraft. The experimental delays were found to be 1.003 ± 0.014 times the theoretical value for Mariner 6 and 1.000 ± 0.012 times the theoretical value for Mariner 7. Unfortunately, these spacecraft are subject to unknown random forces produced by outgassing of the attitude control systems and by fluctuations in the solar radiation pressure. These unknown perturbations increase the uncertainties of the measurement; a realistic estimate of the total uncertainty is ± 0.03.

A comment on just what is meant by the "slowing down" of the speed of propagation of light (Eq. (142)) may be helpful. Although a light signal that passes close to the sun will be delayed, this does not mean

---

* Since the earth is moving, the time dilation effect of special relativity must also be taken into account.

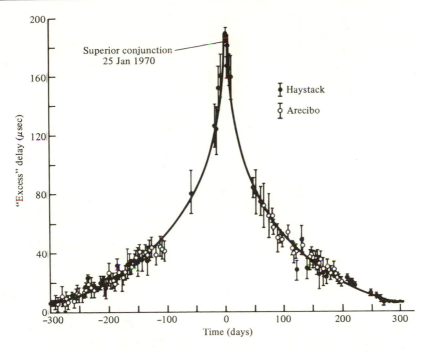

Fig. 3.7 *Earth-Venus time-delay measurements. The solid curve gives the theoretical prediction.* (*From Shapiro et al.,* Phys. Rev. Lett. *26, 1132 (1971).*)

that an observer who measures the speed of the signal at a point near the sun will obtain a result different from the usual speed of light. If this observer uses "good" clocks and "good" meter sticks* he will always find a speed $c = 3 \times 10^{10}$ cm/sec. The velocity given by Eq. (142) is not a locally measured velocity but an effective velocity measured with standards of time and length that are far away, outside of the gravitational field.

An idealized situation is shown in Fig. 3.8a. We have two points P_1, P_2 connected by a light signal. The points P_1, P_2 are outside of the gravitational field (alternatively, the field is so weak at these points that it can be neglected for our purposes), but the light ray connecting them passes through a region where the field is strong. In order to measure the velocity of the light signal along P_1P_2 we could try to lay out our meter sticks along P_1P_2; but since, according to Eq. (109), the gravitational field affects the measurement of space and time, let us not do this. An alternative method is to use surveying instruments to lay out the lines P_1A, AB, BP_2, where all the angles are right angles (the figure is drawn under the simplifying assumption that there is no appreciable bending of the light rays). All these lines are outside of the gravitational field and their length can be measured without trouble. According to Euclidean geometry the distance P_1P_2 equals AB. Furthermore, clocks at P_1, A, B,

* The search for "good" standards of time and length will be our concern in Chapter 5.

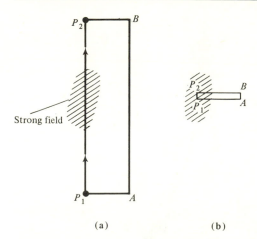

Fig. 3.8 *Light path and survey used for the measurement of the effective velocity.*

(a)   (b)

P_2 can be synchronized by the procedure familiar from special relativity. We may then say that the effective velocity along P_1P_2 is the distance measured along AB divided by the time elapsed between departure from P_1 and arrival at P_2. It is this effective velocity that is smaller than c.

Of course this procedure only gives us the average effective velocity. If we want to obtain the instantaneous effective velocity, then we must make the segments P_1P_2 and AB infinitesimal and then both P_1 and P_2 will be in the region where the gravitational field is appreciable (see Fig. 3.8b). In this case surveying instruments cannot be used directly to lay out the rectangle of Fig. 3.8b because the light rays on which these instruments rely will be bent in the gravitational field. If we assume that the formula for the amount of bending is known, we can of course correct for it and lay out the necessary straight lines. Furthermore, when measuring the time elapsed between departure of a light signal from P_1 and arrival at P_2 we will not want to use the local clocks of Fig. 3.8b, but rather a clock outside of the gravitational field. This can be done by signaling the departure and arrival to a clock at a large distance and using that clock to measure the time interval. Note that essentially this is an operational procedure for setting up the coordinates x, y, z, t by reference to the coordinates at large distance, the velocity v (Eq. (142)) being defined in terms of the coordinate changes Δt, Δx, Δy, Δz.

The disagreement between local velocity (c) measured along P_1P_2 and the effective velocity measured indirectly by the construction P_1ABP_2 shows quite clearly that something is foul with flat spacetime, or our measurement instruments, or both. In a curved spacetime, the disagreement is not at all surprising because then the distance P_1P_2 need not equal AB. In fact, the effective velocity of propagation can be obtained immediately from the curved spacetime interval (111). The propagation of light is characterized by a lightlike interval, that is,

$$0 = ds^2 = g_{\mu\nu}dx^\mu dx^\nu$$

[145]

(This is the obvious generalization of what we know to be the case in flat spacetime; as we will see in Chapter 5, Eq. (145) is a direct consequence of the definition of spacetime measurements.) With $g_{\mu\nu} = \eta_{\mu\nu} + h_{\mu\nu}$ and with the sun's gravitational field given by (126), we obtain

$$0 = \left(1 - \frac{2GM_\odot}{r}\right)(dx^0)^2 - \left(1 + \frac{2GM_\odot}{r}\right)[(dx^1)^2 + (dx^2)^2 + (dx^3)^2]$$

Hence the velocity of propagation of the light signal is

$$v = \frac{[(dx^1)^2 + (dx^2)^2 + (dx^3)^2]^{1/2}}{dx^0} = \left[\frac{1 - 2GM_\odot/r}{1 + 2GM_\odot/r}\right]^{1/2}$$

$$\simeq (1 - 2GM_\odot/r) \qquad\qquad [146]$$

To first order in G this agrees with Eq. (142).

Since both the light deflection and the time delay can be obtained from the velocity formula (142) or (146), we could have derived these effects without using the equation of motion (127) explicitly. Note, however, that this alternative derivation is not entirely independent of the equation of motion because the interpretation of $g_{\mu\nu}$ as metric hinges on that equation (see Section 3.6).

Finally, let us see what our general equation (128) for the change in the wave vector has to say about the redshift of light. We have

$$dk_0 = \frac{\kappa}{2} h^\alpha{}_{\beta,0} k_\alpha k^\beta$$

Hence, in any static gravitational field ($h_{\alpha\beta,0} = 0$) light propagates *without change in frequency*. This result should not cause any surprise: it is well known that when light enters a dielectric substance with an inhomogeneous (but time-independent) index of refraction it propagates without change in frequency; only the *wavelength* changes and produces the deflection of the wavefronts.

As we will see later, the famous "redshift" of light is due to a slowing down of clocks when placed in a gravitational field. The redshift is not due to a gradual change of frequency* as the light propagates through the gravitation field; rather, it is due to a reduction of the frequency of oscillation of the emitter when placed deep in a gravitational potential. Of course, such statements are meaningless if no careful operational definition is given as to how clocks are to be constructed and how their relative rates are to be adjusted when they are separated by some distance. The whole question of measurement of time and length will have to be faced in Chapter 5.

---

* In this context "frequency" means frequency as reckoned by a clock placed at large distance from the gravitational field, and therefore not affected by the field.

3.10   THE FIELD OF A ROTATING MASS

The gravitational field surrounding a rotating mass differs from that surrounding a nonrotating mass. We can understand this by analogy with the case of a rotating, uniformly charged sphere; such a sphere produces both electric and magnetic fields while a nonrotating sphere produces only an electric field.

Since the rotations of the sun and of the earth are quite slow (nonrelativistic), the effects of the rotation on the gravitational fields and on the motion of particles in these fields are very small. However, a high-precision experiment now under construction will attempt to detect the effect of the rotation of the earth on a gyroscope placed near the earth. We will study this effect in Section 8.3, and we will here lay the groundwork by solving for the gravitational field surrounding a nonrelativistic rotating mass. The general solution with relativistic rotation will be given in Section 9.4.

We will assume that the rotation is steady, so that the field is time independent. It is convenient to begin with the general solution of Eq. (55). In the time-independent case, this equation reads

$$- \partial_k \partial_k \phi^{\mu\nu}(\mathbf{x}) = - \kappa T^{\mu\nu}(\mathbf{x}) \qquad [147]$$

and has the solution

$$\phi^{\mu\nu}(\mathbf{x}) = - \frac{\kappa}{4\pi} \int \frac{T^{\mu\nu}(\mathbf{x}')}{|\mathbf{x} - \mathbf{x}'|} \, d^3x' \qquad [148]$$

EXERCISE 22. Show that (148) solves Eq. (147).

In order to evaluate the integral appearing in Eq. (148), we use the Taylor series expansion for $1/|\mathbf{x} - \mathbf{x}'|$ (see Eq. (1.14)) and keep only the first two terms:

$$\frac{1}{|\mathbf{x} - \mathbf{x}'|} \simeq \frac{1}{r} + \frac{x^k x'^k}{r^3} \qquad [149]$$

This gives

$$\phi^{\mu\nu}(\mathbf{x}) \simeq - \frac{\kappa}{4\pi r} \int T^{\mu\nu}(\mathbf{x}')d^3x' - \frac{\kappa x^k}{4\pi r^3} \int x'^k T^{\mu\nu}(\mathbf{x}')d^3x' \qquad [150]$$

The first integral in Eq. (150) has the value

$$\int T^{00}(\mathbf{x}')d^3x' = M \qquad [151]$$

$$\int T^{k\mu}(\mathbf{x}')d^3x' = 0 \qquad [152]$$

where M is the mass of the system. Eq. (151) is obvious and requires no comment. To derive Eq. (152), we begin with the conservation law

$$\frac{\partial}{\partial x'^0} T^{0\mu}(\mathbf{x}') + \frac{\partial}{\partial x'^l} T^{l\mu}(\mathbf{x}') = 0 \qquad [153]$$

Since the matter distribution is assumed to be time independent, this reduces to

$$\frac{\partial}{\partial x'^l} T^{l\mu}(\mathbf{x}') = 0 \qquad [154]$$

Multiply Eq. (154) by x'^k and integrate over the volume of the system:

$$\int x'^k \frac{\partial}{\partial x'^l} T^{l\mu}d^3x' = 0 \qquad [155]$$

An integration by parts then gives Eq. (152).

EXERCISE 23. Check this.

The second integral appearing in Eq. (150) has the value

$$\int x'^k T^{00}(\mathbf{x}')d^3x' = 0 \qquad [156]$$

$$\int x'^k T^{l0}(\mathbf{x}')d^3x' = \tfrac{1}{2}\epsilon^{kln}S^n \qquad [157]$$

$$\int x'^k T^{ln}(\mathbf{x}')d^3x' = 0 \qquad [158]$$

where S^n is the spin angular momentum of the system and ϵ^{kln} the object defined in Eq. (1.53).

Eq. (156) simply states that the origin of coordinates is at the center of mass; this is an *assumption* which we make for the sake of simplicity.

The derivation of Eq. (157) proceeds as follows: Multiply Eq. (154) by $x'^k x'^n$ and integrate:

$$\int x'^k x'^n \frac{\partial}{\partial x'^l} T^{l\mu}d^3x' = 0 \qquad [159]$$

Upon integration by parts, we obtain the identity

$$-\int x'^n T^{k\mu}d^3x' - \int x'^k T^{n\mu}d^3x' = 0 \qquad [160]$$

Before we can proceed, we must write down the definition of the spin angular momentum S^n. Since T^{l0} is the density of l-momentum, the definition of, say, S^1 must be

$$S^1 = \int (x'^2 T^{30} - x'^3 T^{20}) d^3 x' \qquad [161]$$

The integrand is simply the density of the x-component of angular momentum.

EXERCISE 24. Show that in general

$$S^n = \int \epsilon^{nkl} x'^k T^{l0} d^3 x' \qquad [162]$$

In view of the identity (160), each of the two terms in the integrand of Eq. (161) contributes the same amount and therefore

$$S^1 = 2 \int x'^2 T^{30} d^3 x' \qquad [163]$$

This equation is exactly the same as the $k = 2$, $l = 3$ component of Eq. (157).

EXERCISE 25. Supply the proof for the other components of Eq. (157).

Finally, to derive Eq. (158) we begin with the identity

$$T^{ln}(\mathbf{x}') = \frac{1}{2} \frac{\partial}{\partial x'^m} (x'^n T^{lm}(\mathbf{x}') + x'^l T^{nm}(\mathbf{x}')) \qquad [164]$$

EXERCISE 26. Derive this identity. (Hint: use the conservation law and the symmetry of the energy-momentum tensor.)

Hence, with an integration by parts,

$$\int x'^k T^{ln}(\mathbf{x}') d^3 x' = \frac{1}{2} \int x'^k \frac{\partial}{\partial x'^m} (x'^n T^{lm} + x'^l T^{nm}) d^3 x'$$

$$= -\frac{1}{2} \int \delta_m^{\ k} (x'^n T^{lm} + x'^l T^{nm}) d^3 x'$$

$$= -\frac{1}{2} \int (x'^n T^{lk} + x'^l T^{nk}) d^3 x' \qquad [165]$$

The last expression vanishes as a consequence of the identity (160).

If we insert the results (151), (152), and (156)–(158) into Eq. (150), we obtain

$$\phi^{00}(\mathbf{x}) = -\frac{\kappa M}{4\pi r}$$

$$\phi^{l0}(\mathbf{x}) = \phi^{0l}(\mathbf{x}) = -\frac{\kappa}{8\pi r^3}\, \epsilon^{kln}x^k S^n$$

$$\phi^{ln}(\mathbf{x}) = 0 \qquad\qquad\qquad\qquad\qquad\qquad [166]$$

The corresponding expressions for $h_{\mu\nu}$ are

$$h^{00}(\mathbf{x}) = -2GM/\kappa r$$

$$h_{l0}(\mathbf{x}) = h_{0l}(\mathbf{x}) = 2G\epsilon^{kln}x^k S^n/\kappa r^3$$

$$h_{ln}(\mathbf{x}) = -(2GM/\kappa r)\delta_l^n \qquad\qquad\qquad [167]$$

If the direction of the spin angular momentum coincides with the z-axis of our coordinate system, we obtain

$$h_{\mu\nu} = \begin{pmatrix} -2GM/\kappa r & -2GS_z y/\kappa r^3 & 2GS_z x/\kappa r^3 & 0 \\ -2GS_z y/\kappa r^3 & -2GM/\kappa r & 0 & 0 \\ 2GS_z x/\kappa r^3 & 0 & -2GM/\kappa r & 0 \\ 0 & 0 & 0 & -2GM/\kappa r \end{pmatrix} \quad [168]$$

which may be compared with Eq. (126). The field of a rotating mass differs from that of a nonrotating mass by the presence of the off-diagonal components h_{0l}.*

The spins of the sun and earth are relatively small, and the effects of the off-diagonal component on the orbits of planets and satellites are insignificant. However, in the case of the earth it may be possible to detect the off-diagonal components of the gravitational field by their effect on a gyroscope. Calculation shows that a gyroscope placed in the field (168) will precess. This is known as the Lense-Thirring effect and will be further discussed in Section 8.3.

This concludes our discussion of the time-independent solutions of the linear field equations. In the next chapter we will investigate the time-dependent fields in the radiation zone, that is, the fields at large distance from some time-dependent mass distribution. We could also investigate the fields in the convection zone, that is, the fields in the immediate vicinity of a time-dependent mass distribution. In fact, we could work out a set of "Lienard-Wiechert" potentials for gravitation entirely in analogy with electromagnetism. But although the close similarity between the linear gravitation and electromagnetism may tempt us to solve the same mathematical problems, we must keep in mind that there are few gravitational problems that are accessible to experiment. We cannot manufacture noticeable gravitational fields as easily as we manufacture electric and magnetic fields. With the one exception of the Cavendish

---

* If the rotation is very fast then there is a further difference: the total mass M of Eq. (168) includes the kinetic energy of rotation.

experiment, the gravitational fields that we use in our experiments have been generated by processes that are beyond our control.

FURTHER READING

Lagrangian field theory provides an alternative, and very elegant, derivation of the linear field equations. This approach is presented in detail in Thirring, "An Alternative Approach to the Theory of Gravitation" (see ref. 16), and more briefly in Misner, Thorne, and Wheeler, *Gravitation* (Chapter 7).

The most fundamental justification of the linear equations emerges from the analysis of the representations of the Lorentz group. This approach is discussed by Weinberg in the incisive article "The Quantum Theory of Massless Particles" (see ref. 20).

Nonlinear corrections to the field equations, and the method of successive approximations that can be used to find them, are discussed in the article by Thirring (ref. 16) and in the article "Einstein's and Other Theories of Gravitation" by Gupta (see ref. 17). Incidentally, both of these articles also explain what is wrong with scalar and vector theories of gravitation.

The most common approach is to derive the linear approximation from the exact Einstein equations, rather than vice versa.* This approach may be found in Landau and Lifshitz, *The Classical Theory of Fields;* Misner, Thorne, and Wheeler, *Gravitation;* and Weinberg, *Gravitation and Cosmology.* The article "On Alternative Approaches to Gravitation" by Halpern (ref. 18) compares the merits of the alternatives.

Klüber, "The Determination of Einstein's Light-Deflection in the Gravitational Field of the Sun," is a thorough and interesting review of the light deflection observations (up to 1952), including a discussion of the experimental difficulties in obtaining reliable results, by an astronomer who participated in several expeditions. Eddington, *Space, Time, and Gravitation,* tells the story of the first expedition that set out to measure the light deflection.

REFERENCES

1. I. I. Shapiro, M. E. Ash, and M. J. Tausner, Phys. Rev. Lett. **17,** 933 (1966).
2. H. von Klüber, *in* A. Beer, ed., *Vistas in Astronomy* (Pergamon Press, London, 1960).
3. B. F. Jones, University of Texas, Austin. Private communication.
4. G. A. Seielstadt, R. A. Sramek, and K. W. Weiler, Phys. Rev. Lett. **24,** 1373 (1970).
5. D. O. Muhleman, R. D. Ekers, and E. B. Fomalont, Phys. Rev. Lett. **24,** 1377 (1970).
6. R. A. Sramek, *Proc. of the International School of Physics,* "Enrico Fermi" (Varenna, 1972).

---

* In Chapter 7 we will use a very simple method to derive the exact nonlinear Einstein equations from the linear equations.

7. J. M. Hill, Mon. Not. Roy. Astron. Soc. **153**, 7P (1971).
8. J. M. Riley, Mon. Not. Roy. Astron. Soc. **161**, 11P (1973).
9. K. W. Weiler, R. D. Ekers, E. Raimond, and K. J. Wellington, Astron. and Astrophys. **30**, 241 (1974).
10. C. C. Counselman, S. M. Kent, C. A. Knight, I. I. Shapiro, T. A. Clark, H. F. Hinteregger, A. E. E. Rogers, and A. R. Whitney, Phys. Rev. Lett. **33**, 1617 (1974).
11. E. B. Fomalont and R. A. Sramek, Ap. J., in press.
12. K. W. Weiler, R. D. Ekers, E. Raimond, and K. J. Wellington, Phys. Rev. Lett., in press.
13. I. I. Shapiro, G. M. Pettengill, M. E. Ash, M. L. Stone, W. B. Smith, R. P. Ingalls, and R. A. Brockelman, Phys. Rev. Lett. **20**, 1265 (1968).
14. I. I. Shapiro, M. E. Ash, R. P. Ingalls, W. B. Smith, D. B. Campbell, R. B. Dyce, R. F. Jurgens, G. M. Pettengill, Phys. Rev. Lett. **26**, 1132 (1971).
15. J. D. Anderson, P. B. Esposito, W. Martin, C. L. Thornton, and D. O. Muhleman, Ap. J. **200**, 221 (1975).
16. W. E. Thirring, Ann. Phys. **16**, 96 (1961).
17. S. N. Gupta, Rev. Mod. Phys. **29**, 334 (1957).
18. L. Halpern, Ann. Phys. **25**, 387 (1963).
19. A. Eddington, *Space, Time, and Gravitation* (Cambridge University Press, Cambridge, 1920).
20. S. Weinberg, *in* S. Deser and K. W. Ford, eds., *Lectures on Particles and Fields* (Prentice-Hall, Englewood Cliffs, 1965). See also S. Weinberg, Phys. Rev. **138**, B988 (1965).
21. P. Jordan, *Schwerkraft und Weltall* (Vieweg, Braunschweig, 1951).
22. C. Brans and R. H. Dicke, Phys. Rev. **124**, 925 (1961).

PROBLEMS

1. Show that if the gravitational field $h_{\mu\nu}$ is constant in some region, then this gravitational field can be reduced to zero by a gauge transformation in that region.

2. Find a gauge transformation that leaves invariant the equation (43). Show that in the limiting case $a \to 1$, your gauge transformation reduces to that given by Eq. (49).

3. Prove that the energy density $t_{(1)}^{00}$ given by Eq. (62) is never negative.

4. Show that for the static, spherically symmetric field given by Eq. (125), the energy density $t_{(1)}^{00}$ (see Eq. (62)) agrees with the energy density of Newtonian theory (see the first term in Eq. (1.12)).

5. In Eq. (36) it has been assumed that $p_\mu = m u_\mu$. To construct a proof of this, first show that Eq. (36) can be written as

$$\frac{d}{d\tau} P_\mu = -q A_{\nu,\mu} u^\nu$$

where $P_\mu = p_\mu - qA_\mu$. Then check that this equation of motion can be obtained from the Lagrangian

$$L = m \sqrt{u^\mu u_\mu} - qA_\mu u^u$$

if and only if $p_\mu = mu_\mu$. This *proves* that the momentum p_μ which appears in Eq. (36) is in fact the ordinary momentum mu_μ.

6. Show that in the nonrelativistic limit ($v \ll 1$), the particle energy given by Eq. (99) agrees with the Newtonian energy $\frac{1}{2}mv^2 + m\Phi$.

7. A small spherical mass (spherical in its own rest frame) moves with uniform velocity v along the z-axis, passing the origin at time $t = 0$. Find the gravitational field $h_{\mu\nu}$ at the point $x = b$ on the x-axis as a function of time. Plot h_{00}, h_{11}, and h_{22} as a function of time for $v = 0.5$ and for $v = 0.95$. (Hint: Perform a Lorentz transformation on Eq. (126)).

8. *In vacuo*, the nonlinear field equation (64) is

$$\partial_\lambda \partial^\lambda \phi^{\mu\nu} = -\kappa t^{\mu\nu}_{(1)}$$

where $t^{\mu\nu}_{(1)}$ is given by Eq. (62). Check that this nonlinear equation has the approximate solution

$$\phi^{\mu\nu} = \begin{pmatrix} -\dfrac{4GM}{\kappa r} + \dfrac{G^2M^2}{\kappa r^2} & 0 & 0 & 0 \\[2ex] 0 & -\dfrac{G^2M^2}{\kappa}\dfrac{x^2}{r^4} & -\dfrac{G^2M^2}{\kappa}\dfrac{xy}{r^4} & -\dfrac{G^2M^2}{\kappa}\dfrac{xz}{r} \\[2ex] 0 & -\dfrac{G^2M^2}{\kappa}\dfrac{xy}{r} & -\dfrac{G^2M^2}{\kappa}\dfrac{y^2}{r^4} & -\dfrac{G^2M^2}{\kappa}\dfrac{yz}{r^4} \\[2ex] 0 & -\dfrac{G^2M^2}{\kappa}\dfrac{xz}{r} & -\dfrac{G^2M^2}{\kappa}\dfrac{yz}{r^4} & -\dfrac{G^2M^2}{\kappa}\dfrac{z^2}{r^4} \end{pmatrix}$$

Take into account terms of order M and M^2, but neglect terms of order M^3. This solution gives the second-order corrections for the field surrounding a spherically symmetric mass; the perihelion precession can be calculated from this (see Problem 8.8).

9. A *relativistic* particle of mass m and velocity v passes by the sun with an impact parameter b. Find the angular deflection suffered by the particle. Make the same approximations as were used in the calculation of the deflection of light (weak gravitational field, small deflection angle).

10. A spherical mass distribution of uniform density has a total mass M and a radius R.

(a) Find the solution of Eqs. (119) and (120) in the exterior and the *interior* of the sphere. (At the boundary $r = R$, the field $\phi^{\mu\nu}$ should be continuous).

(b) Show that the deflection angle of a light ray with impact parameter $b < R$ is

$$\theta = \frac{4GM}{b} \left[1 - \frac{(R^2 - b^2)^{3/2}}{R^3} \right]$$

(c) Show that the deflection can also be expressed as $\theta = 4GM(b)/b$ where $M(b)$ is that part of the mass which is contained in a cylinder of radius b centered on the mass distribution. In this form the result is quite general and applies to any spherically symmetric mass distribution. (Since matter absorbs light, this calculation is of little interest for the propagation of light. However, matter is quite transparent to gravitational waves and neutrinos. In principle, the interior of a star can act as a "lens" to focus gravitational radiation.)

11. According to Eq. (146) [or (142)], the gravitational field of the sun may be regarded as a medium with an effective index of refraction

$$n = (1 + 2GM_\odot/r)$$

Use this expression for n and Huygens's principle for the propagation of wavefronts to derive the deflection formula (135).

12. Find an expression for the deflection angle of a light ray in the field of a *rotating* mass. Assume that the light ray moves in the equatorial plane of the rotating mass. Consider both the case of a light ray whose orbital angular momentum is parallel to the spin of the central mass and the case of angular momentum antiparallel to spin. Numerically estimate the magnitude of the effect produced by the spin of the sun for an impact parameter $b = R_\odot$.

4. GRAVITATIONAL WAVES

The earth is just a silly ball
To them, through which they simply pass,
Like dustmaids down a drafty hall
Or photons through a sheet of glass.

John Updike

Gravitational effects cannot propagate with infinite speed. This is obvious both from the lack of Lorentz invariance of infinite speed, and from the causality violations that are associated with signal speeds in excess of the speed of light. Since the speed of light is the only Lorentz-invariant speed, we expect that gravitational effects propagate in the form of waves at the speed of light.

As a concrete example, consider an apple that hangs on a tree. At some time, the stem of the apple breaks and the apple falls to the ground, i.e., there is a sudden change in the terrestrial mass distribution. The gravitational field surrounding the earth must then adapt itself to this new mass distribution. The change in the field will not occur simultaneously throughout the universe; at any given point of space the change will be delayed by a time equal to the time needed for a light signal to travel from the earth to that point. Hence the disturbance in the gravitational field propagates outwards at the speed of light. Such a propagating disturbance is a gravitational wave.

Since the existence of gravitational waves is such an obvious consequence of (special) relativity, why would the experimental discovery of these waves cause much excitement? Although the *existence* of waves is assured by general arguments, the strength and type of wave depends on the details of the gravitational theory and hence investigation of the properties of the waves would serve as a test of gravitational theory. Even more important: gravitational-wave astronomy would be a useful supplement to optical, radio, and X-ray astronomy. The energy, pulse shape, and polarization of bursts of gravitational radiation could tell us a great deal about the astrophysical processes in which these bursts are generated. Table 4.1, adapted from a review by Press and Thorne,[1] lists gravitational-wave frequency bands and their likely origins.

TABLE 4.1   FREQUENCY BANDS FOR GRAVITATIONAL WAVES

Designation	Frequency	Typical Sources
Extremely low frequency	10^{-7}/sec to 10^{-4}/sec	cosmological? explosions in quasars and galactic nuclei binaries
Very low frequency	10^{-4}/sec to 10^{-1}/sec	short-period binaries huge black holes ($\sim 10^5$ to $10^8\ M_\odot$)
Low frequency	10^{-1}/sec to 10^2/sec	pulsars
Medium frequency	10^2/sec to 10^5/sec	black holes (1–$10^3\ M_\odot$) collapse of stars supernovae, birth of neutron stars
High frequency	10^5/sec to 10^8/sec	man-made?
Very high frequency	10^8/sec to 10^{11}/sec	black-body cosmological?

The most promising frequency band is that of medium frequency. There are several possible sources and, fortunately, detectors that respond in this frequency range can be built. There is little doubt that gravitational waves are incident on the earth; the question is, can we build a detector sufficiently sensitive to feel them?

4.1   PLANE WAVES

According to Eqs. (3.55) and (3.56), the linear field equation *in vacuo* is

$$\partial_\lambda \partial^\lambda \phi^{\mu\nu} = 0 \qquad\qquad [1]$$

with the gauge condition

$$\partial_\mu \phi^{\mu\nu} = 0 \tag{2}$$

Let us look for plane wave solutions of the form

$$\phi^{\mu\nu} = \epsilon^{\mu\nu} \cos k_\alpha x^\alpha \tag{3}$$

where $\epsilon^{\mu\nu}$ is a constant tensor and k_α is a constant vector; these are called, respectively, the *polarization tensor* and the *wave vector*. This plane wave will satisfy Eqs. (1) and (2) provided

$$k_\alpha k^\alpha = 0 \tag{4}$$

$$\epsilon^{\mu\nu} k_\mu = 0 \tag{5}$$

EXERCISE 1. Show this.

The frequency of the wave is $\omega = k^0$. According to Eq. (4),

$$k^0 = (k_x{}^2 + k_y{}^2 + k_z{}^2)^{1/2} = |\mathbf{k}| \tag{6}$$

and hence $\omega/|\mathbf{k}| = 1$, so that the velocity of the wave is that of light.

The tensor $\epsilon^{\mu\nu}$ is symmetric. A general symmetric second-rank tensor has ten independent components, but Eq. (5) imposes an extra four conditions on these ten components. Hence $\epsilon^{\mu\nu}$ has only six independent components, i.e., there are *six* linearly independent tensors that will solve Eq. (5). Let us suppose that the wave propagates in the z-direction, so that

$$k^\alpha = (\omega, 0, 0, \omega) \tag{7}$$

For this special case, the linearly independent solutions of Eq. (5) may be taken to be the following:

$$\epsilon^\mu{}_{(1)} \epsilon^\nu{}_{(2)} + \epsilon^\nu{}_{(1)} \epsilon^\mu{}_{(2)} \tag{8}$$

$$\epsilon^\mu{}_{(1)} \epsilon^\nu{}_{(1)} - \epsilon^\mu{}_{(2)} \epsilon^\nu{}_{(2)} \tag{9}$$

$$\epsilon^\mu{}_{(1)} k^\nu + \epsilon^\nu{}_{(1)} k^\mu \tag{10}$$

$$\epsilon^\mu{}_{(2)} k^\nu + \epsilon^\nu{}_{(2)} k^\mu \tag{11}$$

$$k^\mu k^\nu \tag{12}$$

$$\epsilon^\mu{}_{(4)} k^\nu + \epsilon^\nu{}_{(4)} k^\mu - \eta^{\mu\nu} \epsilon^\alpha{}_{(4)} k_\alpha \tag{13}$$

where the vectors $\epsilon^{\mu}{}_{(1)}$, $\epsilon^{\mu}{}_{(2)}$, and $\epsilon^{\mu}{}_{(4)}$ are defined as follows:

$$\epsilon^{\mu}{}_{(1)} = (0,1,0,0) \qquad [14]$$

$$\epsilon^{\mu}{}_{(2)} = (0,0,1,0) \qquad [15]$$

$$\epsilon^{\mu}{}_{(4)} = (1,0,0,-1) \qquad [16]$$

EXERCISE 2. Show that (8)–(13) satisfy Eq. (5). Show that (8)–(13) are linearly independent.

EXERCISE 3. Show that the tensors (8)–(12) are traceless –

$$\epsilon_{\mu}{}^{\mu} = 0 \qquad [17]$$

– and show that (13) is *not* traceless.

The polarization tensors (8) and (9) are called *transverse*. They have the form

$$\epsilon^{\mu\nu}{}_{I} = \epsilon^{\mu}{}_{(1)}\epsilon^{\nu}{}_{(2)} + \epsilon^{\nu}{}_{(1)}\epsilon^{\mu}{}_{(2)} = \begin{pmatrix} 0 & 0 & 0 & 0 \\ 0 & 0 & 1 & 0 \\ 0 & 1 & 0 & 0 \\ 0 & 0 & 0 & 0 \end{pmatrix} \qquad [18]$$

and

$$\epsilon^{\mu\nu}{}_{II} = \epsilon^{\mu}{}_{(1)}\epsilon^{\nu}{}_{(1)} - \epsilon^{\mu}{}_{(2)}\epsilon^{\nu}{}_{(2)} = \begin{pmatrix} 0 & 0 & 0 & 0 \\ 0 & 1 & 0 & 0 \\ 0 & 0 & -1 & 0 \\ 0 & 0 & 0 & 0 \end{pmatrix} \qquad [19]$$

These are the only polarizations that correspond to physical gravitational waves. The other four polarizations, (10)–(13), carry no energy and no momentum. In fact, waves of the other four types can always be completely canceled by a gauge transformation. For example, consider a wave of type (10),

$$\phi^{\mu\nu} = (\epsilon^{\mu}{}_{(1)}k^{\nu} + \epsilon^{\nu}{}_{(1)}k^{\mu}) \cos k_{\alpha}x^{\alpha} \qquad [20]$$

The gauge transformation (3.49) for $h^{\mu\nu}$ implies the gauge transformation

$$\phi^{\mu\nu} \rightarrow \phi'^{\mu\nu} = \phi^{\mu\nu} + \Lambda^{(\mu,\nu)} - \tfrac{1}{2}\eta^{\mu\nu}\Lambda^{\alpha}{}_{,\alpha} \qquad [21]$$

for $\phi^{\mu\nu}$. It is easy to see that this gives $\phi'^{\mu\nu} = 0$ if we take

$$\Lambda^{\mu} = -2\epsilon^{\mu}{}_{(1)} \sin k_{\alpha}x^{\alpha} \qquad [22]$$

EXERCISE 4. Find similar gauge transformations that cancel waves of the types (11)–(13).

If we substitute the wave solution (3) into the energy-momentum tensor $t^{\mu\nu}{}_{(1)}$ of Eq. (3.62), we obtain

$$t^{\mu\nu}{}_{(1)} = \tfrac{1}{2}(\epsilon^{\alpha\beta}\epsilon_{\alpha\beta} - \tfrac{1}{2}\epsilon^{\alpha}{}_{\alpha}\epsilon^{\beta}{}_{\beta})k^{\mu}k^{\nu}\sin^2 k_{\sigma}x^{\sigma} \qquad [23]$$

The transverse waves, with amplitudes A_I, A_{II},

$$\phi_I^{\mu\nu} = A_I\epsilon_I^{\mu\nu}\cos k_{\alpha}x^{\alpha} \qquad [24]$$

$$\phi_{II}^{\mu\nu} = A_{II}\epsilon_{II}^{\mu\nu}\cos k_{\alpha}x^{\alpha} \qquad [25]$$

give, respectively, energy-momentum tensors

$$(A_I)^2 k^{\mu}k^{\nu}\sin^2 k_{\alpha}x^{\alpha} \qquad [26]$$

$$(A_{II})^2 k^{\mu}k^{\nu}\sin^2 k_{\alpha}x^{\alpha} \qquad [27]$$

The $\mu = 3$, $\nu = 0$ components of these tensors give the energy flux in the z-direction. For the average energy flux, we must replace $\cos^2 k_{\alpha}x^{\alpha}$ by $\tfrac{1}{2}$ and we obtain, respectively,

$$\tfrac{1}{2}(A_I)^2\omega^2 \qquad [28]$$

$$\tfrac{1}{2}(A_{II})^2\omega^2 \qquad [29]$$

To get the energy flux in cgs units, each of these expressions must be multiplied by c^{-1}.

Each of the "ghost" waves of types (10)–(13) has $t^{\mu\nu}_{(1)} = 0$. Furthermore, if $\epsilon^{\mu\nu}$ in Eq. (23) is an arbitrary superposition of (8)–(13), then the value of $t^{\mu\nu}_{(1)}$ is exactly what you get if the terms involving (10)–(13) are omitted from the superposition; that is, the value of $t^{\mu\nu}_{(1)}$ depends on the amplitudes A_I and A_{II} only.

EXERCISE 5. Show this.

EXERCISE 6. Show that gravitational waves, just as electromagnetic waves, carry not only energy but also momentum. Show that the flux of momentum is equal to the energy flux (in our units).

It is often convenient to make use of the complex waves

$$\epsilon_I^{\mu\nu}\, e^{ik_{\alpha}x^{\alpha}} \qquad [30]$$

$$\epsilon_{II}^{\mu\nu} e^{ik_\alpha x^\alpha} \qquad\qquad\qquad\qquad [31]$$

in which case it is understood that the real part is to be regarded as the physical wave.

We can also define circularly polarized waves as

$$(\epsilon_I^{\mu\nu} + i\epsilon_{II}^{\mu\nu}) e^{ik_\alpha x^\alpha} \qquad\qquad\qquad\qquad [32]$$

$$(\epsilon_I^{\mu\nu} - i\epsilon_{II}^{\mu\nu}) e^{ik_\alpha x^\alpha} \qquad\qquad\qquad\qquad [33]$$

It can be shown that such circularly polarized waves carry angular momentum. The amount of angular momentum is proportional to the amount of energy carried by the wave:

$$\text{(angular momentum in wave)} = \frac{2}{\omega} \text{ (energy in wave)} \qquad [34]$$

Waves with the polarization (32) have angular momentum parallel to the energy flux and are said to have *positive helicity;* those with the polarization (33) have angular momentum antiparallel to the energy flux and are said to have *negative helicity*. The result (34) cannot be obtained directly from our solution (3); this solution ignores the boundaries of the wave in the transverse direction and it is precisely the boundary region which is crucial for the transport of angular momentum.

The quantum mechanical interpretation of Eq. (34) is that the quanta of the gravitational field, or gravitons, have spin $2\hbar;$ in that case the ratio of angular momentum to energy per quantum is

$$2\hbar/\hbar\omega = 2/\omega$$

in agreement with Eq. (34). Incidentally, our result that there exist only two physical states of polarization for a gravitational wave of given momentum corresponds to the well-known result of relativistic quantum theory that a particle of mass zero must have its spin along the direction of motion (positive helicity) or opposite to the direction of motion (negative helicity); all other directions for the spin are forbidden.

What is the effect of a gravitational wave on a particle? The equation of motion of a particle is (see Eq. (3.93))

$$\frac{du_\mu}{dt} = -\kappa(h_{\mu\alpha,\beta} - \tfrac{1}{2}h_{\alpha\beta,\mu})u^\alpha u^\beta \qquad\qquad [35]$$

For a particle initially at rest, or moving at low speed, $u^0 \simeq 1$, $u^k \simeq 0$ and hence

$$\frac{du_\mu}{dt} = -\kappa(h_{\mu 0,0} - \tfrac{1}{2}h_{00,\mu}) \tag{36}$$

But for each of the physical gravitational waves (24) and (25) we have $h_{\mu 0} = 0$; hence the right side of Eq. (36) vanishes. Although this means that the position of the particle, as measured by our coordinates, does not change, we must be careful not to jump to the conclusion that there is no physical effect at all. The point is this: Suppose we are dealing with a wave of type (25). Consider two particles placed on the x-axis, one at $x = x_0$ and the other at $x = -x_0$. According to Eq. (36) the coordinate interval between these particles remains fixed at the value $\Delta x = 2x_0$. But the physical distance, as measured by a meter stick, depends not only on Δx, but also on the *metric* of spacetime. We have already argued (see Section 3.6) that the metric is not $\eta_{\mu\nu}$, but rather

$$g_{\mu\nu} = \eta_{\mu\nu} + \kappa h_{\mu\nu} \tag{37}$$

If this is so, then the measured distance between the two particles is

$$\Delta l^2 = -g_{11}\Delta x^2 = (1 - \kappa h_{11})(2x_0)^2$$
$$= (1 - \kappa A_{II} \cos \omega t)(2x_0)^2$$

In the context of the linear approximation, the amplitude A_{II} is small and hence

$$\Delta l \approx \left(1 - \frac{\kappa}{2} A_{II} \cos \omega t\right)(2x_0) \tag{38}$$

Thus the distance between the particles oscillates. Expressed in another way, the particles remain at rest relative to the coordinates, but the coordinates oscillate relative to meter sticks.

EXERCISE 7. Show that the distance between two particles on the y-axis, at $y = \pm y_0$, exposed to the gravitational wave (25), varies as follows:

$$\Delta l = \left(1 + \frac{\kappa}{2} A_{II} \cos \omega t\right)(2y_0) \tag{39}$$

Comparison of (38) and (39) shows that the distance between the first pair of particles is at minimum when the distance between the second pair is at maximum, and vice versa.

Fig. 4.2 shows the deformation produced by the gravitational wave (25) on a "necklace" of *free* particles originally placed in a circle in the x-y plane. Fig. 4.1 shows the corresponding result for the gravitational wave (24). It is obvious that the only difference between these figures is

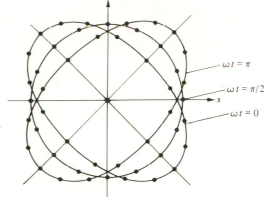

Fig. 4.1 *Deformation of a circular "necklace" of particles by a gravitational wave with polarization of type I.*

a rotation by 45°. In fact, the polarization tensor (18) can be obtained from (19) by a 45°-rotation.

EXERCISE 8. Prove this.

Finally, Figs. 4.3 and 4.4 show the deformation produced by the circularly polarized waves (32) and (33). In these two cases only the deformation, or bulge, rotates in the direction shown; the necklace does not rotate, the particles only oscillate around their initial positions.

The deformations shown in all these figures are essentially *tidal* effects, similar to those of Section 1.6, but time-dependent rather than static.

Let us calculate the tidal force needed to produce the motion given in Eq. (38). The acceleration of each particle towards the origin is

$$\frac{d^2}{dt^2}(\Delta l) = -\frac{\kappa}{2}A_{II}x_0\frac{d^2}{dt^2}\cos\omega t$$

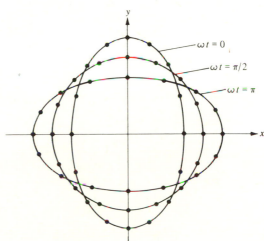

Fig. 4.2 *Deformation by a gravitational wave with polarization of type II.*

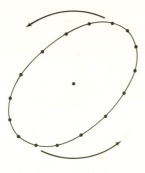

Fig. 4.3   *Deformation by a wave of positive helicity. The deformation rotates in the counterclockwise direction if the wave is propagating towards you.*

$$= \frac{\kappa}{2} A_{II} \omega^2 x_0 \cos \omega t \qquad [40]$$

Hence the tidal force on a particle on the x-axis is

$$f_x = \left(m \frac{\kappa}{2} A_{II} \omega^2 \cos \omega t \right) x_0 \qquad [41]$$

where m is the mass of the particle. Likewise, the tidal force on particles on the y-axis, at $y = y_0$, is

$$f_y = - \left(m \frac{\kappa}{2} A_{II} \omega^2 \cos \omega t \right) y_0 \qquad [42]$$

These expressions should be compared with our earlier results (see Eqs. (1.35)–(1.37)) for the tidal force in a static gravitational field. In both cases the strength of the tidal force increases directly with the distance. Note that in Eqs. (41) and (42) x_0 and y_0 may be regarded, to a sufficient approximation, as the distances of the particles from the origin. The difference between the true distance (Δl) and the coordinate interval (Δx) is very important in (38), but it is not important in (41) and (42) since there it only leads to higher-order corrections.

Incidentally, the forces f_x and f_y associated with a gravitational wave can be written in the form

$$f_x = g(t)x_0 \qquad f_y = -g(t)y_0 \qquad [43]$$

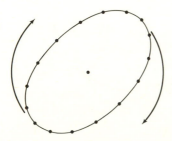

Fig. 4.4   *Deformation by a wave of negative helicity.*

where we are now assuming that both x_0 and y_0 are nonzero so that both forces (41) and (42) are present simultaneously. From this it is easy to see that

$$\frac{\partial f_x}{\partial x_0} + \frac{\partial f_y}{\partial y_0} = 0$$

Hence the divergence of **f** is zero and this implies that the force field can be represented graphically by field lines. These field lines are shown in Fig. 4.5, for $t = 0$. Obviously, such a force field will tend to produce deformations of the type shown in Fig. 4.2.

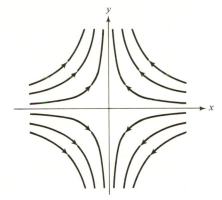

Fig. 4.5 *The tidal field lines for a gravitational wave of type II. The field lines reverse direction every half period.*

EXERCISE 9. Show that the radial component of the tidal force can be written

$$f_r = \left(m \frac{\kappa}{2} A_{II} \omega^2 \cos \omega t \right) r_0 \cos 2\phi \qquad [44]$$

where $r_0 = (x_0^2 + y_0^2)^{1/2}$ is the radial coordinate of the particle and ϕ is the azimuthal coordinate.

Note that our results are only valid for *weak* gravitational waves. To be precise, the wave must have a small amplitude, so that $\kappa A \ll 1$.

As an illustration of these results, consider the effect of a gravitational wave on the earth-moon system. If the frequency of the wave is much larger than the frequency of the orbital motion of the moon, then we can regard the earth and the moon as two free particles. In response to the gravitational wave, incident transversely, the earth-moon distance will oscillate as indicated by Eq. (38). The quantity $2x_0$ may be interpreted as the earth-moon distance measured when the wave is absent ($2x_0 \simeq 3.8 \times 10^{10}$ cm).

For a not too unreasonable astrophysical source of gravitational radiation (see Table 4.3), a wave with a rather large energy flux of $\sim 10^{10}$ ergs/cm^2 sec (at $\omega \sim 10^4$/sec) has $\kappa h_{11} = \kappa A_{II} \sim 10^{-17}$. This means that the earth-moon distance varies by about one part in 10^{17}; this amounts to roughly 10^{-7} cm. By means of a laser pulse sent to the moon and reflected back to the earth, the distance can nowadays be measured to within ~ 10 cm. Clearly, this laser-ranging technique is not sufficiently precise for detection of the gravitational wave effect.

A much more precise measurement of distance can be carried out with an interferometer of laboratory size. Forward and Moss[2] have constructed a gravitational wave detector which consists of a laser interferometer of the Michelson-Morley type. A gravitational wave incident from a direction perpendicular to the two arms of the interferometer, with a suitable polarization, will lengthen one arm and shorten the other. The apparatus is capable of detecting changes in distance of $\sim 10^{-13}$ cm, or roughly one part in 10^{15}.

The expressions for the change in distance (Eqs. (38) and (39)) and for the tidal force (Eqs. (41) and (42)) hinge, of course, on the interpretation of (37) as the metric of spacetime. We will rigorously justify this interpretation in the next chapter, but for now we only note that some of the important results of the present chapter can be derived without relying on this geometric interpretation. In particular, the expressions for the cross section for absorption of gravitational waves (see Eqs. (127), (130)) can be obtained by arguments which rely only on general theorems of scattering theory[3] or on the principle of detailed balance.[4] However, for the sake of simplicity we will base our derivation on the above expressions for the tidal force.

4.2   THE EMISSION OF GRAVITATIONAL RADIATION

If we have some system with an energy-momentum tensor that changes in time, then gravitational waves will be emitted by the system. The calculation of the radiation fields proceeds in much the same way as the calculation of the electromagnetic radiation fields associated with a time-dependent charge distribution. We begin with the (linear) field equation (3.55),

$$\partial_\lambda \partial^\lambda \phi^{\mu\nu}(t,\mathbf{x}) = -\kappa T^{\mu\nu}(t,\mathbf{x}) \qquad [45]$$

where it is understood that $T^{\mu\nu}$ is the energy-momentum tensor of *matter*. The retarded solution of Eq. (45) is

$$\phi^{\mu\nu}(t,\mathbf{x}) = -\frac{\kappa}{4\pi} \int \frac{T^{\mu\nu}(t - |\mathbf{x} - \mathbf{x}'|, \mathbf{x}')}{|\mathbf{x} - \mathbf{x}'|} \, d^3x' \qquad [46]$$

EXERCISE 9. Verify that this is a solution of (45). (Hint: Look at any textbook of electromagnetism.)

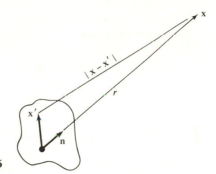

Fig. 4.6

We will assume that the field point **x** is in the radiation zone, far away from the matter system (see Fig. 4.6); this means that we can replace $|\mathbf{x} - \mathbf{x}'|$ by $|\mathbf{x}| \equiv r$ in the denominator of the integrand in Eq. (46). We will further assume that $T^{\mu\nu}$ does not change very quickly in time; if $T^{\mu\nu}$ does not have an excessively strong time dependence, then it will be a good approximation to replace $t - |\mathbf{x} - \mathbf{x}'|$ by $t - |\mathbf{x}|$ in the numerator of the integrand in Eq. (46).* With the notation $r \equiv |\mathbf{x}|$ we then obtain

$$\phi^{\mu\nu}(t,\mathbf{x}) = \frac{-\kappa}{4\pi r} \int T^{\mu\nu}(t - r, \mathbf{x}')d^3x' \qquad [47]$$

In the linear approximation, $T^{\mu\nu}$ satisfies the condition

$$\partial_\nu T^{\mu\nu} = 0 \qquad [48]$$

Let us separate (48) into space and time components:

$$\frac{\partial}{\partial t} T^{k0} = -\partial_l T^{kl} \qquad [49]$$

$$\frac{\partial}{\partial t} T^{00} = -\partial_l T^{0l} \qquad [50]$$

As a consequence of (49), we have the following identity—

$$\int T^{kl}d^3x = \frac{1}{2}\frac{\partial}{\partial t} \int (T^{k0}x^l + T^{l0}x^k)d^3x \qquad [51]$$

where the integration volume includes all of the matter system.

* The condition for the validity of this approximation can be expressed in the alternative forms $\omega b/c \ll 1$, $b/\lambda \ll 1$, or $v/c \ll 1$, where ω is the frequency of the emitted radiation, λ the wavelength, b the typical dimension of the radiating system, and v the typical velocity of the particles in the system. Corrections to our approximation give multipole radiation of higher order.

EXERCISE 10. Derive this identity. (Hint: Use (49) on the right side of (51) and integrate by parts.)

Similarly, as a consequence of (50), we have the identity

$$\int (T^{k0}x^l + T^{l0}x^k)d^3x = \frac{\partial}{\partial t}\int T^{00}x^kx^ld^3x \qquad [52]$$

EXERCISE 11. Derive this identity. (Hint: Use (50) and integration by parts.)

If we combine (51) and (52), we find that

$$\int T^{kl}d^3x = \frac{1}{2}\frac{\partial^2}{\partial t^2}\int T^{00}x^kx^ld^3x \qquad [53]$$

which shows that most of the integrals appearing on the right side of (47) can be expressed in terms of T^{00}. For nonrelativistic matter, we can approximate

$$T^{00} = \text{(energy density)} \simeq \text{(rest mass density)} \equiv \rho \qquad [54]$$

and we obtain

$$\phi^{kl}(t,\mathbf{x}) = -\left[\frac{\kappa}{8\pi r}\frac{\partial^2}{\partial t^2}\int \rho(\mathbf{x}')x'^kx''^ld^3x'\right]_{t-r} \qquad [55]$$

This can also be expressed in terms of the *quadrupole moment tensor* (see Eq. (1.17))

$$Q^{kl} = \int (3x'^kx''^l - r'^2\delta_k{}^l)\rho(\mathbf{x}')d^3x' \qquad [56]$$

in terms of which

$$\phi^{kl}(t,\mathbf{x}) = -\frac{\kappa}{8\pi r}\frac{1}{3}\left[\frac{\partial^2}{\partial t^2}Q^{kl} + \delta_k{}^l\frac{\partial^2}{\partial t^2}\int r'^2\rho d^3x'\right]_{t-r} \qquad [57]$$

For the calculation of the energy flux, we can omit the term proportional to $\delta_k{}^l$ in the bracket of Eq. (57) because this part of the wave carries away no energy. To see this, note that if the distance r is large enough, then ϕ^{kl} can be regarded, to a sufficient approximation, as a plane wave in the vicinity of the point $\mathbf{x}$. According to the preceding section, the only plane wave polarizations that carry energy are $\epsilon_I^{\mu\nu}$ and $\epsilon_{II}^{\mu\nu}$ as given by Eqs. (18) and (19); obviously $\delta_k{}^l$ is not a polarization of this type. If we omit the $\delta_k{}^l$ term, then

$$\phi^{kl}(t,\mathbf{x}) = -\frac{\kappa}{8\pi r}\frac{1}{3}\ddot{Q}^{kl} \qquad [58]$$

where the dots stand for time derivatives and it is understood that the right side is to be evaluated at the retarded time $t - r$.

The energy flux in the direction of the radial direction is given by

$$t^{0s}n^s = \frac{n^s}{4} \left(2\phi^{\alpha\beta,0}\phi_{\alpha\beta},^s - \phi^{,0}\phi^{,s} \right) \tag{59}$$

where $n^s = (n_x, n_y, n_z)$ is the unit vector in the radial direction (see Fig. 4.6). Let us separate $\phi^{\alpha\beta}$ into space and time parts,

$$t^{0s}n^s = \frac{n^s}{4} \left(2\phi^{kl,0}\phi^{kl,s} - 4\phi^{k0,0}\phi^{k0,s} + 2\phi^{00,0}\phi^{00,s} - \phi^{kk,0}\phi^{ll,s} \right.$$

$$\left. + \phi^{kk,0}\phi^{00,s} + \phi^{00,0}\phi^{kk,s} - \phi^{00,0}\phi^{00,s} \right)$$

$$= \frac{n^s}{4} \left(2\phi^{kl,0}\phi^{kl,s} - 4\phi^{k0,0}\phi^{k0,s} + \phi^{00,0}\phi^{00,s} \right) \tag{60}$$

where we have taken into account that, according to (58) and (56),

$$\phi^{kk} \propto Q^{kk} = 0$$

In the evaluation of $\phi_{\alpha\beta,s}$ we can neglect the derivative of the factor $1/r$ appearing in (58) since we are assuming that r is very large.* The only other place where a dependence on r occurs is in the retarded time, $t - r$; this means that r and t *only* appear in the combination $t - r$ and therefore

$$\phi^{\alpha\beta},_s = -\phi^{\alpha\beta},_0 \frac{\partial r}{\partial x^s} = -\phi^{\alpha\beta},_0 n^s \tag{61}$$

In particular, the derivative appearing in the first term in the parentheses in Eq. (60) can be written

$$\phi^{kl,s} = \phi^{kl,0}n^s \tag{62}$$

We can obtain convenient expressions for the other terms appearing in Eq. (60) by beginning with our familiar gauge condition

$$\phi^{\mu 0},_0 = -\phi^{\mu l},_l \tag{63}$$

This gives

$$\phi^{k0,0} = -\phi^{kl},_l = \phi^{kl},_0 n^l = \phi^{kl,0}n^l \tag{64}$$

---

* This approximation is justified whenever $r \gg \lambda$, where λ is the wavelength of the emitted radiation.

and also

$$\phi^{00,0} = -\phi^{0l},_l = \phi^{0l},_0 n^l = \phi^{kl,0} n^k n^l \tag{65}$$

Substituting (62), (64), and (65) into (60), we obtain

$$t^{0s} n^s = \frac{n^s}{4} \left(2\phi^{kl,0}\phi^{kl,0} n^s - 4\phi^{kl,0} n^l \phi^{km,0} n^m n^s \right.$$

$$\left. + \phi^{kl,0} n^k n^l \phi^{mr,0} n^m n^r n^s \right) \tag{66}$$

Using $n^s n^s = 1$ and then Eq. (58), this becomes

$$t^{0s} n^s = \tfrac{1}{4}(2\phi^{kl,0}\phi^{kl,0} - 4\phi^{kl,0}\phi^{km,0} n^l n^m + \phi^{kl,0}\phi^{mr,0} n^k n^l n^m n^r)$$

$$= \frac{1}{9}\left(\frac{\kappa}{8\pi r}\right)^2 (\tfrac{1}{2}\ddot{Q}^{kl}\ddot{Q}^{kl} - \ddot{Q}^{kl}\ddot{Q}^{km} n^l n^m + \tfrac{1}{4}\ddot{Q}^{kl}\ddot{Q}^{mr} n^k n^l n^m n^r) \tag{67}$$

The energy radiated per unit solid angle and unit time in the direction n^s is

$$-\frac{d^2 E}{dt d\Omega} = r^2 t^{0s} n^s \tag{68}$$

Obviously the angular distribution is very complicated; however, the total power radiated can be obtained by integrating (68) over all solid angles and the final expression turns out to be quite simple. We have

$$-\frac{dE}{dt} = \int r^2 t^{0s} n^s d\Omega$$

$$= \frac{1}{9}\left(\frac{\kappa}{8\pi}\right)^2 \left(\tfrac{1}{2}\ddot{Q}^{kl}\ddot{Q}^{kl} \int d\Omega - \ddot{Q}^{kl}\ddot{Q}^{km} \int n^l n^m d\Omega \right. \tag{69}$$

$$\left. + \tfrac{1}{4}\ddot{Q}^{kl}\ddot{Q}^{mr} \int n^k n^l n^m n^r d\Omega \right)$$

The average values of $n^l n^m$ and $n^k n^l n^m n^r$ over the surface of a sphere are, respectively,

$$\frac{1}{4\pi}\int n^l n^m d\Omega = \frac{1}{3}\,\delta_l^{\,m} \tag{70}$$

$$\frac{1}{4\pi}\int n^k n^l n^m n^r d\Omega = \frac{1}{15}\,(\delta_k^{\,l}\delta_m^{\,r} + \delta_k^{\,m}\delta_l^{\,r} + \delta_k^{\,r}\delta_l^{\,m}) \tag{71}$$

EXERCISE 12. Show this. (Hint: Unless the indices appearing in (70) and (71) are equal in pairs, the integral vanishes because the integrand is odd. This shows that the right and left sides of these equations must be *proportional;* determine the constant of proportionality.)

Substituting (70) and (71) into (69), we obtain

$$-\frac{dE}{dt} = \frac{4\pi}{45}\left(\frac{\kappa}{8\pi}\right)^2 \dddot{Q}^{kl}\dddot{Q}^{kl} \qquad [72]$$

EXERCISE 13. Check this.

If we insert the value $\kappa^2 = 16\pi G$ and return to cgs units, the radiated power becomes

$$-\frac{dE}{dt} = \frac{G}{45c^5}\dddot{Q}^{kl}\dddot{Q}^{kl} \qquad [73]$$

In Section 1.3 we have seen that a shift of the origin of coordinates changes the quadrupole moment by an additive constant (see Eq. (1.18)). Since the emission of radiation only involves the time-dependent part of Q^{kl}, it follows that changes in the origin will not affect our results; we can choose any convenient point as origin.

The radiation we have calculated above is often called *quadrupole radiation*. Since our results all involve the quadrupole moment tensor, this is as good a name as any. However, it should be kept in mind that if we were to make in the electromagnetic case the same kind of approximations that we have made above in the gravitational case, we would get electromagnetic dipole radiation—that is, the analog of gravitational "quadrupole" radiation is electromagnetic dipole radiation.

4.3 EMISSION BY A VIBRATING QUADRUPOLE

We will now discuss some examples of systems that can radiate gravitational waves. One of the simplest such systems is the linear quadrupole consisting of two equal masses connected by a spring (see Fig. 4.7). The positions of the two masses are

$$z = \pm (b + a \sin \omega t) \qquad [74]$$

respectively, and we assume that $a << b$. Hence we can *approximate*

$$z^2 \simeq b^2 + 2ba \sin \omega t \qquad [75]$$

The retarded values of the quadrupole moment are

$$Q^{kl} \simeq \left[1 + \frac{2a}{b}\sin \omega(t-r)\right]Q(0)^{kl} \qquad [76]$$

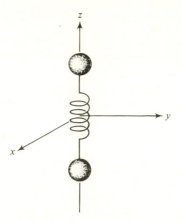

Fig. 4.7 *Oscillating quadrupole.*

where

$$Q(0)^{kl} = \begin{pmatrix} -2mb^2 & 0 & 0 \\ 0 & -2mb^2 & 0 \\ 0 & 0 & 4mb^2 \end{pmatrix}$$

[77]

is the static quadrupole moment.

EXERCISE 14. Derive (76).

EXERCISE 15. Show that the radiated field is

$$\phi^{kl} = + \frac{\kappa}{8\pi r} \frac{2a}{3b} \, \omega^2 \sin \omega(t - r) Q(0)^{kl}$$

[78]

The angular distribution of radiation given by Eq. (68) is

$$\frac{d^2E}{dt\,d\Omega} = \frac{1}{9}\left(\frac{\kappa}{8\pi}\right)^2 \left[\tfrac{1}{2}(\ddot{Q}^{11})^2 + \tfrac{1}{2}(\ddot{Q}^{22})^2 + \tfrac{1}{2}(\ddot{Q}^{33})^2 \right.$$
$$- (\ddot{Q}^{11}n^1)^2 - (\ddot{Q}^{22}n^2)^2 - (\ddot{Q}^{33}n^3)^2$$
$$\left. + \tfrac{1}{4}(\ddot{Q}^{11}n^1n^1 + \ddot{Q}^{22}n^2n^2 + \ddot{Q}^{33}n^3n^3)^2\right]$$

[79]

In terms of the usual polar angles, shown in Fig. 4.8, we have

$$n^1 = \sin \theta \cos \phi$$
$$n^2 = \sin \theta \sin \phi$$
$$n^3 = \cos \theta$$

[80]

and Eq. (79) reduces to

$$-\frac{d^2E}{dt\,d\Omega} = \left(\frac{\kappa}{8\pi}\right)^2 \left[2mab\omega^3 \cos \omega(t - r)\right]^2 \sin^4 \theta$$

[81]

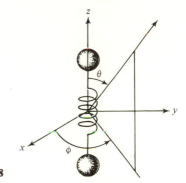

Fig. 4.8

EXERCISE 16. Derive (81) from (79).

The radiation pattern is shown in Fig. 4.9. Note that this radiation pattern in no way resembles the electromagnetic quadrupole pattern; in the latter case the flux is zero in at least three perpendicular directions (and their opposite directions). If anything, Fig. 4.9 reminds one vaguely of an electromagnetic dipole radiation pattern.

The total emitted power is

$$-\frac{dE}{dt} = \frac{G}{45c^5}\, \dddot{Q}^{kl}\dddot{Q}^{kl} = \frac{32G}{15c^5}\, [mab\ \omega^3 \cos \omega(t - r)]^2 \qquad [82]$$

with a time average

$$-\frac{\overline{dE}}{dt} = \frac{16G}{15c^5}\, (mab)^2\ \omega^6 \qquad [83]$$

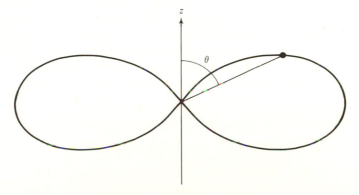

Fig. 4.9 *The radiation pattern for the emission of gravitational radiation by a quadrupole oscillating along the z-axis. This is a polar plot of the function* sin⁴ θ. *The distance between the origin and a point on the curve is proportional to the flux emitted in that direction. The pattern has cylindrical symmetry about the z-axis.*

Let us define the *damping rate* of the oscillator as the ratio of rate of loss of energy to energy

$$\gamma_{rad} = -\frac{1}{E}\frac{\overline{dE}}{dt} \tag{84}$$

Since the energy of each mass is $\frac{1}{2}m\omega^2a^2$, we obtain

$$\gamma_{rad} = \frac{16G}{15c^5}\,mb^2\omega^4 \tag{85}$$

The quantity $1/\gamma_{rad}$ is called the *damping time;* it is the time the oscillator takes to lose a fraction $1/e$ of its energy to radiation.

Although Eq. (83) was derived for a system of two point masses connected by a spring, it gives an order of magnitude estimate for the gravitational wave energy emitted in the vibration of any elastic body provided the vibration is, roughly, in one direction.* For example, we can use (83) to estimate the power radiated by a steel rod in longitudinal vibration; a must be the amplitude of oscillation, m the mass of the rod, b its length, and ω the frequency at which it is oscillating. A more precise calculation of the radiation emitted by a vibrating elastic body can be carried out by replacing the amplitude of oscillation by an effective amplitude (see Eq. (131)).

Radiation emitted by quadrupole vibrations can be very important in the case of pulsating neutron stars. Fig. 9.21 shows the structure of a typical neutron star with $m \simeq 0.7\,M_\odot$ and $R \sim 10$ km. Roughly, the star may be described as a giant nucleus made of neutrons; the average density is $\sim 10^{14}$ g/cm^3, which is comparable to the density of nuclear matter. If the body of the neutron star is disturbed from its equilibrium shape, the star will pulsate. The frequency of the lowest quadrupole mode of vibration is of the order of $\sim 3 \times 10^3$ cycles/sec.

EXERCISE 17. Show that the pulsation frequencies of an (incompressible) fluid sphere held together by gravitation are of the order

$$\omega \sim \sqrt{G\rho} \tag{86}$$

where ρ is the density. Insert relevant numbers into (86). (Hint: The velocity of sound waves in a fluid is $v \sim \sqrt{p/\rho}$ where p is the pressure. The typical pressure for a sphere of radius R is Gm^2/R^4 and $\omega \sim v/R$.)

If a neutron star is created in a supernova implosion, it is likely that the star will initially be strongly deformed and have available a large amount of vibrational energy. This vibrational energy is dissipated by

---

* Eq. (83) *cannot* be applied to radial pulsations of a sphere; in the case of spherical symmetry the radiation is exactly zero.

gravitational radiation. If the deformation has an amplitude δR, then the corresponding vibrational energy is $\frac{1}{2}m\omega^2(\delta R)^2$; for $m \sim 0.7M_\odot$, $\omega \sim 2\pi \times (3 \times 10^3)$/sec, $R \sim 10$ km, this energy can be expressed as

$$\sim \left(\frac{\delta R}{R}\right)^2 \times 10^{53} \text{ ergs} \tag{87}$$

If $\delta R/R \sim 10^{-1}$, the energy available for gravitational radiation is $\sim 10^{51}$ ergs; expressed in mass units this corresponds to $10^{-3}M_\odot$! The radiation rate is quite large and most of the energy is radiated in a time $\tau_{\text{rad}} \lesssim 1$ sec; this means that gravitational radiation is a very efficient dissipation mechanism for the vibrational energy of neutron stars.[5]

EXERCISE 18. Calculate the damping time τ_{rad} from the formula (85). Explain why the result disagrees with $\tau_{\text{rad}} \sim 1$ sec.

If we assume that the birth of such a neutron star took place at a distance of 10^4 light years from the earth, we might expect a pulse of gravitational radiation at a frequency of $\sim 3 \times 10^3$ cycles/sec, lasting for somewhat less than 1 sec and carrying a flux of $\sim 10^6$ ergs/cm^2 sec. Such a pulse would be barely detectable with presently available equipment.

"Starquakes" on a neutron star can also excite vibrations and emit gravitational radiation long after the star has exhausted the vibrational energy supply given it at birth. However, the amplitude of starquake vibrations is likely to be quite small and detection of the emitted radiation will be impossible unless the star is very near the earth.

4.4 EMISSION BY A ROTATING QUADRUPOLE

Another simple radiating system is the quadrupole rotator, consisting of two spherical masses moving in circular orbits about their common center of mass (see Fig. 4.10). The radiated power is

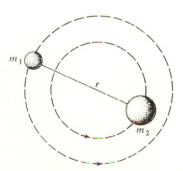

Fig. 4.10 *Two masses in circular orbits about their center of mass.*

$$-\frac{dE}{dt} = \frac{32G}{5c^5} \left(\frac{m_1 m_2}{m_1 + m_2}\right)^2 r^4 \omega^6$$ [88]

where m_1, m_2 are the masses, r is the distance between them, and ω is the frequency of rotation. The gravitational waves emitted by this system have a frequency 2ω; this is because the quadrupole moment repeats when the masses move through one-half of their orbit.

EXERCISE 19. Derive (88). Show that the result is independent of the dimension of the masses. (Hint: Show that the quadrupole moment of any arrangement of spherical masses is the same as that for the corresponding arrangement of point masses.)

EXERCISE 20. Show that the radiation emitted in the direction perpendicular to the plane of the orbit is circularly polarized; show that the radiation emitted in the plane of the orbit is "linearly" polarized.

In the case of a binary star system the force holding the masses in their orbits is gravitational, and then r and ω will be related by Kepler's law,

$$\omega^2 = G(m_1 + m_2)/r^3$$ [89]

so that Eq. (88) becomes

$$-\frac{dE}{dt} = \frac{32G^4}{5c^5 r^5}(m_1 m_2)^2(m_1 + m_2)$$ [90]

As the gravitating system loses energy by radiation, the distance between the masses decreases at a rate

$$\frac{dr}{dt} = -\frac{64G^3}{5c^5 r^3} m_1 m_2 (m_1 + m_2)$$ [91]

EXERCISE 21. Show this. (Hint: Use $E = -\frac{1}{2} G m_1 m_2/r$ and (90)).

Appreciable amounts of gravitational radiation can be emitted by a binary star system provided the distance between the components is small. For example, the system W Ursae Majoris consists of two stars in a very tight orbit with a period of only 8 hr. The characteristics of the system are listed in Table 4.2. According to Eq. (90), the emission of gravitational radiation amounts to 5×10^{29} erg/sec; the frequency of the radiation is 0.25 cycle/hr, twice the frequency of the orbital motion. The distance of this system to the earth is ~ 360 light years and the flux produced at the earth is only 10^{-13} erg/cm^2, quite undetectable. As the binary loses energy, the separation between the components decreases; Eq. (91) gives $dr/dt \simeq 10$ cm/year.

TABLE 4.2 CLOSE BINARY SYSTEMS

System	m_1	m_2	r	Period
W UMa	$0.76 M_\odot$	$0.57 M_\odot$	1.5×10^{11}cm	8.0 hr
AM CVn	0.29 to $1.07 M_\odot$	$0.041 M_\odot$	1.07 to 1.60×10^{10}cm	17.5 min
Neutron stars				
(hypothetical)	$1.0 M_\odot$	$1.0 M_\odot$	1.0×10^7cm	0.0122 sec

In the case of W UMa, the distance between the stars is comparable to their size (see Fig. 4.11). Closer binary systems are only possible if one (or both) of the components is an extremely compact object, such as a white dwarf or neutron star. The system AM Canum Venaticorum appears to be a binary with a very short period of 17.5 min. According to one model (proposed by Faulkner et al.[6]), this system consists of two white dwarf stars with the characteristics given in Table 4.2. The emission rate of gravitational radiation is around 10^{32} erg/sec. The rate of change of the distance between the components cannot be calculated from Eq. (91) because there is a significant mass transfer from the less compact (and less massive) to the more compact (and more massive) component (see Fig. 4.12).

Fig. 4.11 *The binary system W Ursae Majoris. (After Struve,* Stellar Evolution.)

Although the gravitational radiation from AM CVn is too weak to be detected at the earth, it has been argued that the existence of this system is in itself evidence for the existence of gravitational radiation. The reason is that loss of energy and angular momentum by gravitational radiation would seem to be the only available mechanism by which a binary system can evolve into an orbit of such very small dimensions.

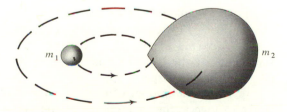

Fig. 4.12 *The binary system AM CVn. The compact star has the larger mass (m_1) and is a blue white dwarf; the other star (m_2) is a low-mass white dwarf. There is mass transfer from the latter to the former. (After Faulkner, Flannery, and Warner, Ap. J.* **175**, L79 *(1972).)*

As a final and extreme example, we consider the hypothetical case of a close binary system consisting of two neutron stars, each of a mass $\sim 1 M_\odot$ (see Table 4.2).[7] As the stars lose energy, they spiral towards each other faster and faster, emitting a crescendo of gravitational radiation. The swan song of the binary system comes to a sudden end when the distance reaches $r \sim 20$ km and the neutron stars collide.

The collapse of a binary system that begins at, say, $r_0 = 100$ km proceeds rather quickly. Initially, the radiated power is 3.3×10^{51} erg/sec and the rate of decrease of the distance is $[-dr/dt]_0 = 2.5 \times 10^6$ cm/sec. If we integrate Eq. (91), we obtain

$$\tfrac{1}{4}(r^4 - r_0^4) = - \frac{64G^3}{5c^5} m_1 m_2 (m_1 + m_2)(t - t_0) \tag{92}$$

which can also be written as

$$t - t_0 = \left(1 - \frac{r^4}{r_0^4}\right) \frac{r_0}{4} \frac{1}{[-dr/dt]_0} \tag{93}$$

where the subscripts $_0$ indicate quantities evaluated at the initial time t_0. The time needed for a decrease in the distance from r_0 to $r \ll r_0$ is therefore

$$\Delta t \simeq \frac{r_0}{4} \frac{1}{[-dr/dt]_0} \tag{94}$$

The result is (approximately) independent of the final distance r because the system spends most of its time at the larger distances, and passes through the smaller distances very quickly. If we insert the values of r_0 and $dr/dt|_0$ given above, we obtain

$$\Delta t \sim 1.0 \text{ sec} \tag{95}$$

A simple calculation shows that in this time interval the binary system completes ~ 130 revolutions; the shape of the orbit is therefore that of a tightly wound spiral. The frequency of the radiation increases with time; initially the frequency is ~ 160 cycles/sec, but towards the end it is ~ 10 times as large.

The total amount of energy emitted in this time interval is equal to the change in the total energy of the system,

$$\Delta E = \frac{G}{2} m_1 m_2 \left(\frac{1}{r_0} - \frac{1}{r}\right) \tag{96}$$

With $r_0 = 100$ km, $r = 20$ km, this amounts to

$$\Delta E = -5.3 \times 10^{52} \text{ erg} \tag{97}$$

which is 1% of the rest energy of the system! Most of this energy is emitted during the last stages of the collapse of the binary system.* The pulse of radiation emitted by a collapsing binary neutron star system somewhere in our galaxy would be of sufficient intensity to be detected with presently available equipment.

Note that the preceding calculation is not exact; relativistic corrections have been ignored. Even at $r = 100$ km, the orbital velocity is $v/c \sim \frac{1}{10}$ and by the time the distance reaches $r = 20$ km, the system is quite relativistic, with $v/c \sim \frac{2}{10}$. It is also necessary to take into account corrections arising from the nonlinear behavior of the gravitational fields. Our calculation must be regarded as an order of magnitude estimate only.

There exists another mechanism by which a neutron star can emit radiation. Most neutron stars rotate (they are pulsars) with periods that are typically in the range from 0.03 to ~ 3 sec. If such a rotating neutron star does not have exact cylindrical symmetry about the axis of rotation, then it constitutes a rotating quadrupole which will emit gravitational radiation. For example, the Crab pulsar rotates with a period of 0.033 sec and if the dimensions in the equatorial plane differ by one part in 1,200 then gravitational radiation would be emitted at a rate of 10^{38} erg/sec.[8] This energy loss would account for the observed slowing down of the rotation observed for the Crab pulsar. However, since it is known that the Crab emits large amounts of electromagnetic energy, most of the loss of rotational kinetic energy is already accounted for, and the emission rate of gravitational energy (and ellipticity in the equatorial plane) had better be smaller.

4.5 EMISSION BY AN ACCELERATED PARTICLE

The sources of gravitational radiation we have mentioned so far have all been periodic (or quasiperiodic) and are based on vibrating or rotating quadrupoles. Since any quadrupole with nonzero third time derivative will radiate, there are many other possible sources. For example, take a particle that is accelerated by falling directly towards a star along a radial line (orbit of zero angular momentum). For simplicity, assume that the mass M of the star is large compared to the mass m of the particle. This implies that the star remains at rest and only the particle contributes a time-dependent quadrupole moment. If we take the z-axis along the direction of incidence and the origin at the center of the star (see Fig. 4.13), then (73) gives

$$-\frac{dE}{dt} = \frac{2Gm^2}{15c^5} (6\dot{z}\,\ddot{z} + 2z\,\dddot{z})^2 \tag{98}$$

---

* An extra pulse of radiation will be emitted in the collision. Also, the system that forms after collision (large neutron star? black hole?) may emit more gravitational radiation.

Fig. 4.13 *Particle falling radially towards a mass.*

EXERCISE 22. Derive Eq. (98).

If the particle starts out at infinity with zero velocity (orbit of zero energy), then

$$\tfrac{1}{2}m\dot{z}^2 = \frac{GmM}{|z|} \qquad\qquad [99]$$

and

$$\dot{z} = \frac{1}{|z|^{1/2}}\,(2GM)^{1/2}$$

$$\ddot{z} = \frac{GM}{z^2}$$

$$\dddot{z} = \frac{2GM}{|z|^{7/2}}\,(2GM)^{1/2} \qquad\qquad [100]$$

Eqs. (98) and (100) give

$$-\,dE = \frac{1}{|z|^{9/2}}\,\frac{2Gm^2}{15c^5}\,(2GM)^{5/2}dz \qquad\qquad [101]$$

The energy radiated during the plunge from $z=-\infty$ to $z=-R$ is obtained by integrating (101)

$$\Delta E = -\,\frac{1}{R^{7/2}}\,\frac{4Gm^2}{105c^5}\,(2GM)^{5/2} \qquad\qquad [102]$$

Obviously, the energy radiated is large when the lower limit R is small. It is therefore advantageous to consider the case of a very compact star or other object. The most compact object that one can have is a black hole. We will discuss these objects later on; for now we only state that a black hole of mass M has a radius

$$R = \frac{2GM}{c^2} \qquad\qquad [103]$$

The gravitational fields at this radius are so intense that any particle (or light, or whatever) that approaches the object nearer than $2GM/c^2$ is unavoidably pulled into the black hole and can nevermore escape.

For the case of a particle falling directly into such a black hole, the lower limit R is that given by Eq. (103) since any gravitational radiation emitted afterwards will remain within the black hole. Substitution into Eq. (102) gives

$$\Delta E = -\frac{2}{105}\, mc^2\left(\frac{m}{M}\right)$$

$$= -0.019\, mc^2\left(\frac{m}{M}\right) \qquad\qquad [104]$$

Our calculation is defective in that we have failed to take into account that the motion becomes *relativistic* as the particle approaches $R = 2GM/c^2$. Also, nonlinear gravitational effects must be considered. A precise calculation by Davis, Ruffini, Press, and Price[9] gives

$$\Delta E = -0.0104\, mc^2\left(\frac{m}{M}\right) \qquad\qquad [105]$$

Note that the amount of gravitational wave energy radiated decreases with M. To get a large pulse of radiation we must choose m large and M as small as we dare (remember: we assumed $M >> m$ in our calculation). If we take a "particle" of mass $m = M_\odot$ and a black hole of mass $M = 10\, M_\odot$, radius $R = 2GM/c^2 \sim 30$ km, then

$$\Delta E = -2 \times 10^{51} \text{ erg} \qquad\qquad [106]$$

Most of this radiation is emitted during the final stages of the plunge — say, when the particle falls from $2R$ to R. The radiation therefore appears as a pulse of length

$$\begin{aligned}
\Delta t &\sim R/\text{(particle velocity)} \\
&\sim R/c \qquad\qquad\qquad\qquad\qquad [107] \\
&\sim 30 \text{ km}/c \sim 10^{-4} \text{ sec}
\end{aligned}$$

Such a pulse will contain frequencies up to $\sim 10^4$ cycles per second. The results of a calculation of the frequency spectrum are shown in Fig. 4.14.

As a final example of a source of gravitational radiation we mention gravitational collapse. When a star evolves into a state of high density, the gravitational field can become so intense that the interior pressure cannot support the weight of the outer layers of the star. The star

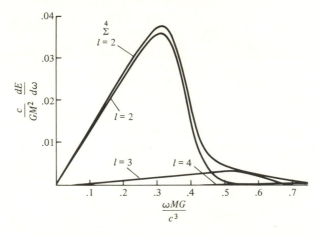

Fig. 4.14 *Spectrum of gravitational radiation emitted by a particle of mass* m *falling radially into a black hole of mass* M. *The quantity* dE/dω *gives the amount of energy radiated per unit frequency interval. The curve marked* l = 2 *corresponds to quadrupole radiation. The other curves (*l = 3, l = 4*) correspond to multipole radiation of higher order. Note that most of the radiation is emitted with frequencies below* ω ~ 0.5c³/GM. *(From Davis, Ruffini, Press, and Price,* Phys. Rev. Lett. *27, 1446 (1971).)*

crushes itself in its own gravitational field. As the material falls inwards faster and faster, the central density increases and at some stage, mass M will have collapsed into a radius less than $R = 2GM/c^2$. A black hole forms. During the collapse process, gravitational radiation can be emitted. Since it is useless to pretend that the linear approximation is applicable to this case, we only quote an estimate given by Misner, Thorne, and Wheeler.[10] In the collapse of a star of $10M_\odot$, the energy radiated may be as much as

$$\sim 2 \times 10^{54} \text{ erg} \qquad\qquad [108]$$

This is emitted as a pulse lasting $\sim 10^{-3}$ sec, and therefore contains a wide band of frequencies up to $\sim 10^3$ cps.

Table 4.3 lists conceivable astrophysical sources of gravitational radiation, and the corresponding frequency spectra and energies of gravitational radiation.

4.6 THE QUADRUPOLE DETECTOR AND ITS CROSS SECTION

A simple quadrupole oscillator, consisting of two masses connected by a spring, can be used as a detector of gravitational waves. Since we know

TABLE 4.3 ASTROPHYSICAL SOURCES OF GRAVITATIONAL RADIATION[a]

Source	Spectrum	Energy received[b]	Comments
Binary star system (of the AM CVn type)	discrete, $\nu = 2 \times 10^{-3}$/sec	10^{-9} erg/cm² sec	not detectable
Collapse of neutron binary system	glissando, $\nu \sim 200$/sec increasing to $\nu \sim 2 \times 10^3$/sec	10^{11} erg/cm²	detectable
Pulsating neutron star	discrete, $\nu = 10^3 - 10^4$/sec	10^9 erg/cm²	detectable with tuned detector
Rotating neutron star (with rigid deformation)	discrete, $\nu = 3 \times 10^2$/sec	10^{-1} erg/cm² sec	not detectable
Rapidly rotating neutron star (with rotation-induced deformation)	discrete, $\nu = 1.5 \times 10^3$/sec (slight drift to higher frequency)	10^9 erg/cm²	detectable with tuned detector
Neutron star falling into black hole (10 $M_\odot$)	continuous, peaked near $\nu \sim 10^4$/sec	10^{10} erg/cm²	detectable
Gravitational collapse of star (10 $M_\odot$) to form a black hole	continuous, peaked near $\nu \sim 10^3$/sec	10^{13} erg/cm²	detectable

[a] Based, in part, on the summary given by Rees, Ruffini, and Wheeler.[4]

[b] The energy is given in terms of the flux (erg/cm² sec) for sources that radiate continuously or in terms of the integrated flux (erg/cm²) for sources that radiate only for a short time (≤ 1 sec).

It has been assumed that the distance to the source has the standard value of 100 light years; if the distance is r, the energy must be multiplied by $(100/r)^2$.

that oscillations of this system radiate gravitational waves, we expect that a wave incident on the system will excite the oscillations. We want to calculate what oscillation amplitude the incident gravitational wave can excite and how much energy is absorbed from the wave. However, there is a difficulty: to the extent that both masses accelerate in the same way, no oscillations of the system will be excited. The gravitational force is not directly observable; only the *difference* between the forces acting on the masses produces an observable effect. Even if the experimenter looks for net translational motion of his apparatus, he will not see any because the experimenter also is accelerated in the gravitational field. Only a relative motion between the two masses, or between them and the experimenter, is noticeable. This means that the observable driving force on the oscillator is the *tidal* force.

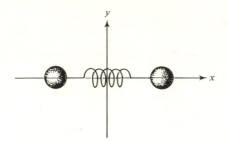

Fig. 4.15   *Quadrupole on the x-axis.*

Suppose that a plane wave, such as, for example, (25), is incident on a simple quadrupole oscillator consisting of two masses placed on the x-axis with equilibrium positions $x = \pm b$ (see Fig. 4.15). In complex notation, we can write the wave as

$$h^{\mu\nu} = A_{\mathrm{II}}\epsilon_{\mathrm{II}}^{\mu\nu}e^{i\omega t - ikz} \tag{110}$$

If the displacement of the two masses always remains small ($|x - b|$ $<< b$), then the tidal force on each mass is given by Eq. (41) as

$$f_x = \left(m\,\frac{\kappa}{2}A_{\mathrm{II}}\omega^2 e^{i\omega t}\right)x \cong \left(m\,\frac{\kappa}{2}A_{\mathrm{II}}\omega^2 e^{i\omega t}\right)b \tag{111}$$

The equation of motion of one of the masses, say, the mass on the positive x-axis, is then

$$m\ddot{x} + m\gamma\dot{x} + m\omega_0^2(x - b) = m\,\frac{\kappa}{2}A_{\mathrm{II}}\omega^2 b e^{i\omega t} \tag{112}$$

Here ω_0^2 is the natural frequency for the free oscillations and γ is the *damping rate* associated with the frictional forces acting on the oscillator. In terms of the mean (average over one cycle) rate of loss of energy to friction,

$$\gamma = -\frac{1}{E}\left(\overline{\frac{dE}{dt}}\right)_{\text{frictional}} \tag{113}$$

The time $\tau_0 \equiv \gamma^{-1}$ is the *relaxation time*, i.e., it is the time interval in which the energy of the free (driving force absent) oscillations decreases by a factor of e. The gravitational tidal force (111) plays the role of a driving force in Eq. (112).

The steady-state solution of the equation of motion is

$$x - b = \frac{\frac{1}{2}\kappa A_{\mathrm{II}}\omega^2 b e^{i\omega t}}{-\omega^2 + \omega_0^2 + i\gamma\omega} \tag{114}$$

EXERCISE 23. Check this.

If the friction in the oscillator is relatively small, so that

$$\gamma << \omega_o \qquad [115]$$

then the steady-state response of the oscillator has a sharp maximum at the frequency $\omega \simeq \omega_0$. Gravitational waves of this frequency are in resonance with the natural oscillations of the system. For $\omega = \omega_0$, Eq. (114) gives an amplitude of oscillation

$$\tfrac{1}{2}\kappa A_{\mathrm{II}}\left(\frac{\omega_0}{\gamma}\right)b \qquad [116]$$

This is larger than the amplitude of motion for a free particle* by a factor ω_0/γ. According to Eq. (115) this factor is large compared to one.

As a measure of the sensitivity of the detector it is often convenient to use a parameter called the *cross section*. Let us begin by defining the *scattering cross section*. As the gravitational wave excites oscillations in the quadrupole detector these oscillations in turn radiate gravitational waves; that is, some of the energy absorbed from the incident gravitational wave is reradiated as a new gravitational wave. The scattering cross section is defined as the ratio of power reradiated to incident flux.

$$\sigma_{\mathrm{scatt}} = \frac{\text{(power reradiated)}}{\text{(incident flux)}} \qquad [117]$$

It is understood that both power and flux are to be averaged over a cycle. The cross section defined by (117) has the dimensions of an area. However, it has nothing to do with any geometrical cross sectional area of the oscillating masses; the cross section is nothing but a parameter that measures how efficiently the oscillator scatters the radiation in all directions.

In our case, the incident wave has a flux $\tfrac{1}{2}A_{\mathrm{II}}^2\omega^2/c$. To find the radiated power, we may use Eq. (83). The quantity a that appears in this equation is the amplitude of oscillation; according to Eq. (114),

$$a = \frac{\tfrac{1}{2}\kappa A_{\mathrm{II}}\omega^2 b}{|-\omega^2 + \omega_0^2 + i\gamma\omega|} \qquad [118]$$

Hence

$$-\left(\frac{\overline{dE}}{dt}\right)_{\mathrm{rad}} = \frac{16G}{15c^5}\frac{(\tfrac{1}{2}m\kappa A_{\mathrm{II}}\omega^2 b^2)^2}{(\omega^2 - \omega_0^2)^2 + \gamma^2\omega^2}\,\omega^6 \qquad [119]$$

* The free particle result can be obtained from Eq. (114) by taking $\omega_0 = \gamma = 0$.

and

$$\sigma_{\text{scatt}} = -\frac{1}{\frac{1}{2}A_{\text{II}}^2\omega^2/c}\left(\frac{dE}{dt}\right)_{\text{rad}}$$

$$= \frac{128\pi G^2}{15c^8}\frac{(mb^2\omega^4)^2}{(\omega^2 - \omega_0^2)^2 + \gamma^2\omega^2} \qquad [120]$$

$$= \frac{15\pi c^2}{2}(\gamma_{\text{rad}})^2\frac{1}{(\omega^2 - \omega_0^2)^2 + \gamma^2\omega^2} \qquad [121]$$

where, as in Eq. (86),

$$\gamma_{\text{rad}} = \frac{16G}{15c^5}mb^2\omega^4 \qquad [122]$$

Note that σ_{scatt} is independent of the amplitude of the incident wave (amplitudes cancel in (120)); this is what makes the cross section useful as a measure of how much the quadrupole scatters radiation out of the incident wave.

For the purpose of detecting gravitational radiation we are more interested in the energy absorbed from the gravitational wave than in the energy scattered. This means we want to know what power the oscillator delivers to the (mechanical) frictional forces in the oscillator. We can define a corresponding absorption cross section as

$$\sigma_{\text{abs}} = \frac{\text{(power lost to mechanical friction)}}{\text{(incident flux)}} \qquad [123]$$

We can establish a connection between σ_{scatt} and σ_{abs} if we note that the total damping rate γ, given by Eq. (113), is made out of two terms

$$\gamma = -\frac{1}{E}\left(\frac{dE}{dt}\right)_{\text{total}} = -\frac{1}{E}\left(\frac{dE}{dt}\right)_{\substack{\text{mechanical}\\ \text{friction}}} - \frac{1}{E}\left(\frac{dE}{dt}\right)_{\text{radiation}}$$

$$= \gamma_m + \gamma_{\text{rad}} \qquad [124]$$

In words: the total "friction" is due to both mechanical friction and radiation damping. By their definitions, σ_{scatt} and σ_{abs} must stand in the ratio $\gamma_{\text{rad}}/\gamma_m$. Hence,

$$\sigma_{\text{abs}} = \frac{\gamma_m}{\gamma_{\text{rad}}}\sigma_{\text{scatt}} \qquad [125]$$

For any reasonable oscillator one might hope to build, $\gamma_m \gg \gamma_{\text{rad}}$ and hence $\gamma_m \simeq \gamma$. Therefore,

$$\sigma_{\text{abs}} \simeq \frac{\gamma}{\gamma_{\text{rad}}} \sigma_{\text{scatt}} \qquad\qquad [126]$$

$$= \frac{15}{2} \pi c^2 \gamma \gamma_{\text{rad}} \frac{1}{(\omega^2 - \omega_0^2)^2 + \gamma^2 \omega^2} \qquad\qquad [127]$$

The cross sections (121) and (127) have been calculated for a quadrupole oscillator oriented in the most favorable direction in the tidal field of the wave. The line of vibration of the masses was taken both perpendicular to the direction of incidence and parallel to one of the principal axes of the tidal deformation field. For an oscillator whose masses are constrained to vibrate along a line making an angle θ with the direction of incidence (z-axis) and an angle ϕ with one of the principal axes of the tidal field (see Fig. 4.16), the component of the tidal force along this line is reduced by a factor

$$\sin^2 \theta \cos 2\phi \qquad\qquad [128]$$

as compared with the most favorable case. This factor is easy to understand: the magnitude of the tidal force is proportional to the transverse dimension of the system ($\sim \sin \theta$); taking the component of this force along the line of vibration results in another factor of $\sin \theta$; finally, the factor $\cos 2\phi$ simply represents the angular dependence of the (radial) tidal field strength in the transverse plane (see Eq. (44)).

The cross section depends on the square of the component of the tidal force along the line of vibration; hence (121) and (127) must be reduced by a factor $\sin^4 \theta \cos^2 2\phi$. This shows that the cross section of the detector is quite sensitive to the orientation and varies between the upper limit given by Eq. (126) and zero. As a compromise, it is convenient to introduce a cross section for random relative orientation between wave and detector. For this, we simply replace the factor $\sin^4 \theta \cos^2 2\phi$ by its average value over all angles,

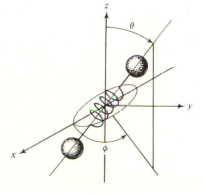

Fig. 4.16 *Orientation of quadrupole detector relative to an incident gravitational wave. The wave propagates along the z-axis. The ellipse shows the polarization of the wave.*

$$\frac{1}{4\pi} \int \sin^4 \theta \cos^2 2\phi d\Omega = \frac{4}{15}$$ [129]

which gives

$$\bar{\sigma}_{abs} = 2\pi c^2 \, \gamma\gamma_{rad} \, \frac{1}{(\omega^2 - \omega_0^2)^2 + \gamma^2\omega^2}$$ [130]

Although our calculations have been made for the special case of a simple linear quadrupole, our results are also valid for an arbitrary mass system vibrating in a mode of cylindrical symmetry. More precisely, since the amplitude of the time-dependent part of the quadrupole tensor is some symmetric matrix Q'^{kl},* it can be diagonalized by a transformation to principal axes. By cylindrical symmetry we mean that two of the diagonal elements of the diagonalized matrix are equal. If we take the axis in the z-direction, then $Q'^{11} = Q'^{22}$ Furthermore, since the trace must be zero, $Q'^{33} = -2Q'^{11}$. Let us now write these components as

$$Q'^{11} = Q'^{22} = -\tfrac{1}{2}Q'^{33} = -2mba'$$ [131]

where m is one-half the mass of the oscillating system, b is one-half the length of the system, and a' is *defined* by Eq. (131). Obviously a' has the dimensions of a length; it can be regarded as an "effective oscillation amplitude" for the masses in the quadrupole. The diagonalized quadrupole matrix is then

$$Q'^{kl} = \begin{pmatrix} -2mba' & 0 & 0 \\ 0 & -2mba' & 0 \\ 0 & 0 & 4mba' \end{pmatrix}$$ [132]

Note that this has essentially the same form as the time-dependent part of the quadrupole moment for a linear quadrupole (see Eqs. (76) and (77)); thus an axially symmetric vibrating system is equivalent to a linear quadrupole. The present-day detectors are cylinders operated in a longitudinal oscillation mode; these detectors satisfy our assumptions. The values of a' are usually of the same order of magnitude as a (Eq. (118)); for example, a long cylinder vibrating in the fundamental longitudinal mode has $a'/a = 2/\pi$.

4.7 EXPERIMENTS WITH DETECTORS OF GRAVITATIONAL RADIATION

Detectors of gravitational radiation were first built by Weber (University of Maryland, 1966).[11] Since then, similar detectors have been built by

---

* The definition of Q'^{kl} is as follows: $Q^{kl} = \text{constant} + Q'^{kl} \sin(\omega t + \delta)$ where Q^{kl} is the total quadrupole moment of the system.

Fig. 4.17 *One of Weber's detectors. A large number of piezo-electric transducers can be seen attached to the waist of the cylinder. When in operation, the detector is placed inside a vacuum tank. (Courtesy of Prof. J. Weber, University of Maryland.)*

Braginskii (U. of Moscow), Tyson (Bell Laboratories), Douglass (U. of Rochester), Levine and Garwin (IBM Research Center), Billing et al. (Max Planck Institute), Bramanti et al. (Frascati), and Drever (U. of Glasgow). Other, more sensitive, detectors are now under construction. A detector of the Weber type consists of a large cylinder of aluminum suspended by a wire inside a vacuum tank. The cylinder support rests on rubber blocks so that external mechanical vibrations cannot disturb it. The oscillations of the cylinder in its fundamental longitudinal mode are measured by piezoelectric strain transducers bonded to the cylinder around its middle (see Fig. 4.17). Table 4.4 summarizes the characteristics of one of Weber's antennas.

Eq. (130) gives a cross section at resonance of[*]

$$\bar{\sigma}(\omega_0)_{abs} = \frac{32\pi G}{15c^3} \left(\frac{\omega_0}{\gamma}\right) \omega_0 mb^2 \simeq 10^{-19} \text{ cm}^2 \qquad [133]$$

[*] *m* and *b* are *one-half* the mass and length given in Table 4.4.

TABLE 4.4 CHARACTERISTICS OF A WEBER DETECTOR

Mass	1.41×10^6 g
Length	153 cm
Diameter	66 cm
Resonant frequency, ν_0	1660/sec
Damping rate, γ	~ 0.05/sec
$Q \equiv \omega_0/\gamma$	$\sim 2 \times 10^5$

This is quite small. And what makes it worse is that unless the radiation has a frequency exactly in tune with the resonance of the cylinder, the cross section will be even smaller. If we assume that the detector reaches steady state, the rate at which the frictional forces absorb energy from the gravitational wave will be

$$-\left(\frac{\overline{dE}}{dt}\right)_{\mathrm{m}} = t^{03}\bar{\sigma}(\omega_0)_{\mathrm{abs}} \qquad [134]$$

and the steady-state energy of the oscillator will be

$$E = -\frac{1}{\gamma}\left(\frac{\overline{dE}}{dt}\right)_{\mathrm{total}} \simeq -\frac{1}{\gamma}\left(\frac{\overline{dE}}{dt}\right)_{\mathrm{m}}$$
$$\simeq \frac{1}{\gamma}\, t^{03}\bar{\sigma}(\omega_0)_{\mathrm{abs}} \qquad [135]$$

What limits our ability to detect this energy is the thermal noise in the oscillations of the cylinder. For an oscillator that is kept at temperature T, the thermal fluctuations in energy will be $\sim kT$ and in order to detect the presence of a gravitational wave, the excitation energy E of Eq. (135) must be at least as large as kT.* Therefore the minimum detectable flux is

$$t^{03} \geq kT\gamma/\bar{\sigma}(\omega_0)_{\mathrm{abs}} \qquad [136]$$

$$\geq 2 \times 10^4 \ \mathrm{erg/cm^2 \ sec} \qquad [137]$$

where we have taken $T = 300°\mathrm{K}$.

The estimate (137) is not realistic since it assumes that steady-state conditions are reached. This requires a gravitational wave that lasts at least for a time $1/\gamma \sim 20$ sec, which is the relaxation time of the detector. Such a long wave train would be hard to produce. Furthermore, we have assumed that the wave frequency coincides with the resonant

---

* The electronic circuitry used to amplify the output of the strain transducers introduces some extra thermal noise. We will ignore this.

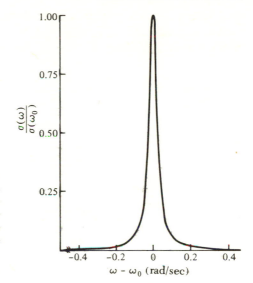

Fig. 4.18 *The cross section of a Weber detector as a function of frequency. The characteristics of this detector are given in Table 4.4. The resonant frequency is $\omega_0 = 10{,}430$ rad/sec (which corresponds to $\nu_0 = 1{,}660$/sec).*

frequency of the detector. Since this resonance has a width of only $\Delta\nu \sim \gamma/2\pi \sim 10^{-2}$/sec (see Fig. 4.18), the frequencies must agree to one part in $\sim 10^5$. This is quite unlikely. Probably the best we can hope for is that the radiation contains a spread of frequencies which overlaps the detector resonance; the spread in frequency of the radiation is likely to be much larger than the width of the resonance.

The relatively short pulses of radiation that are generated by realistic sources are best described by the total energy per unit frequency interval (and unit area) that they carry. If we designate this energy per unit frequency interval and unit area by $t^{03}(\omega)$, then the energy absorbed by the detector from the pulse is

$$\Delta E = \int_0^\infty t^{03}(\omega)\sigma_{\text{abs}}(\omega)d\omega \qquad [138]$$

We will not bother with a formal derivation because Eq. (138) is rather obvious: by Fourier analysis we can express the pulse, of finite duration in time, as a superposition of harmonic waves $e^{i\omega t}$ of many different frequencies. Essentially, $t^{03}(\omega)$ gives the total energy, per unit area, that is associated with the wave of frequency ω. Since this wave is of infinite duration, the oscillator reaches steady state with it. The cross section $\sigma(\omega)$, by its definition, tells us what fraction of the total energy, per unit area, is absorbed from the wave of frequency ω. The integral over all the participating frequencies will then give us the desired absorption for the complete pulse.

If $t^{03}(\omega)$ is a reasonably smooth function, then most of the contribution to the integral (138) comes from the vicinity of the frequency ω_0 where $\sigma(\omega)$ has a very sharp peak (see Fig. 4.18). A pulse of radiation that lasts a time Δt which is short compared to γ^{-1} will contain a wide range of frequencies. Typically, the function $t^{03}(\omega)$ will have a broad maximum of a width

$$\Delta\omega \sim \frac{1}{\Delta t} \gg \gamma \qquad [139]$$

This means that everywhere in the region of the peak of $\sigma(\omega)$, which is the only important region for the integral, the function $t^{03}(\omega)$ has approximately the same value, $t^{03}(\omega) \simeq t^{03}(\omega_0)$. Hence we can factor this function out of the integral and obtain

$$\Delta E \simeq t^{03}(\omega_0) \int \sigma_{\text{abs}}(\omega) d\omega \qquad [140]$$

To evaluate this integral, we will use the cross section $\bar{\sigma}_{\text{abs}}$ for random direction. It is convenient to replace the expression (130)

$$\bar{\sigma}_{\text{abs}}(\omega) = 2\pi c^2 \gamma\gamma_{\text{rad}}(\omega) \frac{1}{(\omega^2 - \omega_0^2)^2 + \gamma^2\omega^2} \qquad [141]$$

by

$$\bar{\sigma}_{\text{abs}}(\omega) = 2\pi c^2 \frac{\gamma\gamma_{\text{rad}}(\omega_0)}{(2\omega_0)^2} \frac{1}{(\omega - \omega_0)^2 + (\gamma/2)^2} \qquad [142]$$

These two expressions are essentially the same in the vicinity of $\omega = \omega_0$; the fact that they differ elsewhere is of no consequence since these other regions contribute little to the integral.

EXERCISE 24. Check that (142) and (141) agree near $\omega = \omega_0$ and do the integral (140).

The final result is

$$\Delta E = \tfrac{1}{2}\pi\gamma\, t^{03}(\omega_0)\bar{\sigma}_{\text{abs}}(\omega_0) \qquad [143]$$

If we demand that the energy deposited by the pulse be at least as large as kT, we are lead to a minimum detectable flux of

$$t^{03}(\omega_0) \geq \frac{2kT}{\pi\gamma\bar{\sigma}_{\text{abs}}(\omega_0)} \qquad [144]$$

With the numbers given in Table 4.4,

$$t^{03}(\omega_0) \geq 5 \times 10^6 \ (\text{erg/cm}^2)/(\text{rad/sec}) \qquad [145]$$

This is the minimum detectable energy per unit frequency interval (at the frequency ω_0). To get the total energy of the pulse, we need to know the shape of the energy spectrum. If we assume that the spectrum is a broad peak of width $\Delta\omega$, the total energy is roughly $t^{03}(\omega_0)\Delta\omega$. Even for a pulse with the rather small bandwidth of $\Delta\omega \sim 2\pi \times 10^3$ rad/sec, the total pulse energy must be $\sim t^{03}(\omega_0)\Delta\omega \sim 10^{10}$ erg/cm²!

To overcome the limit (144) set by thermal noise, several routes of attack are available. Two, or more, detectors may be operated in coincidence; if a signal is observed simultaneously in all of these detectors, then we can be reasonably certain that an external influence is at work, even if these signals are small compared to the thermal noise. Even with a single detector a small signal can be distinguished from the noise if this signal consists of a very short pulse which produces a *sudden* jump in the detector motion. The reason is that thermal fluctuations typically take a time γ^{-1} (relaxation time) to change the oscillation energy by $\sim kT;$ in a time Δt, short compared to γ^{-1}, they only change the energy by $\sim kT(\Delta t\gamma)^{1/2}$. Hence an energy fluctuation of, say, $kT/10$ in a time $10^{-4}\gamma^{-1}$ is very unlikely to be thermal. Finally, it is possible to get rid of the thermal fluctuations by cooling the detector. Several detectors now under construction will use cylinders of $\sim 4 \times 10^6$ gm cooled to 3×10^{-3} °K; this should result in improvements in sensitivity by a factor of 10^5 to 10^7.[12,13]

Weber has reported signals, with an energy in excess of kT, in his detectors.[11] These signals arrive at the rate of a few per day and there is some indication that their direction of arrival correlates with the direction of the galactic center.[14] The bandwidth of the pulses is at least $\Delta\nu = 630$ cps because coincident pulses are observed in cylindrical detectors of a resonant frequency $\nu_0 = 1660$ cps and also in a disc-shaped detector of a resonant frequency $\nu_0 = 1030$ cps.

According to the estimate given above, a gravitational pulse would need an energy of $\sim 10^{10}$ erg/cm² to make such a signal. If these pulses originate at the galactic center, the total energy per pulse is[15]

$$4\pi r^2 \times 10^{10} \text{ erg/cm}^2 = 10^{56} \text{ erg} \sim 100 \, M_\odot \qquad [146]$$

where $r = 3 \times 10^4$ light years is the distance to the galactic center. Eq. (146) makes the plausible assumption that the radiation is emitted more or less uniformly in all directions; some reasonable angular dependence in emission (e.g., $\sim \sin^4 \theta$) would not change the estimate (146) appreciably. With several such pulses per day, the amount of mass converted into gravitational radiation would add up to $\sim 10^5 \, M_\odot$ per year. Such a large rate of loss of mass, if continuous, is in contradiction with astronomical evidence (the reduction of the mass would imply a reduction of the force that holds the galaxy together and the galaxy would expand).[16] Furthermore, it is hard to imagine what kind of monstrous machinery could send out a pulse of gravitational radiation of energy 100 $M_\odot$, and do this in a time of ~ 1 sec or less.

The energy requirements can be reduced if it is assumed that some focusing mechanism sharply concentrates the emission in the galactic plane. But this only helps if the radiation is concentrated in a rather narrow angle ($\sim 10^{-3}$ rad) and no plausible mechanism is known.

A drastic reduction of the energy requirements is also possible if one assumes that the source is not at the galactic center, but very near the earth. For example, a source at a distance of 10 light years would only need to emit 10^{49} erg per pulse. There is enough gravitational and nuclear energy in a neutron star to emit 10^3 such pulses; the emission mechanism might involve a starquake.

Unfortunately, Weber's observation of pulses has received no confirmation from other experimenters using similar detectors. None of the other experimenters has been able to find pulses with the size and abundance of those reported by Weber.[13] For example, Tyson[17] has built a larger (mass 3.6×10^6 gm, length 357 cm) and more sensitive antenna with a resonant frequency of 710 cps. Since Weber finds pulses at 1660 cps and at 1030 cps, it seems reasonable to expect that these pulses will also show up in Tyson's detector. But the latter detector registers no pulses at all—more precisely, no pulses of an energy above $\frac{1}{4}kT$. There may or may not be some pulses of $\frac{1}{8}kT$, but these presumably have nothing to do with Weber's large pulses. According to Tyson, the upper limit on the possible pulse energy is 3×10^6 (erg/cm^2)/(rad/sec). Similar negative results have been published by Levine and Garwin[18] and by Bramanti et al.[19]

The weight of the evidence therefore suggests that it is unlikely that Weber's pulses are caused by gravitational radiation. It seems that the discovery of gravitational waves will have to wait for the next generation of detectors (cooled detectors).

FURTHER READING

A very clear and comprehensive discussion of gravitational radiation is given in Misner, Thorne, and Wheeler, *Gravitation*. Among other things, these authors present a variety of alternative detection schemes and give many details concerning possible sources of radiation.

Much fascinating information about sources of gravitational radiation is contained in the book *Black Holes, Gravitational Waves, and Cosmology* by Rees, Ruffini, and Wheeler (see ref. 4). This book also presents a clever derivation of the cross section for absorption (Eq. (130)) by a method that relies on detailed balance. Yet another derivation is given in Weinberg, *Gravitation and Cosmology;* this relies simply on energy conservation (or the optical theorem).

The article "Gravitational-Wave Astronomy" by Press and Thorne (see ref. 1) speculates about the future use of gravitational antennas to carry out astronomical observations. Table 4.1 is based on their article, which also contains a general analysis of detector sensitivity and a warning against the indiscriminate use of the cross section as a measure of this sensitivity.

Articles by Logan ("Gravitational Waves—a Progress Report," *Physics Today,* March 1973) and Sejnowski ("Sources of Gravity Waves," *Physics Today,* January 1974) provide, respectively, a review of the experimental results obtained with detectors of the Weber type, and a preview of the results that are likely to be obtained with the next generation of detectors (cooled detectors).

A very complete review of all the experiments performed up to 1973 will be found in Papini, "Gravitational Radiation and its Detection" (ref. 13); this also includes a discussion of sources of radiation and a theoretical analysis of detector sensitivity.

REFERENCES

1. W. H. Press and K. S. Thorne, Ann. Rev. Astron. Astrophys. **10**, 335 (1972).
2. G. Moss, L. Miller, and R. Forward, Applied Optics **10**, 2495 (1971).
3. S. Weinberg, *Gravitation and Cosmology* (Wiley, New York, 1972).
4. M. Rees, R. Ruffini, and J. A. Wheeler, *Black Holes, Gravitational Waves, and Cosmology* (Gordon and Breach, New York, 1974).
5. K. S. Thorne, Ap. J. **158**, 1 (1969).
6. J. Faulkner, B. P. Flannery, and B. Warner, Ap. J. **175**, L79 (1972).
7. This example, with slightly different numbers, was given by F. J. Dyson *in* A. G. W. Cameron, ed., *Interstellar Communication* (Benjamin, New York, 1963).
8. A. Ferrari and R. Ruffini, Ap. J. **158**, L71 (1969).
9. M. Davis, R. Ruffini, W. H. Press, and R. H. Price, Phys. Rev. Lett. **27**, 1466 (1971).
10. C. W. Misner, K. S. Thorne, and J. A. Wheeler, *Gravitation* (Freeman, San Francisco, 1973), p. 1041.
11. For a review of Weber's experiments, see J. Weber in B. Bertotti, ed., *Proceedings of Course 56 of the International School of Physics "Enrico Fermi"* (Academic, New York, 1973); see also V. Trimble and J. Weber, Ann. N.Y. Acad. Sci. **224**, 93 (1973).
12. D. Douglass, Fifth Cambridge Conference on Relativity, 1974.
13. G. Papini, Can. J. Phys. **52**, 880 (1973).
14. J. Weber, Phys. Rev. Lett. **25**, 180 (1970).
15. G. W. Gibbons and S. W. Hawking, Phys. Rev. D **4**, 2191 (1971).
16. D. W. Sciama, G. F. Field, and M. J. Rees, Phys. Rev. Lett. **23**, 1514 (1969).
17. J. A. Tyson, Ann. N.Y. Acad. Sci. **224**, 74 (1973); Phys. Rev. Lett. **31**, 326 (1973).
18. J. L. Levine and R. L. Garwin, Phys. Rev. Lett. **33**, 794 (1974).
19. D. Bramanti, K. Maischberger, and D. Parkinson, Lett. Nuov. Cim. **7**, 665 (1973).

20. O. Struve, *Stellar Evolution* (Princeton University Press, Princeton, 1950), p. 234.

PROBLEMS

1. Show that the gravitational waves $\phi_I^{\mu\nu}$ and $\phi_{II}^{\mu\nu}$ (see Eqs. (24) and (25)) carry a momentum flux equal to their energy flux.

2. Suppose that the wave $\phi_I^{\mu\nu}$ (see Eq. (24)) is observed from a new reference frame moving at velocity v along the z-axis of the old reference frame. (a) Find the polarization tensor of the wave in the new frame and express it as a superposition of the standard polarizations given by Eqs. (8)–(13). (b) Find the amplitude A_I' of the wave in the new frame. (c) Find the energy flux in the new frame and compare it with the flux in the old frame.

3. Suppose that a gravitational wave has the form

$$\phi^{\mu\nu} = A\,\eta^{\mu\nu}\,\cos\,\omega(t - z)$$

Express this as a superposition of waves with the standard polarizations. Find a gauge transformation that reduces the wave to zero.

4. An extremely low frequency gravitational wave is incident on the earth. Assuming that the frequency is so low that quasistatic conditions prevail, use the methods of Section 1.6 to find an expression for the height of the tide raised by this wave on the earth. Describe how the shape of the tidal surface differs from that generated by the tidal forces of the moon.

5. Suppose that a plane gravitational wave of frequency $\omega = 10^{-5}$/sec with an energy flux of 10^7 erg/cm²sec is incident on the earth. The tidal torques will change the direction of the spin of the earth. Estimate the maximum rate of change of direction that this wave can generate.

6. Use Eq. (83) to estimate the radiation rate for a "rod" of steel of radius 1 m, length 40 m, oscillating longitudinally with the maximum amplitude permitted by the tensile strength of steel. Assume that the rod is oscillating in its lowest acoustic mode, i.e., $\frac{1}{2}\lambda_{\text{sound}} = 40$ m.

7. Prove that spherically symmetric oscillations ("breathing mode") of a spherical star cannot emit any gravitational radiation.

8. Calculate the rate at which the earth radiates gravitational energy as it moves around the sun. What is the corresponding rate of change of the orbital radius?

9. For each of the close binaries listed in Table 4.2, calculate the rate of change of the orbital period. Are these changes observable? (For this calculation ignore the mass transfer in *AM CVn*.)

10. A meteorite with a mass of 2×10^7 kg hits the surface of the earth with a vertical velocity of 11 km/sec. The meteorite comes to rest after penetrating the earth to a depth of 200 m. Use Eq. (98) to estimate the total gravitational energy radiated during the impact; assume constant deceleration. Describe the polarization of the pulse of gravitational radiation that reaches an observer on the surface of the earth at some short distance from the impact

point. If an antenna of the Weber type is to detect the gravitational radiation, how near must it be to the impact point?

11. Two stars, similar to the sun, collide head on with a relative velocity of $v/c = 10^{-3}$. Roughly estimate the gravitational energy radiated during the impact. Treat the stars as mass points that come to rest after moving through a distance of about one solar radius; pretend that the deceleration is constant. If such a collision takes place at the center of our galaxy, can we detect the gravitational radiation with available detectors?

12. A long thin rod of iron lies along the z-axis. When in equilibrium, the rod occupies the interval $-b \leq z \leq b$. The rod is made to oscillate in a longitudinal acoustic mode with

$$\psi(z,t) = a \sin \frac{\pi}{2} \frac{z}{b} \cos \frac{\pi v t}{2 b}$$

where $\psi(z,t)$ is the longitudinal displacement suffered by a point whose equilibrium position is z, and where v is the velocity of sound waves along the rod. Find the "effective oscillation amplitude" a' defined by Eq. (131). Repeat the calculation for the mode

$$\psi(z,t) = a \sin \frac{3\pi}{2} \frac{z}{b} \cos \frac{3\pi v t}{2 b}$$

Which of these modes has the shorter (gravitational) damping rate?

5. THE MEASUREMENT OF SPACETIME

> I heard once from a learned man, that
> the motion of the sun, moon, and stars
> constituted time, and I assented not.
> For why should not the motions of all
> bodies rather be times? Or, if the lights
> of heaven should cease, and a potter's
> wheel run round, should there be no
> time by which we might measure those
> whirlings . . . ?
>
> St. Augustine, *Confessions*

Our discussion of the relativistic equation of motion for a particle in the gravitational tensor field led to the conjecture that spacetime is curved. The investigation of the geometry of spacetime must begin with a very careful definition of exactly what the units of distance and time are, and exactly what operational procedure is to be used in making measurements. It is good to remember the primordial role that the synchronization procedure for clocks (by means of light signals) plays in special relativity; in the case of "general relativity" we must take an even harder look at our clocks because the problem of synchronization shrinks to insignificance compared with the much more serious problem of adjusting the *relative rates* of clocks placed in different positions in a gravitational field.

    The following simple example, due to Robertson, illustrates the difficulties one may run into if one is careless about the definition of length.[1] Suppose one has a large plane surface, different points of which are kept at different temperatures ("hot plate"). If lengths are measured on this plane by means of a (small) metallic ruler held in thermal contact with the "hot plate," then the corresponding geometry will be that of a two-dimensional curved space. An intelligent physicist would experiment with rulers of different coefficients of thermal expansion and be quick to discover that the fault lies with the rulers; the physicist would then conclude that the two-dimensional curved geometry is spurious. This example shows very clearly that the question of curvature of spacetime is intimately tied in with the definition of measurement of length and time. Before we jump to the conclusion that spacetime is curved, we must carefully study our standards of length and time and eliminate the possibility that these standards are changing.

The arguments presented in this chapter will show that when the standards of length and time are defined in the best possible way, then the measured spacetime geometry is curved. This conclusion is a quite direct consequence of the Galileo principle of equivalence. Hence our arguments apply not only to the tensor theory of gravitation, but also to other theories that ingenuity may contrive.

5.1 THE PROBLEM OF CHANGING STANDARDS

The measurement of any physical quantity is carried out in two steps: we define a standard, and then specify an operational procedure for comparing the standard with the quantity that is to be measured. The popular fundamental standards are three: the *meter* for the unit of length, the *second* for time, and the *kilogram* for mass.

Why do we need no more nor less than *three* units? We need at *least* three because the simplest possible physical system, a classical (point) particle, has a particular position (meters) at a given time (seconds) and, furthermore, different particles may have different masses (kilograms). With these three units one can give a complete description of a (point) particle. Everything that can be measured about a particle can be expressed in terms of length, time, and mass. Roughly, the first two units are needed to describe the *kinematics* in space and time (position, velocity, acceleration) and the last is needed to describe the *dynamics* (momentum, equation of motion). Why do we need no more than three units? Because any complex measurable quantity can be expressed in terms of *how it affects the motion of particles*. For example, the electric charge can be expressed in terms of the acceleration produced on two particles separated a standard distance, etc. Ultimately, physical measurements always refer back to particle properties.

Our standard of length is nowadays defined by the statement that 1 meter equals 1,650,763.73 wavelengths of the orange line of krypton. What we might worry about is then this: How can we be sure that the krypton atoms emit the same wavelength from one day to the next? How can we be sure that two atoms emit the same wavelength when at different places? To see how such changes could come about, we note that the wavelength of light emitted in an atomic transition is proportional to (m_e is the electron mass):

$$\lambda \propto \frac{\hslash}{m_e c} \frac{1}{(e^2/\hslash c)^2} \tag{1}$$

The constant of proportionality, omitted in (1), involves some dimensionless number which characterizes the transition, and it involves the atomic number; neither of these numbers can change in time or space. A change in the wavelength can result from a change in $\hslash$, c, e, or m_e.

Of course there are also other ways in which the wavelength emitted by an atom might change. External perturbations, gravitational or other, can affect the atom. But we will for now take the optimistic view that such external perturbations can be detected and identified because they affect different spectral lines in different ways. We will only worry about those changes in wavelength that might easily be overlooked and that therefore pose a serious threat to our attempts to set up a standard of length.

Although any fool can take the equations of physics and replace the constants that appear in them by whatever functions of space or time that strike his (or her) fancy, nothing is easier than to make a serious mistake. The presence of a "constant" that changes will violate the principle of relativity and energy conservation unless this changing "constant" is regarded as a scalar field with a field equation of its own. There must be an energy density associated with this field, there must be waves of changing "constant," etc. To replace, say, $\hslash$ by a function $\hslash(x)$ certainly makes no sense unless one also constructs some relativistic field equation for $\hslash(x)$ (and, as we will see, makes little sense even then).

General arguments rule out changes of some of the constants appearing in Eq. (1). Thus, the possibility of changes in c and $\hslash$ can be dismissed. The reason is that talk of a "changing constant" is nothing but a lazy way of approximately describing certain interactions taking place with some field that is penetrating the laboratory. Such talk is useful, provided it does not conflict with fundamental physical laws. For example, the effective electron mass used by solid-state physicists to describe some of the interaction between electron and lattice is a useful concept because consistent equations of motion can be written with this effective mass. However, a "changing velocity of light" or a "changing Planck constant" would not be useful. The constancy of c is an essential feature of special relativity and the constancy of $\hslash$ is an essential feature of quantum mechanics; no consistent equation can be written if these quantities are taken as variable.* Incidentally, this suggests

---

* For example, in $\int p\,dq = n\hslash$, the left side is an adiabatic invariant (roughly, it changes either suddenly or not at all) and hence we cannot have any gradual change in the $\hslash$ on the right side.

that rather than regard length as a fundamental unit, we should regard velocity as a fundamental unit and length as a derived unit, defined in terms of velocity and time; the velocity of light would serve as the standard of velocity. Likewise, we should regard action (or angular momentum) as a fundamental unit and mass as a derived unit; the quantum of action $\hbar$ (or the spin of the electron) would serve as the standard of action. Thus the conventions $c = 1$ and $\hbar = 1$, often adopted in theoretical physics, take on a meaning that is much deeper than that of mere computational convenience.

As concerns changes in e, Maxwell's equations forbid this because they demand the conservation of charge. However, Maxwell's equations permit a changing dielectric constant of the vacuum; it is therefore preferable to interpret any, say, decrease in the electric force between two charges as an increase in the dielectric constant. In Eq. (1) this will have the same effect as a decreasing e. To be more precise, we should include the dielectric constant explicit in Eq. (1) by writing

$$\lambda \propto \frac{\hbar}{m_e c} \frac{1}{\alpha^2} \qquad [2]$$

where

$$\alpha = \frac{e^2}{\epsilon_0 \hbar c} \qquad [3]$$

is the *fine structure constant*, modified so as to include the dielectric constant ϵ_0 of the vacuum.* The quantity α is a dimensionless number which characterizes the strength of the electromagnetic interaction between charges; its (present) value is $\alpha \simeq 1/137.036$.

If we suspect that changes in constants produce a change in the wavelength of light, we can check this by comparing the wavelength with some other standard of length. The obvious thing to try is a comparison with the old standard meter defined by the length of the old platinum bar. The fundamental length involved in the platinum bar is the size of an atom, i.e., some multiple of the Bohr radius

$$a_0 = \frac{\hbar}{m_e c} \frac{1}{\alpha} \qquad [4]$$

The ratio of the new to the old standard of length is therefore proportional to

$$\frac{\lambda}{a_0} \propto \frac{1}{\alpha}$$

---

* In our Gaussian units, the accepted value of ϵ_0 is of course $\epsilon_0 = 1$. Note that any change in the dielectric constant must be matched by a change in the magnetic permeability of the vacuum so that the speed of light, $c \propto 1/\sqrt{\epsilon_0 \mu_0}$, remains constant.

In principle, comparison of the two standards of length would allow us to detect changes in α.

Alternatively, we can compare the wavelength of light with the Compton wavelength

$$\lambdabar_C = \frac{\hbar}{m_e c} \qquad\qquad [5]$$

The ratio of the standards of length defined by Eqs. (2) and (5) is

$$\frac{\lambda}{\lambdabar_C} = \frac{1}{\alpha^2} \qquad\qquad [6$$

Note that $\lambdabar_C$ would seem to be the best of these standards of length since it does not depend on α. In fact, we will later give some strong theoretical arguments that the Compton wavelength is an absolute constant both in space and in time (see Section 5.5).

EXERCISE 1. Show that $2\pi\lambdabar_C$ may be operationally defined as the wavelength of the two photons produced in the annihilation of an electron-positron pair (initially at rest).

As regards our standard of time and our standard of mass, the situation is quite similar. These standards are some functions of the atomic constants, and if the constants change, then so will the standards. Comparisons between different standards can be used to detect changes in the constants.

The tightest limits on possible changes in physical constants are set by comparison of the rates of different time standards, rather than length or mass standards. Among the many types of clocks that we can compare, nuclear "clocks" are particularly helpful. These clocks measure time by the decay of a radioactive isotope, such as e.g., α-decay of U^{238} or U^{235}, β-decay of Rb^{87}, Re^{187}, or I^{129}, electron capture in K^{40}, spontaneous fission of U^{238}, etc. Although these clocks are very inaccurate for the measurement of short time intervals, they are very good for long time intervals and are widely used by geologists in the determination of ages of rocks and meteorites (for ages up to the age of the earth, 4.6×10^9 years). Radioactive decay rates are very sensitive to changes in the physical constants. If the values of the physical constants were different in the past from what they are now, the nuclear "clock" ages would disagree. In fact, the ages of rocks as measured by these different clocks agree quite well, typically to within 5% or 10%. This sets the limits given in Table 5.1 on possible changes in the physical constants.[3]

Here M_n and M_p are, respectively, the neutron and proton masses. For the purpose of Table 5.1 it has been assumed that c, $\hbar$, m_e serve as standards and are therefore constant by definition; this particular choice

TABLE 5.1 EMPIRICAL LIMITS ON THE TIME VARIATIONS OF CONSTANTS

$$\left|\frac{1}{\alpha}\frac{d\alpha}{dt}\right| < 5 \times 10^{-15}/\text{year}$$

$$\left|\frac{1}{M_n}\frac{dM_n}{dt}\right| < 5 \times 10^{-13}/\text{year}$$

$$\left|\frac{1}{M_p}\frac{dM_p}{dt}\right| < 5 \times 10^{-13}/\text{year}$$

of standards will be justified in Section 5.5. The rates of change quoted in the table are of course averages for the last $\sim 10^9$ years.

Changes in constants presumably have something to do with cosmology. We expect that *if* the constants change at all, then they should change on the time scale set by the universe; since the age of the universe is $\sim 10^{10}$ years (see Section 10.4), we expect changes of $\sim 10^{-10}$/year. The limits given in Table 5.1 are much smaller than this and it is therefore very likely that these constants are absolutely independent of the age of the universe.

Another constant which might possibly change is the gravitational constant G. Theories with a changing value of G have been constructed by Jordan and Brans-Dicke (see Section 3.2).* If the value of G decreases then the periods of planetary motion would increase; hence a limit on changes of G with time may be set by comparing the rate of an atomic clock with the celestial (or ephemeris) clock defined by the motion of the planets. Such a comparison was carried out by Shapiro et al.,[4] who measured planetary positions (by radar-echo time delays) as a function of the atomic time given by the cesium beam standard of the U.S. Naval Observatory. The deviation between the two clocks was found to be less than one part in 5×10^9 per year. The corresponding limit on changes in G is

$$\left|\frac{1}{G}\frac{dG}{dt}\right| < 4 \times 10^{-10}/\text{year}$$

A different comparison was carried out by Van Flandern,[19] who analyzed measurements of the position of the moon as a function of atomic time. Since the mean motion of the moon suffers from a deceleration produced by tidal interaction with the earth, such measurements do not yield a direct comparison of the atomic clock with the ephemeris clock. Two steps are needed: first, atomic time is compared to the motion of the moon, and then the motion of the moon is compared

---

* In view of the possibility of changes in G it would be somewhat misleading to adopt the convention $G = 1$ in the choice of units. This is not a mere convention; it makes the implicit assumption that G is in fact constant.

to ephemeris time; the net result is a comparison of atomic and ephemeris time. According to Van Flandern's analysis:

$$\frac{1}{G}\frac{dG}{dt} = (-7.5 \pm 2.7) \times 10^{-11}/\text{year} \qquad [7]$$

However, according to a more recent analysis[19] the rate of change of G comes out to be only half as big as the value given by (7), the difference being due to a newly discovered systematic error. This, of course, raises the suspicion that there may be further systematic errors in the analysis.

Changes in G can also be detected by comparing the sidereal time, defined by the rotation of the earth, with ephemeris time. The comparison can be carried out over an interval of more than 3,000 years by means of ancient records of eclipses from Babylonian, Chinese, Greek, etc., sources; a shift in the angle of rotation of the earth (or in the position of the moon and sun) would shift the locality at which the eclipse is visible. In order to make this comparison meaningful, the ordinary sidereal time must be corrected for changes in the rotation rate of the earth due to causes other than changing constants. The main long-term effects that must be considered are tidal friction and variation in the sea level (variation in the moment of inertia); the corrected rotation rate then corresponds essentially to that of a rigid body. After carefully taking into account such effects, Dicke found[20]

$$\frac{1}{G}\frac{dG}{dt} = (-3.8 \pm 0.4) \times 10^{-11}/\text{year} \qquad [8]$$

Both (7) and (8) suggest that G is decreasing. However, both these results hinge on determinations of the rate of increase of the period of the motion of the moon (deceleration of the mean motion of the moon). There are several determinations of this parameter[19,21] and there are inconsistencies among these determinations (the differences between them are much in excess of the quoted standard errors). In view of this, the reliability of (7) and (8) is questionable. In any case, we can regard $\sim 10^{-10}/\text{year}$ as an upper limit on $(1/G)dG/dt$ and, in the absence of firm evidence to the contrary, we will take the optimistic and conservative stand that G remains unchanging.

Although we have quite good evidence that constants do not change in time, it is also important to find out if they change in space. For example, if the wavelength (2) of our standard of length changes as the Kr atom is moved away from the surface of the earth, then the geometry derived by means of this standard of length would be very misleading. The results of the Eötvös experiment can be used to set limits on changes of constants with height above the earth. If we suppose that α changes with height, then the internal energy of an atom must change with height. Since the internal Coulomb energy (nuclear and electronic) of an atom depends on α, any dependence of α on position would imply that the energy of an atom is a function of position. This means that the "α-field" must do work on the atom as the latter is moved about; hence the "α-field" would exert a force on the atom and a modification of the gravitational acceleration would result.

Let us discuss this quantitatively.[5] The Coulomb energy U is proportional to α; hence,

$$\frac{dU}{U} = \frac{d\alpha}{\alpha}$$

i.e.,

$$dU = U\frac{1}{\alpha}d\alpha$$

This implies that the "α-field" exerts a vertical force

$$-\frac{dU}{dh} = -U\frac{1}{\alpha}\frac{d\alpha}{dh}$$

on the atom. This force must be added to the gravitational force $-Mg$, and hence the acceleration of a falling atom is

$$a = \frac{1}{M}\left(-Mg - U\frac{1}{\alpha}\frac{d\alpha}{dh}\right)$$
$$= -g\left(1 + \frac{c^2}{g}\frac{U}{Mc^2}\frac{1}{\alpha}\frac{d\alpha}{dh}\right) \qquad [9]$$

The quantity U/Mc^2 is the ratio of electrostatic energy to total rest mass energy of the atom. For a platinum atom this ratio is $\sim 4 \times 10^{-3}$, for carbon it is about 6×10^{-5} (the main contributions to the electrostatic energy are of nuclear origin). Hence the acceleration of a platinum atom differs by about one part in

$$\sim 4 \times 10^{-3}\frac{c^2}{g}\frac{1}{\alpha}\frac{d\alpha}{dh}$$

from that of carbon. Since the Eötvös experiment shows that the accelerations in the gravitational field of the earth are in fact equal to 1 part in $\sim 10^{10}$, we can place the limit

$$\left|\frac{1}{\alpha}\frac{d\alpha}{dh}\right| < 3 \times 10^{-26}/cm$$

on changes in α with height.

We may also consider changes in α with distance from the sun. The Dicke-Braginskii experiments (masses falling towards the sun) then set the limit

$$\left| \frac{1}{\alpha} \frac{d\alpha}{dr} \right| < 10^{-31}/\text{cm}$$

on the rate of change of α with radial distance to the sun.

EXERCISE 2. Show this.

The Eötvös experiment can also be used to set limits on changes in (rest) mass with position. If the mass of, say, an electron changes with height, then an argument similar to that leading to Eq. (9) tells us that the electron has an acceleration

$$a = -g \left(1 + \frac{c^2}{gm_e} \frac{dm_e}{dh} \right)$$

According to the Eötvös experiment, the accelerations of different particles are equal or nearly equal. Hence, the values of $(1/m)(dm/dh)$ must be the same for all particles. The maximum deviation between the values of this quantity for electrons, protons, and neutrons permitted by the experimental results is

$$\left| \Delta \frac{1}{m} \frac{dm}{dh} \right| < 2 \times 10^{-24}/\text{cm}$$

If we consider changes in mass with distance to the sun, then the corresponding limit is

$$\left| \Delta \frac{1}{m} \frac{dm}{dr} \right| < 10^{-29}/\text{cm}$$

We may therefore safely conclude that if the masses change with position, then they all change in the same way. But a change in all masses by a common factor amounts to no change at all: only relative mass changes are observable. This statement is perfectly obvious if we use a piece of mass, such as, e.g., the platinum cylinder of the MKS system, as our standard of mass, and define other masses relative to it. We will see in Section 5.5 that the statement remains true even if we do not use mass as a fundamental unit and instead use a fundamental unit of action ($\hbar$).

There is at present no empirical evidence for changes in any of the fundamental constants of nature. Nevertheless, the theoretical possibility of such changes is a serious defect of the popular system of units. For example, it is conceivable that a measurement of the speed of light with the MKS standards of length and time gives the result that this speed varies in space and time. The physics based on this system of units would then conflict with special relativity! Obviously, it is preferable to define an ideal system of units in such a way that those quantities that are required to be constant by basic theoretical considerations are constant by definition.

In our ideal system of units, we regard the velocity of light as the standard of velocity. Thus, the meter/second is a fundamental unit.* As our next fundamental unit, we may use a unit of time. Note that this immediately fixes the unit of length in terms of the unit of time. The relation $(2.99792458 \times 10^8$ m) = (distance that light travels in one second) defines meters in terms of seconds. Alternatively, we could take the unit of distance as fundamental and the unit of time as derived. Which alternative one prefers is a matter of taste; it depends on whether one likes meter sticks or clocks better. In what follows we will often treat units of length and time interchangeably and not bother to state which is fundamental, which derived. Incidentally, the Comité International de Poids et Mesures is now considering the possibility of defining the meter in terms of the unit of time and the velocity of light.[2]

For the definition of the standard of time, we need a standard clock. The essential requirement that this clock must satisfy is that the worldlines of free particles are straight lines when measured with this clock. It is obvious that this requirement fixes the rate of the clock to within a constant factor. Mathematically: The equation of a straight worldline is $x = vt +$ const. This worldline remains straight if and only if we perform a linear transformation $t \rightarrow a_0 t + b_0 x + c_0, x \rightarrow a_1 t + b_1 x + c_1$. Here, nonzero values of c_0, c_1 give a shift of origin, and nonzero values of b_0, b_1 indicate a change in inertial frame (Lorentz transformation); neither of these possibilities is relevant in the present context. Thus, we may assume that only a_0 and a_1 are nonzero (since the standard of length is defined in terms of the standard of time, we will actually have $a_0 = a_1$). This corresponds to a change in the rate of the clock by a constant factor; that is, we have the freedom to adjust the duration of the first tick of the clock, but the duration of the succeeding ticks is then uniquely determined.

The preceding does not, of course, tell us how to construct the standard clock; it only tells us how to recognize whether a given clock is or is not satisfactory. In the next section, we will present an ingenious procedure for the actual construction of a satisfactory standard clock. This clock is called the geometrodynamic clock; it relies on nothing but free particles and light rays.

---

* Perhaps it would be good to devise a clever name for this unit.
** We ignore the possibility of projective transformations since these have singularities.

As our third and final unit, we take angular momentum (or action) with $\hbar$ as the standard. More concretely, we may take the spin of an electron as the standard of angular momentum. This choice is justified by the theoretical arguments given above which show that $\hbar$ is an unchangeable constant of nature. Note that in our ideal system of units, mass is a derived unit which must be expressed in terms of velocity, time, and angular momentum. Since the definitions of our standards incorporate the conditions that c and $\hbar$ are constant, it would be sensible to choose the units in such a way that $c = 1$ and $\hbar = 1$. We have already made implicit use of $c = 1$ in many of our equations by omitting factors of c. (For dealing with numerical examples, we will however retain the familiar MKS or cgs values of c and $\hbar$.)

The ideal system of units, based on the velocity of light, the geometrodynamic clock, and Plank's constant, is not very realistic. The catch is that the geometrodynamic clock, for all its theoretical elegance, is not workable in practice. However, as we will see, the standard defined by the geometrodynamic clock is equivalent to a standard based on the Compton wavelength. The ideal system of units may therefore be alternatively based on the velocity of light, the Compton wavelength, and Planck's constant. In these terms, the system could be achieved in practice although it would require some effort.

5.2 THE GEOMETRODYNAMIC CLOCK

The experimental evidence for "constant constants" is very good. However, this evidence does involve finite error bars and, furthermore, the limits given in Table 5.1 apply only to the average rate of change over the last $\sim 10^9$ years. Short-term changes in constants would be much harder to detect. Hence it would certainly be preferable if we could set up an *absolute* standard of length of time, i.e., one that does not depend on any of the physical constants e, m_e, M_n, M_p, G, etc. Since all clocks must be made of matter, it would seem impossible to construct a clock which is completely independent of the structure of matter. Nevertheless, a clock fulfilling this condition has been designed by Marzke and Wheeler.[6] It is known as the *geometrodynamic clock*.

The existence of an absolute standard of length is of crucial importance for the investigation of the geometry of spacetime. The assertion that spacetime is curved becomes unambiguous only if the measurements of the geometry are carried out with an absolute standard. If the measurements are carried out with some arbitrary atomic standard, one will always wonder to what extent the apparent curvature is the result of changes in the standard as it is moved about in the gravitational field.

To describe the operation of the geometrodynamic clock, let us start with the simplifying assumption that spacetime is free of gravitational fields (flat spacetime); this assumption must be, and will be, dropped later on. The geometrodynamic clock is an extremely simple device. Take a free particle with some worldline *PF*. Take a second free particle with worldline *AB parallel* to *PF* (see Fig. 5.1). Reflect a light signal back and forth between these particles. The "ticks" of the clock will be taken as the arrivals of the light pulse back at the first particle. This clock in no way depends on the structure of the matter contained in the particles. One might perhaps worry that there are time delays in reflection that depend on the structure of matter, but in principle we can replace the reflection of light by a somewhat more complicated process with zero time delay; for example, an observer riding on each particle checks when the worldlines of light signal and particle coincide, and he sends out a new signal towards the other particle at the instant the coincidence occurs.

Fig. 5.1 *The geometrodynamic clock.*

A concrete picture of the geometrodynamic clock is the following: take two mirrors (or corner reflectors) in free motion with a constant distance between them and let light bounce back and forth.

To complete the construction, a crucial question must be answered. What is meant by "constant distance" or "parallel" worldlines? How can we define "parallel" *without* using a standard of length to check whether the worldlines are keeping their distance?

Fortunately, there is a way to construct parallels which does not involve any length measurements. Fig. 5.2a shows the worldline *PF* of a free particle (particle I). To construct a parallel to *PF*, introduce a second particle with a worldline *QF* that intersects *PF* at some point *F* in the future. Such a worldline *QF* can be found by simply shooting the second particle towards the first in such a way that they collide. Arrange for particle I to emit a light signal at *P*, and let this light signal be returned to particle I after reflection by particle II (see dashed line in Fig. 5.2a). Introduce a third particle with worldline *PG* that passes through *P*. Next, arrange for particle I to emit a light signal towards particle III in such a way that this light signal passes exactly through *F* after reflection by particle III (see Fig. 5.2b). Obviously it will be difficult to arrange for

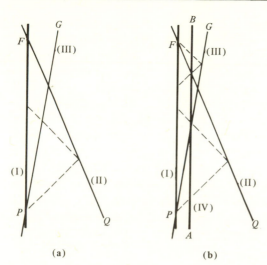

(a)   (b)

Fig. 5.2 *Step-by-step construction of the parallel.*

this. It is probably best to use a trial and error procedure: let particle I emit many light signals (continuous emission) and discard all of them except for the one signal which does reach *F*. Consider now the intersection of *PG* with the first light signal and of *QF* with the second light signal. If a fourth free particle has a worldline *AB* that passes through both of these points (see Fig. 5.2b), then *AB is parallel to PF*.

To prove this, we can use similarity of the triangles appearing in Fig. 5.3. Triangle *PQF* is similar to *HBF:*

$$\frac{PQ}{HB} = \frac{PF}{HF}$$

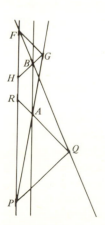

Fig. 5.3

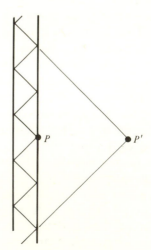

Fig. 5.4 *Measurement of the space-like interval* PP'.

Triangle *FGP* is similar to *RAP:*

$$\frac{FG}{RA} = \frac{FP}{RP}$$

Finally, triangle *FGH* is similar to *PQR:*

$$\frac{FG}{PQ} = \frac{FH}{RP}$$

Combining these equations, one finds $HB = RA$, which implies that *HR* and *BA* are parallel.

The assumptions that went into this proof are that the worldlines are straight and that light rays have worldlines that slope at 45° in the spacetime diagram. The former is true because straight worldlines are *defined* as free-particle worldlines,* and the latter is true because we are using the velocity of light as a standard of velocity.

We can now take as the unit of time the interval between two "ticks" of the geometrodynamic clock. We can take as the unit of length the distance that light travels in one unit of time; this means we take as the unit of distance twice the separation between the two parallel world-lines.

The geometry of spacetime measured by our geometrodynamic clock (in the absence of gravitational fields!) is necessarily the Lorentz geometry with the flat spacetime metric $\eta_{\mu\nu}$. Whether or not the geometry measured with an atomic standard of time is also flat depends on whether the rates of the atomic and geodesic clocks agree. In case of disagreement we will trust the latter.

Note that the construction of the geometrodynamic clock guarantees that the time measured by this clock coincides with the time variable *t* that appears in the equations of motion of a particle, in the Maxwell equations, in the Lorentz transformation equations, etc. We may express this by saying that the geometrodynamic time scale agrees with the inertial time scale. On the other hand, the agreement between an atomic or nuclear time scale and the inertial time scale is open to some doubts because of the possibility of changes in "constants."

Before we put our geometrodynamic clock to some use, one more important remark: nothing in the prescription for the manufacture of the clock tells us just what initial separation to pick for the worldlines. This means that if we want to define a unit of time, we must build *one standard clock* to be kept at some safe place; all other clocks must have their

---

* This statement is somewhat circular because we have given no operational procedure for checking when a particle is free. The thing to do is to impose a self-consistency condition: we say that particles are free if worldlines that are initially parallel (according to above construction) *remain* parallel.

rates adjusted to agree with the standard clock. This is, of course, exactly what used to be the case for the meter: the standard meter bar was kept at Sèvres, near Paris. One of the reasons for the change to the Kr wavelength as standard of length is that Kr atoms are universally accessible and therefore everybody can keep a "standard meter" in his attic. In this respect, the use of a single geometrodynamic clock as primary standard is a step backward. However, questions of principle must take precedence over questions of convenience.

5.3 MEASUREMENT OF SPACETIME WITH THE GEOMETRODYNAMIC CLOCK

The geometrodynamic clock measures time intervals directly. In order to measure distances, we use the time that a light signal takes to cover this distance. Fig. 5.4 (p. 194) shows two points, P, P' separated by a purely spatial interval. The distance between these points is measured by sending a light signal from the clock to P' and waiting for the reflected signal. If the number of ticks of the clock between departure of the signal and return of the echo is $2N_1$, then the distance is $\Delta l = N_1$. Obviously this measurement technique amounts to radar ranging.

Let us now apply our geometrodynamic clock to the measurement of a spacetime interval, such as the interval PP' in Fig. 5.5. We could, of course, measure Δx and Δt separately and calculate the interval from $\Delta s^2 = \Delta t^2 - \Delta x^2$. There is a more elegant way of obtaining the answer. Suppose you send a light signal from the clock towards P' and receive a

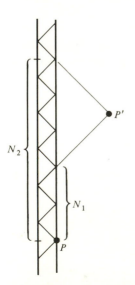

Fig. 5.5 *Measurement of an arbitrary interval* PP'.

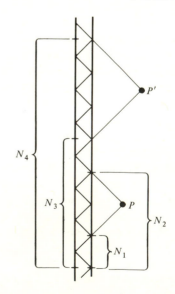

Fig. 5.6 *Measurement of another arbitrary interval* PP'.

reflected signal. If the numbers N_1, N_2 of "ticks" are as indicated in Fig. 5.5, then one can easily show that the spacetime interval PP' is

$$\Delta s^2 = N_1 N_2 \qquad [10]$$

EXERCISE 3. Derive (10).

This result can be generalized. Consider the interval PP' of Fig. 5.6. If the numbers N_1, N_2, N_3, and N_4 are reckoned from some initial instant O as indicated in this figure, then the spacetime interval is

$$\Delta s^2 = (N_4 - N_2)(N_3 - N_1) \qquad [11]$$

This result applies to arbitrary timelike, spacelike, or lightlike intervals. In the general case it must be kept in mind that numbers of ticks previous to the initial instant O must be counted as negative.

EXERCISE 4. Show that the interval between the points PP' of Fig. 5.6 is given by Eq. (11).

The result (10) gives us a nice way of calibrating a clock that is moving with some uniform velocity relative to the standard clock. Since the spacetime interval PP' shown in Fig. 5.7 (p. 198) must be invariant, the moving clock is correctly adjusted if

$$N_1 N_2 = (N_3)^2 \qquad [12]$$

where the left side expresses the interval PP' in terms of ticks of the standard clock (at rest) and the right side expresses the same interval in terms of ticks of the moving clock. When the condition (12) is satisfied, the rates of the two clocks will of course not be "equal," but will differ by the usual time dilation factor.

Note that since a geodesic clock is supposed to consist of *free* particles, a clock that is at rest in some place *cannot* be transported to another place. Hence, if you want a clock in your attic, you must built it there and calibrate it by means of an auxiliary clock which uniformly moves from the standard clock to the attic clock; when the auxiliary is passing by the standard, it can be calibrated, and when it passes through the attic, the calibration can be transferred to your clock.

This procedure seems, and is, rather cumbersome. Why can't we simply use light signals (or WWV radio signals from the U.S. Naval Observatory) to calibrate our clocks at different places? In the absence of gravitational fields it is certainly possible to use light signals. But in the presence of gravitational fields, a calibration by means of light signals will not give the same result as a calibration by means of a traveling

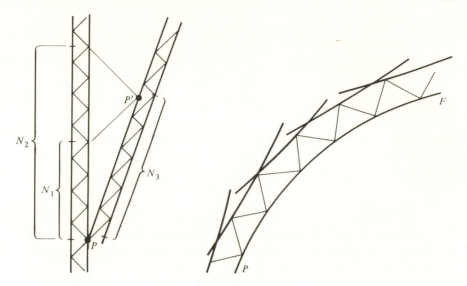

Fig. 5.7 *Calibration of a moving clock.*

Fig. 5.8 *A sequence of geometrodynamic clocks
based on the worldline PF of a freely
falling particle.*

auxiliary clock;* since the latter method is absolutely foolproof we will
stick to it.

5.4 MEASUREMENT IN THE GRAVITATIONAL FIELD

We come now to the crucial question of what happens to our geometrodynamic clock in a gravitational field. We have so far assumed that the
particles constituting the clock were *free,* i.e., shielded from all forces.
But there is no way we can shield the particles from the gravitational
forces.

EXERCISE 5. Why is gravitational shielding impossible? (Hint: How is shielding
of electrostatic forces accomplished?)

Since we cannot get free particles for our clock, we will have to use the
next best thing to a free particle: i.e., a particle in *free fall* in the gravitational field. With this substitution, we then construct the geometrodynamic clock according to the same design used above. Note that since
all (point) particles fall at the same rate in a gravitational field, clocks
made of different kinds of particles will behave in the same way.

There is, however, one modification we must make in the operation
of the clock. The action of the gravitational tidal forces will destroy the
parallelism of the worldlines. To be precise: the worldline *PF* of the first

---

* What enters here is the gravitational time dilation effect, to be discussed later.

particle (see Fig. 5.8) is a curve in spacetime. We can construct a parallel to this curve at one point by giving it the direction of the tangent at that point. But the two worldlines will gradually separate (or approach) because of tidal effects and parallelism will fail. To save our clock, we must subdivide *PF* into many short segments, and construct a parallel to each (see Fig. 5.8). This really means that we construct many clocks; each will be used only for a short time and then the timekeeping duties will be handed over to the next clock. Each change of clock of course involves a calibration procedure.

A concrete picture is helpful: the clock consists of two mirrors in free fall, with light bouncing back and forth between them. If the second mirror acquires an excessive velocity relative to the first by tidal acceleration, we replace it by a new mirror at rest at the proper distance.

One can show that by proceeding to the limit of a sufficiently large number of clocks used in succession, the error in timekeeping can be made arbitrarily small. Suppose we want to keep time accurately for an interval *T*. If we use *N* clocks in succession, each running for only a time *t*, then

$$T = Nt \tag{13}$$

and the error in *T* is smaller than*

$$\Delta T \le N\Delta t \tag{14}$$

where Δt is the largest error introduced by one clock. If the tidal force produces a (maximum) acceleration *a* between two of the freely falling particles, then the deviation in distance is

$$\Delta x \le \tfrac{1}{2}at^2 \tag{15}$$

The error in timekeeping due to this deviation is $2\Delta x$ per tick and hence after *n* ticks of the clock it is

$$\Delta t \le 2\Delta xn \le at^2n \tag{16}$$

so that

$$\Delta T \le Nat^2n \le aT^2n/N \tag{17}$$

We can regard *n* as a fixed number. It then follows that $\Delta T \to 0$ as $N \to \infty$,†

---

* Note: The error ΔT is probably smaller than $N\Delta t$ since the individual clocks do not err in the same direction.

† In this limit it will of course also be necessary to use clocks of a very short ticking period.

The preceding tells us how to measure proper time along any worldline that belongs to a particle in free fall. To measure the interval along any other kind of worldline, we need only subdivide this worldline into short segments; the interval corresponding to such a short segment can then be obtained by using a nearby geometrodynamic clock in free fall and the construction shown in Fig. 5.6. This procedure may be applied to any worldline, whether timelike, spacelike, or lightlike. As is customary in special relativity, the interval measured along a timelike worldline will be called the *proper time*. If the worldline is that of a particle in free fall, then the proper time is simply the number of ticks of a geometrodynamic clock falling with the particle. If the worldline does not correspond to free fall, then the proper time can not be measured with a single geometrodynamic clock—many must be used successively. For example, the proper time for a point at rest on the surface of the earth must be measured by using a geometrodynamic clock which is near the point and in free fall; as soon as this clock has fallen too far away from the point, it must be replaced by a new nearby clock, etc. It is convenient, but not essential, to arrange that each of these clocks be initially at rest relative to the earth. Whenever we speak of the time as measured by a "clock" at rest in a gravitational field, or time measured by a "clock" carried along a worldline, this must always be understood as shorthand for complicated measurement procedures involving many geometrodynamic clocks.

It remains to specify the calibration procedure that is to be used to adjust the rates of geometrodynamic clocks placed at different positions. If we have set up a standard geometrodynamic clock whose ticks define the unit of time, how do we compare the ticks of some other geometrodynamic clock, placed at some distance, with the standard? The procedure is very simple; we use an auxiliary geometrodynamic clock and let it fall from the standard clock to the other clock. While the auxiliary clock is near the standard, we calibrate it using the method indicated by Fig. 5.7. When the auxiliary arrives near the other clock, we calibrate the latter by means of the former. Since the auxiliary clock, like any geometrodynamic clock, is constructed in such a way as to maintain a constant rate, it makes sense to use it to "carry" the unit of time from one place to another.

One immediate consequence of this calibration procedure is that the gravitational field affects the rates of clocks: a clock placed in a deep gravitational potential runs slow. Suppose we have two clocks (I and II) some distance apart in a gravitational field and we want to compare their proper time rates. For example, clock I might be located on the surface of the earth and clock II on the top of a tower.* Suppose that from clock I signals are sent towards clock II at intervals $d\tau_1 = 1$ second proper

* It must be kept in mind that "clock" in this context stands for a measurement procedure (see above).

time (one tick of clock I). These signals could be flashes of light, cannon-balls, or messenger rockets launched at one-second intervals, or a con-tinuous radio wave with a period of one second. The signals are received at clock II at intervals of $d\tau_2$ seconds proper time. We want to find the relation between $d\tau_1$ and $d\tau_2$. It is convenient to concentrate on "cannon-ball" signals because these are particles in free fall with a simple equation of motion. Fig. 5.9 shows the worldlines of clocks I and II and shows the worldlines of several cannonballs which start at I with an initial velocity v_1 at regular intervals and arrive at II with a final velocity v_2 at regular intervals. We may suppose that each cannonball carries a geometro-dynamic clock. If we use the clock on one cannonball to measure the time interval between the departure of that cannonball and the next, we will obtain a time dt which differs from $d\tau_1$ by a dilation factor,

$$dt = d\tau_1 / \sqrt{1 - v_1^2/c^2} \qquad [18]$$

We can likewise measure the time interval between arrival of one cannon-ball and the next at clock II. This arrival interval has the same value as the departure interval dt. The reason is that all cannonballs have worldlines with the same shape; the worldlines are only shifted in time. The, say, second cannonball follows the orbit of the first one and passes each point a time dt later (as measured by its own clock). The arrival interval dt differs from $d\tau_2$ by the corresponding time dilation factor,

$$dt = d\tau_2 / \sqrt{1 - v_2^2/c^2} \qquad [19]$$

Note that (18) and (19) are valid because the prescription for adjusting the relative rates of clocks I and II was based on the use of a freely falling geometrodynamic clock, identical to that carried by the freely falling cannonballs. Since

$$\frac{\sqrt{1 - v_2^2/c^2}}{\sqrt{1 - v_1^2/c^2}} \simeq 1 - \frac{1}{2}(v_2^2 - v_1^2)/c^2 \simeq 1 + \Delta\Phi/c^2$$

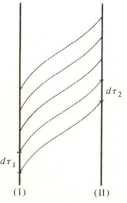

Fig. 5.9 *Worldlines of two clocks at rest in a static gravi-tational field. Timing signals are sent from clock I to clock II.*

where $\Delta\Phi = \Phi_2 - \Phi_1$ is the change in the Newtonian gravitational potential between clocks I and II, it follows from Eqs. (18) and (19) that

$$d\tau_2 = d\tau_1(1 + \Delta\Phi/c^2) \tag{20}$$

This shows that if clock II is at the higher potential ($\Delta\Phi > 0$), then $d\tau_2$ is larger than $d\tau_1$. The signals sent out from clock I at one-second intervals arrive at clock II with intervals larger than one second. Clock I, which is deeper in the gravitational potential, runs slow. This effect does not depend on the type of signal that is used (light flashes, cannonballs, messenger rockets, periodic radio wave, etc.). It is only important that each signal have exactly the same motion as the preceding one (same shape of worldline). Eq. (20) is the gravitational time dilation formula; it will be further discussed in Section 5.5. The equation is valid to lowest order in the gravitational field strength. Furthermore, if the potential Φ is time dependent and changes appreciably in the time needed to send the signal from one clock to the other, then (20) fails.

5.5 CURVED SPACETIME

We now come to a fundamental theorem which establishes the connection between the measured spacetime geometry and the motion of particles in this geometry. In order to measure the metric of some region of spacetime with our geometrodynamic clocks we must begin by setting up a coordinate system so that each point of spacetime is labeled by a set of four numbers $x^\mu = (x^0, x^1, x^2, x^3)$. These could be, but need not be, rectangular coordinates; as we will see in the next chapter, general coordinates are often more convenient. Consider two points P, P' separated by a coordinate interval dx^μ. By means of the geometrodynamic clock and the radar-ranging procedure outlined in Fig. 5.5 we can measure the spacetime interval between these points. We will suppose that the connection between the coordinate interval and the measured spacetime interval is of the general form

$$ds^2 = g_{\mu\nu}(x)dx^\mu dx^\nu \tag{21}$$

where $g_{\mu\nu}(x)$ is a second-rank symmetric tensor. The tensor $g_{\mu\nu}(x)$ is the metric tensor of the spacetime. Any geometry in which the connection between ds^2 and dx^μ is given by a quadratic form (21) is known as a *Riemannian geometry*. The measurement of the metric of spacetime amounts to the measurement of the values of ds^2 that go with different possible choices of dx^μ. It must be emphasized that the general expression (21) encompasses both the case of a flat spacetime geometry and the case of a curved spacetime geometry. Thus, in writing down Eq. (21) we are not prejudging the issue of whether spacetime is flat or curved. The arguments that follow are intended to show that, in the presence of gravitation, spacetime is in fact curved.

We can prove a fundamental theorem: *If $g_{\mu\nu}(x)$ is the metric as measured with the geometrodynamic clock, then the motion of particles in free fall is along the geodesics of $g_{\mu\nu}$.* By geodesics of $g_{\mu\nu}$ we mean those curves that make the integral

$$\int \sqrt{g_{\mu\nu}(x)dx^{\mu}dx^{\nu}}$$

stationary with respect to variations of the worldline; as we have seen in Section 3.6 (Eq. (3.117)), the differential equation for the geodesic curves is

$$\frac{d}{d\tau}\left(g_{\mu\nu}\frac{dx^{\nu}}{d\tau}\right) - \tfrac{1}{2}g_{\alpha\beta,\mu}\frac{dx^{\alpha}}{d\tau}\frac{dx^{\beta}}{d\tau} = 0 \qquad [22]$$

The proof of the theorem is as follows. Suppose that a particle in its free-fall motion is accompanied by a geometrodynamic clock. As this clock falls along worldline I, (see Fig. 5.10), between P and P', it ticks a certain number of times. We compare the number of ticks with the corresponding number for a second clock transported* along a nearby worldline II which has the same initial and final points P, P'. What we must show is that if the difference between the two worldlines is infinitesimal, then the number of ticks is stationary (to first order in the displacement between the worldlines).

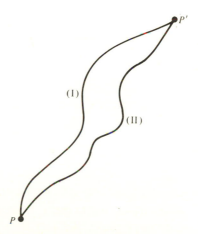

Fig. 5.10 *Worldlines of two clocks. Clock I is in free fall. Clock II is not in free fall.*

To show this, we use a local rectangular reference frame that moves with the clock in free fall along I. Such a reference frame can be very simply set up by using the clock as origin and measuring distances in the vicinity of the origin with the procedure given in Fig. 5.5. Since we are only interested in the effect of infinitesimal variations of the

---

* The second "clock" is not in free fall; it is pushed along its worldline by some nongravitational force. This "clock" is therefore not a geometrodynamic clock, but rather a procedure with a succession of many geometrodynamic clocks.

worldline, we can suppose that II is *very near* to I; as we will see, in this limit the tidal effects can be ignored and the gravitational field can be taken as zero in the freely falling reference frame. Hence the behavior of geometrodynamic clocks in this reference frame is exactly the same as in the absence of gravitational fields. The first clock is at rest and the second has some velocity **v** with respect to the reference frame; therefore the latter suffers the familiar time dilation by a factor $\sqrt{1 - v^2}$ relative to the former. According to this, the time indicated by the first clock is at a maximum and therefore certainly stationary.

To complete the proof, we must show that the tidal gravitational field can in fact be neglected. For this, we need to know something about the effect of a gravitational field on clock rates. The effect is given by Eq. (20),

$$d\tau_2 = d\tau_1(1 + \Delta\Phi/c^2) \qquad [23]$$

Here $\Delta\Phi$ is the potential difference due to the tidal field,

$$\Delta\Phi = \frac{1}{2}\frac{\partial^2\Phi}{\partial x^k\partial x^l}\,\delta x^k\delta x^l \qquad [24]$$

where the second derivative is evaluated at the position of the first clock and δx^k measures the displacement from the first to the second clock.

EXERCISE 6. Check that the tidal forces of Eq. (1.38) are correctly given by

$$f^k = -m\frac{\partial(\Delta\Phi)}{\partial(\delta x^k)}$$

Eq. (23) then leads to the following accumulated time difference between the two clocks as they move from P to P'

$$\int d\tau_2 - \int d\tau_1 = \int \frac{1}{2c^2}\frac{\partial^2\Phi}{\partial x^k\partial x^l}\,\delta x^k\delta x^l d\tau_1 \qquad [25]$$

This is *quadratic* in δx^k and therefore does not affect our result that the time indicated by a freely falling clock is stationary with respect to variations of the worldline.* Note that this conclusion is valid even if Φ is time dependent; we are only interested in the case of an infinitesimal separation δx between the clocks, hence the time $\delta x/c$† needed to carry

---

* Eq. (25) does have an important consequence. Although terms of second order in δx^k have no effect on the stationary nature of the proper time shown by the freely falling clock, such terms will be important in deciding whether the proper time is maximum, minimum, or neither. According to Exercise 3.16, worldlines of free particles in flat spacetime are of maximum proper time; this conclusion fails in the presence of gravitational fields.

† $\delta x/c$ is the time needed to carry out the comparison if light flashes or very fast (relativistic) cannonballs are used as signals.

out the comparison of clocks is an infinitesimal of first order and in the limit $\delta x \to 0$ the dominant contribution to the gravitational time dilation comes from that part of the potential that is constant in time.

It is a corollary of this theorem that spacetime is curved. We know that in a gravitational field the worldlines of particles are certainly *not* straight lines. Hence these worldlines can only be geodesics if the spacetime is curved.

Thus, when we measure distance and time intervals with the best clock that can be conceived, we find that the metric is that of a curved space rather than that of a flat space. Furthermore, the effect of a gravitational field on a particle is *entirely* accounted for by the warping of spacetime: *gravitation is geometry.*

The important role played by the equivalence principle in the construction of the geometrodynamic clock and in the question of curvature of spacetime must be emphasized. If different particles were to fall at different rates in a gravitational field, it would not be possible to construct a clock that is independent of particle properties. Thus, regardless of the details of the gravitational theory, the experimental evidence for equal free-fall rates for all particles must in itself be regarded as experimental evidence for the curved geometry of spacetime, since it shows that geometrodynamic clocks *can* be constructed.

The above conclusion is perfectly general, i.e., all theories of gravitation must be geometric theories. The only difference between diverse theories is how the metric tensor is related to the gravitational field variables, and how the latter are related to the energy-momentum tensor. Let us restate our general conclusion in the particular case of our tensor theory of gravitation. We have seen in Section 3.6 that in the linear approximation particles move as though spacetime were curved; that is, the equation of motion has the appearance of a geodesic equation with a proper time

$$d\tau^2 = (\eta_{\mu\nu} + \kappa h_{\mu\nu})dx^\mu dx^\nu$$

Our fundamental theorem shows that the appearances are not deceiving. To be precise, our theorem establishes that the geodesics of the metric tensor measured by the geometrodynamic clock are exactly the geodesics of the tensor $\eta_{\mu\nu} + \kappa h_{\mu\nu}$. But the geodesic equations constructed from two tensors cannot agree unless the tensors agree.* Hence, we reach the unavoidable conclusion that the metric tensor $g_{\mu\nu}$ measured by the clock must coincide, in the linear approximation, with the tensor $\eta_{\mu\nu} + \kappa h_{\mu\nu}$ which describes the gravitational field:

$$g_{\mu\nu} = \eta_{\mu\nu} + \kappa h_{\mu\nu}$$

* Inspection of Eq. (22) shows that if the two tensors differ by a constant factor, then the corresponding equations of motion will agree. To see that the value of this factor is unity, we need only consider the limiting case of a vanishing gravitational field.

As a further corollary, we can conclude, by means of a powerful argument due to Dicke,[7] that the (rest) masses of particles must be constants independent of space and time. Let us begin by writing down the Lagrangian for a particle with a variable mass, $m = m(x)$. The Lagrangian (3.115) implicitly assumes a constant mass; in general, the mass appears as a factor multiplying the particle Lagrangian (compare Eqs. (3.82) and (3.89)). Hence, to take into account the possibility of a variable mass we need only multiply the right side of (3.115) by $m(x)$:

$$L = m(x) \sqrt{g_{\mu\nu} \frac{dx^\mu}{d\tau} \frac{dx^\nu}{d\tau}} \qquad [26]$$

EXERCISE 7. Show that the Lagrangian equation of motion corresponding to (26) is

$$m \frac{d}{d\tau} \left(g_{\mu\nu} \frac{dx^\nu}{d\tau} \right) - \frac{m}{2} g_{\alpha\beta,\mu} \frac{dx^\alpha}{d\tau} \frac{dx^\beta}{d\tau} = \frac{\partial m}{\partial x^\mu}$$

The quantity $-\partial m/\partial x^\mu$ may be regarded as a force that produces a deviation of the particle from geodesic motion.

But Eq. (26) can also be written as

$$\begin{aligned} L &= m_0 \sqrt{\left[\frac{m(x)}{m_0(x)} \right]^2 g_{\mu\nu} \frac{dx^\mu}{d\tau} \frac{dx^\nu}{d\tau}} \\ &= m_0 \sqrt{\bar{g}_{\mu\nu} \frac{dx^\mu}{d\tau} \frac{dx^\nu}{d\tau}} \end{aligned} \qquad [27]$$

where m_0 is the mass that the particle has when at some reference point (say, at infinity) and where $\bar{g}_{\mu\nu}$ is a new tensor related to the old tensor by $\bar{g}_{\mu\nu} \equiv (m/m_0)^2 g_{\mu\nu}$. Since m_0 is a constant, it follows from (27) that the particle moves along the geodesics of $\bar{g}_{\mu\nu}$. According to our fundamental theorem, $\bar{g}_{\mu\nu}$ will then coincide with the metric tensor measured by the geometrodynamic clock. This means that any space or time dependence of the mass gets included in the definition of the observable metric. The variable mass $m(x)$ disappears from sight and the constant mass m_0 is the only observable mass left in our equations.[7]

Note that this conclusion hinges on the universality of the "mass function" $m(x)$. If different particles have functions $m(x)$ that differ by more than a constant multiplicative factor, then our definition of the metric by means of the geometrodynamic clock would not eliminate the variable masses. At best, we could only eliminate *one* of the variable masses by using one kind of particle (say, electrons) in the construction of the geometrodynamic clock.* Fortunately, we do not have to contend

---

* Thus, for the purposes of Table 5.1, we could assume without loss of generality that the electron mass is constant.

with this difficulty because there is excellent experimental evidence that masses vary in the same way if they vary at all (see Section 5.1).

Finally, how well is the rate of a geometrodynamic clock known to agree with the rate of an atomic clock? This question is of practical importance because experiments, such as the redshift experiment, are always carried out with an atomic or nuclear clock of some kind. Since the construction of a geometrodynamic clock involves complicated maneuvers, difficult to carry out with high precision, this kind of clock will forever remain a theoretician's dream. We cannot, therefore, make any direct comparison between clock rates. But we can get around this difficulty by noting that the ideal system of units based on c as standard of velocity, $\hbar$ as standard of action, and the geometrodynamic clock as standard of time (see Section 5.1) is equivalent to a system based on $c, \hbar$, and the Compton wavelength $\lambda_C = \hbar/m_e c$.* This is an immediate consequence of the result established above: when the spacetime metric is measured with the geometrodynamic clock, the masses of particles are necessarily constant. Hence, the standard of length based on the geometrodynamic clock is equivalent to the standard based on the Compton wavelength. The latter only differs from the atomic standard of length by a factor of α^2 (see Eq. (6)). Since α is known to a very high degree of accuracy to be constant both in space and in time, it follows that the deviation between the geometrodynamic standard and the atomic standard is certainly very small, probably zero.

We might also worry about tidal effects: can the tidal forces disturb an atomic clock? A simple estimate (see Exercise 8) shows that not even the strong tidal forces found on the surface of a neutron star can change the frequency of atomic vibrations by very much; the internal forces of the atom dominate and resist the tidal distortions.

EXERCISE 8. Estimate the change in the frequency of an atomic transition produced by tidal forces. Show that for an atom in free fall at the surface of a neutron star, the change in proper frequency amounts to no more than one part in $\sim 10^{23}$.

(Hint: For a rough estimate, you may regard the electron as bound to the nucleus by a harmonic force with spring constant k_0. Show that the effect of the tidal force (Eq. (1.35)–(1.37)) can be described as a change in spring constant to $\sim (k_0 \pm GMm_e/r^3)$. The frequency of oscillation is then changed from $\omega_0 = \sqrt{k_0/m_e}$ to

$$\sqrt{\frac{k_0 \pm GMm_e/r^3}{m_e}} \simeq \sqrt{\frac{k_0}{m_e}} \left(1 \pm \tfrac{1}{2} \frac{GMm_e}{k_0 r^3}\right)$$

$$\simeq \omega_0 \left(1 \pm \tfrac{1}{2} \frac{GM}{\omega_0^2 r^3}\right)$$

For optical transitions, the value of ω_0 is typically $\omega_0 \simeq 4 \times 10^{15}/\text{sec.}$)

---

* We could of course just as well take $c, \hbar, m_e$, or c, m_e, λ_C, or something else.

Atomic clocks are usually not found in free fall. They are usually at rest in the gravitational field (e.g., at rest on the surface of the earth or a star) and this means that they are accelerated relative to a freely falling clock. We might worry about the disturbances caused by this acceleration. These disturbances can be estimated (see Exercise 9) and they are negligible. Incidentally, there is some direct experimental evidence which confirms that even very large accelerations have an insignificant effect on the frequency of vibration of a nucleus. Nuclei have been placed in an ultracentrifuge and whirled around with an acceleration of 10^8 cm/sec^2 with no effect on their frequency of vibration (other than the ordinary time dilation predicted by special relativity).[18]

EXERCISE 9. An atom, or nucleus, at rest in a gravitational field is held in this position by an electric force generated by the atoms that provide the support. Compare the magnitude of the electric force needed to support an electron against the earth's gravity with the magnitude of the typical internal electric forces in an atom. Compare the magnitude of the electric force needed to support a typical nucleus against the earth's gravity with the magnitude of the electric fields that the nucleus generates at its own surface. Repeat the calculation for the case of an atom or nucleus placed on the surface of a neutron star.

5.6 GRAVITATIONAL TIME DILATION

According to the theorem of the preceding section, the tensor $\eta_{\mu\nu} + \kappa h_{\mu\nu}$ obtained in the study of the equation of motion (Section 3.6) is, in the linear approximation, the metric $g_{\mu\nu}$ that is measured by means of geometrodynamic clocks. In the particular case of a spherically symmetric mass distribution (see Eq. (3.126)) the metric tensor is therefore approximately given by

$$g_{\mu\nu} = \eta_{\mu\nu} + \kappa h_{\mu\nu}$$

$$= \begin{pmatrix} 1 - 2GM/r & 0 & 0 & 0 \\ 0 & -(1 + 2GM/r) & 0 & 0 \\ 0 & 0 & -(1 + 2GM/r) & 0 \\ 0 & 0 & 0 & -(1 + 2GM/r) \end{pmatrix} \quad [28]$$

and the spacetime interval by

$$ds^2 = g_{\mu\nu}dx^\mu dx^\nu = \left(1 - \frac{2GM}{r}\right)dt^2$$

$$- \left(1 + \frac{2GM}{r}\right)(dx^2 + dy^2 + dz^2) \quad [29]$$

This interval corresponds to a curved spacetime; we will calculate the curvature in the next chapter.

Although the arguments of the preceding section prove that (29) describes the geometry surrounding, say, the earth and that therefore we are living in a curved spacetime, it would be good to have some very direct experimental evidence for this curvature. In 1821 Gauss[8] attempted to find a curvature of space by setting up a triangle, about 100 km on one side, on the surface of the earth and measuring the sum of interior angles by means of surveying instruments. He found no deviation from 180°, to within experimental error ($\sim 0.7''$). But even if he had found a deviation, could this not have been interpreted as a simple bending of the path of light in the earth's gravitational field with no geometrical implications? In view of our definition of c as the local standard of velocity, the answer must be in the negative. The deflection of light implies that the effective wave velocity defined by reference to standards of length and time kept outside of the gravitational field (see the discussion at the end of Section 3.9) is a function of position. Hence this effective velocity disagrees with the locally measured velocity, a paradox that can only be resolved if spacetime is curved. Thus, the experimental observations of the deflection of light, or of the extra time delay for light signals, constitute direct evidence for the curvature of spacetime. From a purely phenomenological point of view, these experiments may be regarded as measurements of the ratio g_{00}/g_{kk}* since this is what enters into the velocity formula, Eq. (3.146). The experiments, of course, lend support to the values of g_{00} and g_{kk} given in Eq. (28).

The gravitational time dilation, or slowing down of clocks in a gravitational potential, serves as an even more direct test of the curvature of spacetime. We have already obtained the time dilation formula in our general discussion of measurement procedures (see Eq. 20). If a regular succession of identical signals (messenger rockets, cannonballs, periodic light pulses, or whatever) is sent from one clock to another, the time intervals that these clocks measure between successive signals are related by

$$\frac{d\tau_2}{d\tau_1} = (1 + \Delta\Phi/c^2) \qquad\qquad [30]$$

where it is assumed that the clocks are at rest in a static gravitational field. The experimental evidence for this effect (see below) serves as direct evidence for the curvature of spacetime. The adjustment procedure for geometrodynamic clocks has been carefully designed so that in a flat spacetime it necessarily gives $d\tau_2 = d\tau_1$ (compare Section 5.3); hence the observation of a difference between $d\tau_2$ and $d\tau_1$ indicates that the spacetime is not flat.

---

* Here kk stands for 11, or 22, or 33; the repeated index is not intended to indicate summation.

In terms of the metric of spacetime, the ratio $d\tau_2/d\tau_1$ may be expressed as follows. According to Eq. (20), the proper time for a clock at rest ($dx = dy = dz = 0$) is

$$d\tau^2 = g_{00}dt^2 \tag{31}$$

It is very important to realize that the time *t* which appears in Eq. (31) is *not* the time that clocks placed in the gravitational field give us. Rather, *t* is the time indicated by a clock placed at large distance from the gravitating body. To understand this, we only need to remember that clocks always show *proper time;* in the limit $r \rightarrow \infty$, $g_{00} \rightarrow 1$ and then $d\tau$ coincides with dt. A clock placed at a finite distance shows a time *interval $d\tau$ that is smaller than dt;* that is, the rate of the clock is reduced by the factor $\sqrt{g_{00}}$. This effect is called the *gravitational time dilation.*

If we compare two clocks at different positions, then, for a given interval dt, we have

$$d\tau_1 = \sqrt{g_{00}(1)}\, dt \tag{32}$$
$$d\tau_2 = \sqrt{g_{00}(2)}\, dt$$

and therefore

$$\frac{d\tau_2}{d\tau_1} = \sqrt{\frac{g_{00}(2)}{g_{00}(1)}} \tag{33}$$

According to Eq. (28), $g_{00} = 1 - 2GM/r$ and hence

$$\frac{d\tau_2}{d\tau_1} = \sqrt{\frac{1 - 2GM/r_2}{1 - 2GM/r_1}} \simeq 1 - \frac{GM}{r_2} + \frac{GM}{r_1} \tag{34}$$

Since $\Phi = -GM/r$, this expression agrees with Eq. (30).

The experiments on gravitational time dilation may be regarded as direct measurements of the g_{00} component of metric tensor. The experiments show that in the vicinity of a massive body the metric does not have the flat spacetime value $g_{00} = 1$, but rather the value $g_{00} = 1 - 2GM/r$. We are constrained to accept this purely geometric interpretation of the time dilation effect because it is impossible for the gravitational field to influence the rate of a clock in any other way. In the case of a geometrodynamic clock, this is obvious because of the design of the clock. In the case of an atomic or nuclear clock used in the actual experiments, a simple estimate of the magnitude of the perturbations expected from the gravitational force and the gravitational tidal force shows that there should be no appreciable change in clock rates (see Exercise 8). A modification of the clock rate by a change in the geometry is the only remaining alternative.

For comparison with the experiments it is convenient to express Eq. (34) as a fractional deviation between the times kept by the two clocks:

$$\frac{\Delta T}{T} \simeq \frac{d\tau_2 - d\tau_1}{dt} = GM\left(\frac{1}{r_1} - \frac{1}{r_2}\right) \qquad [35]$$

In terms of the clock rate, or frequency, this deviation amounts to a shift of frequency:

$$\frac{\Delta \nu}{\nu} = -\frac{\Delta T}{T} = -GM\left(\frac{1}{r_1} - \frac{1}{r_2}\right) \qquad [36]$$

If $r_1 < r_2$, then $\Delta \nu / \nu < 0$ and hence a periodic signal emitted in step with the ticking of the clock at r_1 will appear to have a reduced frequency when it arrives at, and is measured by, the clock at r_2. In particular, the spectral lines in the light emitted by an atom placed deep in a gravitational potential will have a *redshift* when compared with the spectral lines emitted by a similar atom placed outside of the potential.

For clocks near the surface of the earth,

$$GM\left(\frac{1}{r_1} - \frac{1}{r_2}\right) \simeq GM\frac{\Delta r}{R^2} = g\Delta r$$

where Δr is the difference in height between the clocks and R is the radius of the earth. With this approximation Eq. (35) becomes, in cgs units,

$$\frac{\Delta T}{T} \simeq g\frac{\Delta r}{c^2} \qquad [37]$$

If $r = 10$ km, this gives

$$\frac{\Delta T}{T} \simeq 10^{-13} \qquad [38]$$

Although the time dilation effect given by Eq. (38) is small, it is not beyond the accuracy of cesium-beam clocks. Hafele and Keating (1972)[9] took cesium-beam clocks on flights around the world in commercial jets. On each flight four cesium clocks were taken and continuously intercompared so that small spontaneous jumps in rate, to which the individual clocks are prone, could be identified and corrected for. The clocks carried to high altitude* by the aircraft were found to have gained time when brought back to a laboratory clock that stayed on the ground. To be precise, the clocks were found to have gained time *after* a correc-

---

* Typical cruising altitude is 30,000 ft.

TABLE 5.2 TIME DILATION WITH CESIUM-BEAM CLOCKS

	Time gain, westward	Time gain, eastward
Observed mean	$(273 \pm 7) \times 10^{-9}$ sec	$(-59 \pm 10) \times 10^{-9}$ sec
Kinematic correction	96 ± 10	-184 ± 18
Remainder (observed dilation)	$177 + 12$	125 ± 21
Predicted dilation	179 ± 18	144 ± 14

tion was subtracted to take into account the special relativistic time dilation ("kinematic correction"). Table 5.2 shows results for clocks that were carried around the earth in an eastward direction, and for clocks carried in a westward direction. The kinematic correction and the predicted redshift were calculated from the flight data of the aircraft.

A somewhat more precise time dilation measurement has been carried out with a nuclear "clock." Pound and Rebka (1959)[10] placed an emitter of γ-rays (Fe^{57}) at ground level and detected the γ-rays with an absorber placed on top of a tower, 22.6 m above the emitter. For $\Delta r = 22.6$ m, Eq. (37) gives a fractional shift in frequency of

$$\frac{\Delta \nu}{\nu} \simeq -\frac{\Delta T}{T} = -g \frac{\Delta r}{c^2} = -2.46 \times 10^{-15} \qquad [39]$$

The measurement of such a small frequency shift is possible because resonant absorption of γ-rays in an absorber of Fe^{57} is very sensitive to the frequency of the incident γ-rays. The experiment relies on the Mössbauer effect to prevent loss of any of the γ-ray energy by recoil of the nucleus during emission or absorption. The most recent version of this experiment (by Pound and Snider, 1964[11]) gave a result of 1.00 ± 0.01 times the predicted value of Eq. (39).

The cesium-beam atomic clock experiment, although less precise than the γ-ray experiment, has one important advantage in that the former experiment shows in the most direct way that clocks in a gravitational potential run slower. The latter experiment does not give quite as direct an indication of whether the frequency shift between the absorbed γ-rays and the natural nuclear oscillations is due to a slower oscillation rate of the emitter, or a loss of frequency suffered by the γ-ray as it climbs upwards in the earth's field. However, we can rule out the possibility of a simple frequency loss during propagation of the light wave by noting that if pulses of light, or whatever, are injected into a static medium (e.g., a gravitational field) at one point and received at another point, then, under steady-state conditions, the rate of injection measured in t-time must be identical with the rate of reception, also measured in t-time. This is obvious because under steady-state conditions, pulses cannot accumulate in the space between the two points. (In optics, this corresponds to the well-known result that if a medium has an index of refraction which varies with position, then light propagates with a changing wavelength but constant frequency.) The fact that the frequency of a nuclear clock disagrees with the frequency of a γ-ray sent to it by another nuclear clock, shows that these clocks do not have the same frequency, as measured in t-time. Thus the rates of clocks depend on the gravitational field, that is, the spacetime geometry depends on the gravitational field.

Gravitational redshifts have also been measured for light emitted by atoms on the surface of stars. For example, oscillations of atoms on the surface of the sun have a frequency shift of

$$\frac{\Delta \nu}{\nu} = -\frac{GM_\odot}{R_\odot c^2} = -2.12 \times 10^{-6} \tag{40}$$

with respect to oscillations of atoms at large distance. It has been difficult to obtain clear-cut results for this redshift because there are strong convection currents in the solar atmosphere and the spectral lines are subjected to Doppler shifts arising from the motion of the gas.[12] However, recent measurements by Brault (1962)[13] and Snider (1971)[14] found redshifts in good agreement with the value given by Eq. (40). These measurements were performed on spectral lines of sodium and potassium which are believed to originate above that region of the photosphere where convection currents produce significant Doppler shifts.

For light emitted by white dwarfs, the redshift is larger than that given by Eq. (40) by a factor of up to ~ 100 because the radius is smaller than $R_\odot$ by approximately this factor, while the mass is about equal to $M_\odot$. The spectral lines of white dwarfs tend to be very broad and diffuse and meaningful measurements have only been possible on the two brightest white dwarfs, Sirius B (the companion of Sirius A) and 40 Eridani B. Both of these white dwarfs are in binary systems and this makes possible the determination of their masses from the observational data on their orbital motion. The radius is determined from the absolute luminosity and surface temperature.

Unfortunately, between ~ 1930 and ~ 1960 the angular separation between Sirius A and Sirius B was too small for carrying out any measurements. The observations made before this period[12] suffered very much from scattering of the light of Sirius A by the atmosphere of Sirius B. Observations made after this period[15] took advantage of a clever diffraction technique to eliminate the undesirable light from Sirius A and gave a redshift, $\Delta\nu/\nu = -(30 \pm 5) \times 10^{-5}$, in excellent agreement with the value $\Delta\nu/\nu = -(28 \pm 1) \times 10^{-5}$ predicted from the measured radius. In the case of 40 Eridani B, the separation from its companion is large and the measured redshift, $\Delta\nu/\nu = -(5.7 \pm 1) \times 10^{-5}$, is in good agreement with the expected value $- (7 \pm 1) \times 10^{-5}$.

Table 5.3 lists the time dilation experiments; the last column gives the ratio of experimental result to theoretical prediction.

TABLE 5.3 TIME DILATION EXPERIMENTS

Experimenter(s)	Year	Method	$\Delta\nu_{ex}/\Delta\nu_{th}$
Adams, Moore[12]	1925, 1928	redshift of H lines on Sirius B	0.2 to 0.5
Popper[15]	1954	redshift of H lines on 40 Eridani B	1.2 ± 0.3
Pound and Rebka[10]	1960	redshift of γ-rays on earth	1.05 ± 0.10
Brault[13]	1962	redshift of Na lines on sun	1.0 ± 0.05
Pound and Snider[11]	1964	redshift of γ-rays on earth	1.00 ± 0.01
Greenstein et al.[15]	1971	redshift of H lines on Sirius B	1.07 ± 0.2
Snider[14]	1971	redshift of K lines on sun	1.01 ± 0.06
Hafele and Keating[9]	1972	time gain of cesium-beam clocks	0.9 ± 0.2

Finally, we present two simple derivations of the redshift. The first argument is based on energy conservation. Suppose an atom in a gravitational field emits a photon and that this photon is absorbed by a second atom at a different position. We assume both atoms are in free fall, but instantaneously at rest. Then, in the rest frame of the first atom, the mass lost by emission of the photon is*

$$\Delta m_1 = h\nu_1 \tag{41}$$

and in the rest frame of the second atom, the mass gained by absorption of the photon is

$$\Delta m_2 = h\nu_2 \tag{42}$$

where ν_2 is the frequency of the photon as seen in this second reference frame. The *total* energy cannot change in the emission-absorption processes:

$$-\Delta m_1 + \Delta m_2 + (\Phi_2\Delta m_2 - \Phi_1\Delta m_1) = 0 \tag{43}$$

The term in parentheses represents the change in gravitational *potential* energy. Using (42) and (43), we obtain

---

* h is Planck's constant.

$$-v_1 + v_2 + \Phi_2 v_2 - \Phi_1 v_1 = 0 \qquad\qquad [44]$$

or

$$\frac{\Delta \nu}{\nu} = \frac{v_2 - v_1}{v_1} = -(\Phi_2 - \Phi_1) \qquad\qquad [45]$$

which agrees with our old result (40).

What is involved in this argument is not just energy conservation, but also the principle of equivalence since in writing the potential energy term in (44) it was assumed that the energy change Δm produces a corresponding change in *gravitational* mass;* it would therefore be wrong to say that energy conservation by itself implies the redshift. Also, the argument should not be construed as a denial of the curved spacetime. Rather, the argument shows that energy conservation and the principle of equivalence imply the redshift and hence, as we have seen above, imply that spacetime is not flat.

Another simple derivation of the redshift makes use of the principle of equivalence to replace the gravitational field by the pseudo-force field existing in an accelerated reference frame. A uniform gravitational field g is simulated by placing the two clocks in a reference frame ("elevator") accelerating upwards at the rate g (in the absence of any gravitational field; see Fig. 5.11). An inertial observer then describes the situation as follows: Clocks I and II, separated by a vertical distance Δr,** are both accelerating upwards with acceleration g. Clock I is

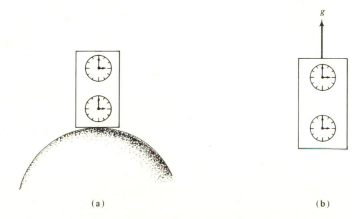

(a)   (b)

Fig. 5.11 (a) *Two clocks at rest in the gravitational field of the earth.*
(b) *Two clocks in an accelerated reference frame.*

* Also, the usual connection $E = h\nu$ was assumed to hold in the freely falling reference frame.
** We ignore length contraction as small; hence, Δr could be measured either with respect to the accelerated reference frame or with respect to the earth.

emitting light signals at the rate ν_1 signals per second. In the time $\sim \Delta r/c$ that these light signals take to reach clock II, the velocity of clock II increases by $\Delta v = g\Delta r/c$. This means that the rate ν_2 at which signals are received by clock II is decreased by the Doppler shift:

$$\nu_2 \simeq \nu_1\left(1 - \frac{\Delta v}{c}\right) = \nu_1(1 - g\Delta r/c^2) \tag{46}$$

i.e.,

$$\Delta\nu/\nu = - g(\Delta r/c^2) \tag{47}$$

which agrees with Eq. (39).

To sum up, Table 5.4 reviews the several lines of direct experimental evidence that converge to the conclusion that spacetime is curved.

TABLE 5.4 DIRECT EVIDENCE FOR CURVED SPACETIME

Experimental Observation	Implication
Equality of inertial and gravitational mass (Newton's principle and Galileo's principle)	Permits the construction of the geometrodynamic clock; particles move along the geodesics measured by this clock ("fundamental theorem"). Hence, orbits in a gravitational field must be geodesics in a curved spacetime (see Section 5.5).
Deflection of light and retardation of light	Shows that velocity of light measured with instruments placed in a gravitational field (c) disagrees with velocity measured by instruments placed at infinity. This contradicts flat spacetime (see Section 3.9).
Gravitational time dilation	Shows that clocks placed in a gravitational field run slow as compared to clocks at infinity. This contradicts flat spacetime (see Section 5.6).

FURTHER READING

The idea that the gravitational constant may decrease with time was proposed by Dirac in a letter to *Nature* (see ref. 16). The effects of a changing G on the earth, the moon, and on the evolution of the sun and other stars have been discussed by Dicke in his book *The Theoretical Significance of Experimental Relativity* and his article "The Significance of a Time-Varying Gravitation" (*in* Chiu and Hoffmann, eds., *Gravitation and Relativity*). The effects on stellar evolution have also been discussed by Teller in the article "Are the Constants Constant?" (*in* Reines, ed., *Cosmology, Fusion and Other Matters*).

A recent analysis of the limits that can be placed on possible changes in the atomic constants may be found in the article "The Fundamental Constants and their Time Variation" by Dyson (see ref. 3).

The important concept of the geometrodynamic clock was first introduced by Marzke and Wheeler in their stimulating article "Gravitation as Geometry" (see ref. 6).

Ehlers, Pirani, and Schild in the article "The Geometry of Free Fall and Light Propagation" (*in* O'Raifeartaigh, ed., *General Relativity: Papers in Honor of J. L. Synge*) carry out a very thorough and sophisticated mathematical study of the assumptions that go into the construction of the spacetime geometry.

The simple derivations of the redshift given at the end of Section 5.6 were found by Einstein long before he developed his theory of general relativity. His article "Über den Einfluss der Schwerkraft auf die Ausbreitung des Lichtes" (see ref. 17; translated in Lorentz, Einstein, Minkowski, and Weyl, *The Principle of Relativity*) is therefore of great historical interest. Note that the light deflection formula derived in this article is off by a factor of two because Einstein did not (yet) take into account the curved spacetime geometry.

REFERENCES

1. H. P. Robertson, *in* P. A. Schilpp, ed., *Albert Einstein: Philosopher-Scientist* (Harper, New York, 1959). See also H. P. Robertson and T. W. Noonan, *Relativity and Cosmology* (Saunders, Philadelphia, 1968), p. 182.
2. Physics Today, January 1974, p. 112.
3. The limit on the rate of change of α was worked out by F. J. Dyson *in* A. Salam and E. P. Wigner, eds., *Aspects of Quantum Theory* (Cambridge University Press, 1972). The limits on dM_n/dt and dM_p/dt have been obtained by a generalization of Dyson's arguments.
4. I. I. Shapiro, W. B. Smith, M. A. Ash, R. P. Ingalls, and G. H. Pettengill, Phys. Rev. Lett. **26**, 27 (1971). Recent preliminary results, presented by I. I. Shapiro at the Fifth Cambridge Conference (1974), indicate $dG/dt = (4 \pm 8) \times 10^{-11}$/year.
5. R. H. Dicke, *in* C. de Witt and B. de Witt, eds., *Relativity, Groups and Topology* (Gordon and Breach, New York, 1964), p. 170.
6. R. F. Marzke and J. A. Wheeler, *in* H.-Y. Chiu and W. F. Hoffman, eds., *Gravitation and Relativity* (Benjamin, New York, 1964).
7. R. H. Dicke, Phys. Rev. Lett. **7**, 359 (1961).
8. An accessible description of the measurements of Gauss may be found in C. Kittel, W. D. Knight, and M. A. Ruderman, *Mechanics* (McGraw-Hill, New York, 1962).
9. J. C. Hafele and R. E. Keating, Science **177**, 168 (1972).
10. R. V. Pound and G. A. Rebka, Phys. Rev. Lett. **4**, 337 (1960).

11. R. V. Pound and J. L. Snider, Phys. Rev. Lett. **13**, 539 (1964); Phys. Rev. B **140**, 788 (1965).
12. A review of the older redshift measurements is given by B. Bertotti, D. Brill, and R. Krotkov, *in* L. Witten, ed., *Gravitation: An Introduction to Current Research* (Wiley, New York, 1962).
13. J. W. Brault, Princeton University Thesis, 1962. A brief description of this measurement is given in ref. 5.
14. J. L. Snider, Phys. Rev. Lett. **28**, 853 (1972).
15. J. L. Greenstein, J. B. Oke, and H. L. Shipman, Ap. J. **169**, 563 (1971).
16. P. A. M. Dirac, Nature **139**, 323 (1937).
17. A. Einstein, Ann. d. Physik **35**, 898 (1911).
18. H. J. Hay, J. P. Schiffer, T. Cranshaw, and P. A. Egelstaff, Phys. Rev. Lett. **4**, 165 (1960); W. Kündig, Phys. Rev. **129**, 2371 (1963); D. C. Champeney, G. R. Isaak, and A. M. Khan, Phys. Lett. **7**, 241 (1963).
19. T. C. Van Flandern, Mon. Not. R. Astr. Soc. **170**, 333 (1975); Bull. Am. Phys. Soc. **20**, 543 (1975); and private communication.
20. R. H. Dicke, *in* B. G. Marsden and A. G. W. Cameron, eds., *The Earth-Moon System* (Plenum Press, New York, 1966).
21. R. R. Newton, *Ancient Astronomical Observations and the Accelerations of the Earth and the Moon* (Johns Hopkins Press, Baltimore, 1970).

PROBLEMS

1. A small (infinitesimal) metallic ruler is used to measure the geometry on a two-dimensional flat hot plate whose temperature is $T = T(x,y)$. The measured geometry is described by an interval $dl^2 = g_{11}dx^2 + g_{22}dy^2 + 2g_{12}dxdy$. Express the functions g_{11}, g_{22}, and g_{12} in terms of $T(x,y)$ and the coefficient of thermal expansion of the ruler.

2. Suppose that both G and α are slowly changing in time. Express the rate of change of the frequency of a pendulum (at the surface of the earth) in terms of the rate of change of G and α. Neglect the effect of the change in G on the size of the earth, but take into account the effect of a change in α on the size of the earth and on the length of the pendulum.

3. Show that if G changes gradually, then the angular velocity ω and the radius r of a planetary orbit change as follows:

$$\frac{1}{\omega}\frac{d\omega}{dt} = \frac{2}{G}\frac{dG}{dt}$$

$$\frac{1}{r}\frac{dr}{dt} = -\frac{1}{G}\frac{dG}{dt}$$

(Hint: The angular momentum, $L = mr^2\omega$, remains constant as G changes because the force always remains central.)

4. In the Skylab satellite, orbiting the earth at a height of 400 km, an astronaut sets up two geometrodynamic clocks consisting of a pair of mirrors with light bouncing between. The first clock is a corrected (or true) clock in which

the distance between the mirrors is permanently kept at exactly 1 cm by means of the procedure shown in Fig. 5.8. The second clock is an uncorrected clock in which the distance between mirrors is initially 1 cm, but no further adjustments are made and the mirrors are permitted to drift under the influence of the tidal force. How long does it take for one of these clocks to lag behind the other by one tick? Assume that the mirrors are separated by a purely radial distance.

Show that in general the time needed to produce a lag of one tick is

$$t = \alpha \left(\frac{3r_0^3}{GM} \Delta t\right)^{1/3}$$

where α is a numerical factor which depends on the orientation of the mirrors, r_0 is the radius of the orbit, M the mass of the earth, and Δt the period of a tick of the clock. (The above formula assumes that the change of orientation of the clock relative to the radial line can be neglected; this will be a good approximation if Δt and t are much smaller than the orbital period of the satellite. $\Delta t << (r_0^3/GM)^{1/2}$.)

5. By what factor does the rate of a clock on the earth differ from that of a clock in intergalactic space (assumed to be flat)? Take into account the potential of the earth, sun, and galaxy.

6. A satellite in a circular orbit of radius r around the earth carries a radio transmitter which emits a signal of constant frequency ω_0 (measured in the rest frame of the satellite). Find the frequency ω measured by a receiver on the surface of the earth when the satellite passes overhead and its radio waves come to the earth along the vertical (no *longitudinal* Doppler shift!). Take the redshift into account and also the orbital velocity of the satellite and the velocity of rotation of the surface of the earth. Express your answer in terms of r (radius of satellite orbit), ω_0 (transmitter frequency), M (mass of earth), R (radius of earth), Ω (angular velocity of rotation of the earth). Assume that both transmitter and receiver are in the earth's equatorial plane.

7. Consider a particle of mass m which moves towards the sun along a straight line of impact parameter b. Assume that the velocity of the particle is nonrelativistic, yet large enough so that the orbit is nearly a straight line. Show that the increment in the proper time of the particle can be expressed (approximately) as

$$d\tau \simeq dt + \left(-\frac{E}{m} - \frac{2GM_\odot}{r}\right)dt$$

where $E \equiv \frac{1}{2}mv^2 - GmM_\odot/r$ is the Newtonian energy. Since E is a constant of the motion, the effect of the gravitational field of the sun is entirely contained in the term $-(2GM_\odot/r)dt$. Integrate this term from $r = r_1$ to $r = b$ to $r = r_2$ and find the total time dilation that may be attributed to the sun. Evaluate numerically for a particle that moves from $r_1 = 1$ light year to $r_2 = 1$ light year with an initial velocity $v = 10^9$ cm/sec and an impact parameter $b = R_\odot$.

8. The earth is in an elliptical orbit around the sun and therefore has a (small) radial velocity. Use the orbital parameters of the earth to calculate the maxi-

mum Doppler shift of the light emitted by the sun. Compare this Doppler shift with the gravitational redshift. Is it necessary to take the Doppler shift into account when making a redshift measurement?

9. The famous twin "paradox" of special relativity can be stated as follows: Romulus and Remus are initially on the earth (regarded as an inertial frame). Romulus stays on the earth. Remus boards a spaceship and flies away at speed u. After a while, he decelerates the spaceship, reverses the velocity, and returns to earth. If we assume that the accelerated portions of the trip last a very short time compared to the unaccelerated portions, then in the inertial reference frame of Romulus, the usual time dilation formula tells us that the elapsed proper times of Romulus and Remus are related by

$$\Delta t_{Ro} = \frac{\Delta t_{Re}}{\sqrt{1 - u^2}} \cong \Delta t_{Re} \left(1 + \frac{u^2}{2} \right)$$

Thus Romulus has aged more than Remus. If we naïvely apply the time dilation formula in the reference frame of Remus, we would reach the opposite conclusion. However, it is not legitimate to apply the time dilation formula in a noninertial reference frame and hence there actually is no paradox.

If one wants to derive the relationship between Δt_{Ro} and Δt_{Re} in the reference frame of Remus, then one must take into account the gravitational time dilation effect. When Remus decelerates, there is an apparent gravitational field in his reference frame, such that Romulus is at the higher gravitational potential and Remus at the lower. Show that the consequent advance of the clocks of Romulus more than compensates for the time dilation that these clocks suffer during the unaccelerated portion of the trip and show that the net result is again given by the above formula. Assume that the accelerated portion of Remus's trip lasts a very short time compared to the unaccelerated portion so that the acceleration essentially occurs at a fixed distance from the earth. (Ignore the accelerations that occur near the earth at the beginning and at the end of the trip.)

6. RIEMANNIAN GEOMETRY

I know why there are so many people who love chopping wood. In this activity one immediately sees the results.

Albert Einstein

The linear tensor theory of gravitation that we developed by analogy with electrodynamics started out as the theory of a tensor field in a flat space-time background. The geometrical interpretation of this tensor field only emerged as an afterthought. However, the analysis of spacetime measurements has shown us that the flat spacetime background is purely fictitious; in a gravitational field, the real geometry measured by our instruments (geometrodynamic clock) is a curved Riemannian geometry.

Before we can proceed with the study of "gravitation as geometry" we need to develop some mathematical concepts, in particular tensors and tensor fields in curved spacetime. To describe tensors, we will use the formalism of the absolute differential calculus of Ricci and Levi-Civita.[1,2] In this formalism tensors are represented by their components relative to a set of coordinates; the tensor calculus is said to be absolute because it is independent of the details of the choice of coordinates, that is, the equations have the same form in all coordinate frames. The tensor calculus for Lorentz tensors given in Chapter 2 is a special case of the general tensor calculus.

Tensors can also be described by the formalism of the exterior

221

differential calculus of Cartan. Vectors, tensors, and differentials are then represented by abstract symbols. For example, the momentum vector of a particle in this abstract notation is represented by the symbol **p**; in component notation the vector is represented by p^α and the connection between these representations is given by $\mathbf{p} = p^\alpha \mathbf{e}_{(\alpha)}$, where the objects $\mathbf{e}_{(\alpha)}$ are a set of four unit vectors. In many instances, the proofs of theorems of the differential calculus take a very elegant and concise form when expressed in terms of the abstract notation. But in other instances this notation results in excessive "orgies of formalism"[3] and component notation is our salvation. Furthermore, we must not forget that the physicist who wishes to measure a tensor has no choice but to set up a coordinate system and then measure the numerical values of the components. Thus, to carry out the comparison of theory and experiment the physicist cannot ultimately avoid the language of components; only a pure mathematician can adhere exclusively to the abstract, coordinate-free language. We will represent tensors by their components because this is physically natural and mathematically adequate.

6.1 GENERAL COORDINATES AND TENSORS

It is quite obvious that unless the space is flat, there is not much reason to try to use rectangular coordinates. In flat spacetime rectangular coordinates are preferred because both the metric ($g_{\mu\nu} = \eta_{\mu\nu}$) and the Lorentz transformations take their simplest form when expressed in these coordinates. In a curved spacetime the use of "rectangular" coordinates does not necessarily result in any simplification of the metric, except in those regions that are far away from all gravitational fields where spacetime becomes asymptotically flat. We will use *general coordinates* to describe points in spacetime. By this is meant that to each point we attach a set of four numbers which serve to identify the point. These numbers can be chosen according to any arbitrary, but well-defined, rule. The coordinates should be assigned in such a way that they vary more or less continuously as one moves about in spacetime; a finite number of discontinuities, such as one finds in polar coordinates at $\phi = 2\pi$, will of course be acceptable.

On a sheet of paper it is easy to label points by coordinates: one simply marks the points with an X and writes the coordinate values next to the X. In empty spacetime this will not do. One way to label spacetime points is to imagine all of space to be filled with point particles, each point particle carrying a dog tag with the values of the coordinates printed on it. There is no need that these particles be at "rest"; a continuous streaming about will keep the coordinate values continuous and acceptable. Of course we must imagine that all the particles, and their dog tags, are massless so that they do not disturb the geometry. A somewhat more realistic proposal for giving coordinates to spacetime is the following. Suppose that at some sufficiently large distance from the region where we want to introduce coordinates, there exist three fixed stars (red, white, and blue) which are not collinear. The space coordinates of a given point in space can then be taken to be the three angles between the stars as seen from the given point (see Fig. 6.1). The time coordinate may be taken as the angle between, say, the red star and a fourth star (yellow) which is in motion with respect to red, white, and blue.* Note that the bending of light by gravitational fields does no harm to these coordinates; all we are trying to find is a well-defined procedure for associating four numbers with each given point of spacetime. We will call these numbers *astrogator coordinates* since an astro-navigator would probably use coordinates of this kind. Incidentally, this second proposal for coordinates is related to the first since it assumes that all of space is filled with photons (but they need not be labelled with dog tags).

These astrogator coordinates are only intended to serve as a demonstration that an operational definition for spacetime coordinates is possible. In practice, our choice of coordinates will be controlled by our desire to keep the solution of the field equations as simple as possible. In the theory of the (electrostatic) potential, the advantages of a clever choice of coordinates are very well known. Unfortunately, in a curved spacetime the coordinates that lead to the simplest mathematics often lack any direct physical interpretation and it will be possible to give an operational definition of the coordinate points only *after* one has solved the field equation for the spacetime metric and after one has calculated just how much the light rays used to locate coordinate points are bent. We have already discussed one instance of this difficulty at the end of Section 3.9.

We will now use the notation $x^\mu = (x^0, x^1, x^2, x^3)$ for general coordinates. It is important to remember that the index on the coordinates is always an *upper* (contravariant) index. Although there exists a general

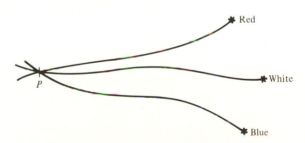

Fig. 6.1 *Astrogator coordinates of a point P are obtained by measuring the angles between rays of light reaching the point from three stars.*

* It is not really necessary that red, white, and blue be fixed. Even if they are in motion, the coordinates will be well defined.

procedure for raising and lowering indices (it involves the *metric* tensor of the space, see below), this procedure gives nothing very useful when applied to x^μ. Thus, we will avoid using x_μ except when dealing, as we were in earlier chapters, with simple rectangular coordinates in flat spacetime.

Let us see what happens if we transform from one kind of general coordinates to another. Suppose that the old coordinates are x^μ and the new x'^μ; the new coordinates are then some functions of the old,

$$x'^\mu = x'^\mu(x) \tag{1}$$

and the old, some functions of the new,

$$x^\mu = x^\mu(x') \tag{2}$$

Small changes in the coordinates obey the transformation law

$$dx'^\mu = \frac{\partial x'^\mu}{\partial x^\nu} dx^\nu \tag{3}$$

$$dx^\mu = \frac{\partial x^\mu}{\partial x'^\nu} dx'^\nu \tag{4}$$

In the case of Lorentz transformation, the partial derivatives $\partial x^\mu / \partial x'^\nu$ were constants. In the present case of general coordinate transformations, these partials are functions of x^μ (or x'^μ). Note that the transformations (3) and (4) are the inverses of one another and therefore

$$\frac{\partial x'^\mu}{\partial x^\beta} \frac{\partial x^\beta}{\partial x'^\nu} = \delta_\nu{}^\mu \tag{5}$$

This also is obvious from the chain rule for derivatives.

The transformation law for the differential operator ∂_μ is obviously

$$\frac{\partial}{\partial x^\mu} = \frac{\partial x'^\nu}{\partial x^\mu} \frac{\partial}{\partial x'^\nu} \tag{6}$$

$$\frac{\partial}{\partial x'^\mu} = \frac{\partial x^\nu}{\partial x'^\mu} \frac{\partial}{\partial x^\nu} \tag{7}$$

The definitions of vectors are natural generalizations of those we gave in the case of Lorentz transformations:

A four-component object A^μ is a contravariant vector under general coordinate transformations if it transforms according to

$$A'^{\mu} = \frac{\partial x'^{\mu}}{\partial x^{\nu}} A^{\nu} \qquad\qquad [8]$$

A four-component object B_{μ} is a covariant vector if

$$B'_{\mu} = \frac{\partial x^{\nu}}{\partial x'^{\mu}} B_{\nu} \qquad\qquad [9]$$

Contravariant vectors have the transformation law of dx^{μ}, covariant vectors that of ∂_{μ}.

Note that x^{μ} is *not a vector* with respect to general coordinate transformations: the transformation law for x^{μ} is given by Eq. (1) and this does not agree with (8). In the exceptional case of a *linear* transformation, Eq. (1) becomes

$$x'^{\mu} = b^{\mu}{}_{\nu} x^{\nu} \qquad\qquad [10]$$

where $b^{\mu}{}_{\nu}$ is a constant matrix. In this case x^{μ} does have the transformation law (8), i.e., (10) can also be written as

$$x'^{\mu} = \frac{\partial x'^{\mu}}{\partial x^{\nu}} x^{\nu} \qquad\qquad [11]$$

This means that x^{μ} is a vector with respect to all *linear* transformations — e.g., it is a vector with respect to Lorentz transformations.

Tensors of higher rank have the transformation laws

$$A'^{\alpha\beta\ldots\lambda} = \frac{\partial x'^{\alpha}}{\partial x^{\mu}} \frac{\partial x'^{\beta}}{\partial x^{\nu}} \cdots \frac{\partial x'^{\lambda}}{\partial x^{\kappa}} A^{\mu\nu\ldots\kappa} \qquad\qquad [12]$$

and

$$B'_{\alpha\beta\ldots\lambda} = \frac{\partial x^{\mu}}{\partial x'^{\alpha}} \frac{\partial x^{\nu}}{\partial x'^{\beta}} \cdots \frac{\partial x^{\kappa}}{\partial x'^{\lambda}} B_{\mu\nu\ldots\kappa} \qquad\qquad [13]$$

if they are contravariant and covariant, respectively.

EXERCISE 1. Show that $\delta_{\mu}{}^{\nu}$ transforms as a tensor covariant in μ, contravariant in ν.

It must be kept in mind that a multicomponent object is (or is not) a tensor *with respect to a given transformation*. Thus, it is quite possible for a quantity to be a tensor with respect to Lorentz transformations, but not a tensor with respect to general coordinate transformations; the converse is of course not possible. By "tensor" we will from here on mean "tensor with respect to general transformations."

Using Eq. (5), it is now easy to show that

$$A^\mu B_\mu \tag{14}$$

is a scalar invariant if A^μ and B_ν are vectors, and that

$$A^{\mu\nu\alpha}{}_\alpha \tag{15}$$

is a second-rank contravariant tensor if $A^{\mu\nu\alpha}{}_\beta$ is a tensor contravariant in $\mu\nu\alpha$, covariant in β, etc.

EXERCISE 2. Show this.

It is also easy to show that the derivative $\partial_\mu\phi$ of a scalar function is a vector. However, it *does not follow* from the transformation laws that the derivative $\partial_\mu A_\nu$ of a vector is a tensor. To construct tensors by differentiating vectors will require a bit more care.

EXERCISE 3. Show that the transformation law of the derivative of a vector is

$$\frac{\partial A'^\mu}{\partial x'^\nu} = \frac{\partial x'^\mu}{\partial x^\alpha}\frac{\partial x^\beta}{\partial x'^\nu}\frac{\partial A^\alpha}{\partial x^\beta} + A^\alpha \frac{\partial x^\beta}{\partial x'^\nu}\frac{\partial^2 x'^\mu}{\partial x^\beta \partial x^\alpha} \tag{16}$$

This is not the transformation law of a tensor. (Hint: Differentiate Eq. (8).)

The metric tensor $g_{\mu\nu}(x)$ was defined in terms of the spacetime interval ds^2 between two points separated by a coordinate interval dx^μ (see Eq. (5.21)):

$$ds^2 = g_{\mu\nu}(x)dx^\mu dx^\nu \tag{17}$$

Without any loss of generality we may assume that the tensor is symmetric: $g_{\mu\nu} = g_{\nu\mu}$. Any antisymmetric part in $g_{\mu\nu}$ would simply drop out of Eq. (17) because $dx^\mu dx^\nu$ is symmetric.

To see that $g_{\mu\nu}$ is indeed a tensor, we recall that ds^2 is an invariant so that

$$ds^2 = ds'^2 = g'_{\alpha\beta}dx'^\alpha dx'^\beta \tag{18}$$

from which it follows that

$$g'_{\alpha\beta}dx'^\alpha dx'^\beta = g_{\mu\nu}dx^\mu dx^\nu \tag{19}$$

By the definition of partial derivative with respect to dx'^μ, dx'^ν we get

$$g'_{\alpha\beta} = g_{\mu\nu} \frac{\partial x^\mu}{\partial x'^\alpha} \frac{\partial x^\nu}{\partial x'^\beta} \qquad [20]$$

exactly what is required for a second-rank, covariant tensor.

Although our main concern is with curved spacetime, the transformation equation (20) is perfectly general and can be used to obtain the metric tensor $g'_{\mu\nu}$ that describes flat spacetime when curvilinear coordinates are used. For example, if $x'^0 \equiv t, x'^1 \equiv r, x'^2 \equiv \theta, x'^3 \equiv \phi$ are the usual polar coordinates, then Eq. (20) with $x^0 = t, x^1 = x, x^2 = y, x^3 = z$, and $g_{\mu\nu} = \eta_{\mu\nu}$ gives

$$g'_{\alpha\beta} = \begin{pmatrix} 1 & 0 & 0 & 0 \\ 0 & -1 & 0 & 0 \\ 0 & 0 & -r^2 & 0 \\ 0 & 0 & 0 & -r^2 \sin^2\theta \end{pmatrix} \qquad [21]$$

The corresponding expression for the interval is

$$\begin{aligned} ds^2 &= g'_{\alpha\beta} dx'^\alpha dx'^\beta \\ &= dt^2 - dr^2 - r^2 d\theta^2 - r^2 \sin^2\theta \, d\phi^2 \end{aligned} \qquad [22]$$

EXERCISE 4. Use Eq. (20) to derive (21).

Note that the expression (22) can also be obtained directly by expressing the differentials $dt^2 - dx^2 - dy^2 - dz^2$ in terms of $dt, dr, d\theta, d\phi$; the metric $g'_{\alpha\beta}$ can then be read off from (22). This procedure is mathematically equivalent to using the transformation law (20), but perhaps somewhat quicker.

The *inverse* $g^{\mu\nu}$ of the metric tensor $g_{\mu\nu}$ is defined by

$$g_{\mu\alpha} g^{\alpha\nu} = \delta_\mu^{\ \nu} \qquad [23]$$

EXERCISE 5. Find the tensor inverse to (21).

We will use the tensors $g^{\mu\nu}$ and $g_{\mu\nu}$ to raise and lower indices, respectively:*

$$dx_\mu \equiv g_{\mu\nu} dx^\nu \qquad [24]$$

$$\partial^\mu \equiv g^{\mu\nu} \partial_\nu \qquad [25]$$

$$A^{\nu\alpha\beta\cdots} = g^{\mu\nu} A_\mu^{\ \alpha\beta\cdots} \qquad [26]$$

$$B_\mu^{\ \mu\alpha} = g^{\alpha\beta} B_\mu^{\ \mu}_{\ \beta} \qquad [27]$$

* Although we will not use general coordinates with lower indices (x_μ), we will use differentials with lower indices (dx_μ).

EXERCISE 6. Show that the quantities appearing on the left sides of Eqs. (24)–(27) have the transformation properties indicated by the position of their indices if it is assumed that the quantities on the right side transform as indicated by their indices.

6.2 THE CHRISTOFFEL SYMBOL AND THE COVARIANT DERIVATIVE

A useful quantity is the *Christoffel symbol*, defined by

$$\Gamma^{\mu}{}_{\alpha\beta} = \tfrac{1}{2}g^{\mu\nu}(g_{\nu\alpha,\beta} + g_{\beta\nu,\alpha} - g_{\alpha\beta,\nu}) \qquad [28]$$

This is a $4 \times 4 \times 4$ component object, symmetric in $\alpha\beta$. The object $\Gamma^{\mu}{}_{\alpha\beta}$ is *not* a tensor under general coordinate transformations.* The transformation law of the Christoffel symbol is

$$\Gamma'^{\mu}{}_{\alpha\beta} = \frac{\partial x'^{\mu}}{\partial x^{\nu}} \frac{\partial x^{\sigma}}{\partial x'^{\alpha}} \frac{\partial x^{\tau}}{\partial x'^{\beta}} \Gamma^{\nu}{}_{\sigma\tau} + \frac{\partial x'^{\mu}}{\partial x^{\nu}} \frac{\partial^2 x^{\nu}}{\partial x'^{\alpha}\partial x'^{\beta}} \qquad [29]$$

This follows from the transformation law of $g^{\mu\nu}$ and $g_{\mu\nu}$. Eq. (29) can also be written as

$$\Gamma'^{\mu}{}_{\alpha\beta} = \frac{\partial x'^{\mu}}{\partial x^{\nu}} \frac{\partial x^{\sigma}}{\partial x'^{\alpha}} \frac{\partial x^{\tau}}{\partial x'^{\beta}} \Gamma^{\nu}{}_{\sigma\tau} - \frac{\partial x^{\nu}}{\partial x'^{\alpha}} \frac{\partial x^{\sigma}}{\partial x'^{\beta}} \frac{\partial^2 x'^{\mu}}{\partial x^{\sigma}\partial x^{\nu}} \qquad [30]$$

EXERCISE 7. Derive Eq. (29).

EXERCISE 8. Derive Eq. (30). (Hint: Differentiate

$$\frac{\partial x'^{\mu}}{\partial x^{\nu}} \frac{\partial x^{\nu}}{\partial x'^{\alpha}} = \delta_{\alpha}{}^{\mu}$$

with respect to x'^{β} and obtain the identity

$$\frac{\partial x'^{\mu}}{\partial x^{\nu}} \frac{\partial^2 x^{\nu}}{\partial x'^{\alpha}\partial x'^{\beta}} + \frac{\partial x^{\nu}}{\partial x'^{\alpha}} \frac{\partial x^{\sigma}}{\partial x'^{\beta}} \frac{\partial^2 x'^{\mu}}{\partial x^{\sigma}\partial x^{\nu}} = 0.)$$

Note that the difference between (29) and the transformation law of a third-rank tensor is due to the presence of the second term on the right side. In the special case of a *linear* transformation, say

$$x'^{\mu} = b^{\mu}{}_{\nu}x^{\nu}$$

---

* We should therefore not use tensor notation for it. The notation $\{^{\mu}_{\alpha\beta}\}$ is sometimes used instead of $\Gamma^{\mu}{}_{\alpha\beta}$, but the latter seems to be more popular.

where $b^\mu{}_\nu$ is a constant, this second term is absent and then $\Gamma^\mu{}_{\alpha\beta}$ does behave as a tensor. $\Gamma^\mu{}_{\alpha\beta}$ is an example of an object that is a tensor with respect to Lorentz transformations (linear) but not a tensor with respect to general nonlinear coordinate transformations.

EXERCISE 9. In a later chapter (see Section 8.1), the following spacetime interval for a curved spacetime will play an important role:

$$ds^2 = e^{N(r)}dt^2 - e^{L(r)}dr^2 - r^2 d\theta^2 - r^2 \sin^2 \theta d\phi^2 \qquad [31]$$

where N, L are some functions of r, but not functions of t, θ, or ϕ (we write these functions as exponentials e^N, e^L because this simplifies some calculations). With $x^0 \equiv t$, $x^1 = r$, $x^2 = \theta$, $x^3 = \phi$ the metric tensor is

$$g_{\mu\nu} = \begin{pmatrix} e^{N(r)} & 0 & 0 & 0 \\ 0 & -e^{L(r)} & 0 & 0 \\ 0 & 0 & -r^2 & 0 \\ 0 & 0 & 0 & -r^2 \sin^2 \theta \end{pmatrix} \qquad [32]$$

Derive the following expressions for the Christoffel symbols (the primes indicate derivatives: $L' = \partial L/\partial r$, $N' = \partial N/\partial r$):

$$\begin{aligned}
\Gamma^0{}_{01} &= \Gamma^0{}_{10} = \tfrac{1}{2}N' & \Gamma^2{}_{12} &= \Gamma^2{}_{21} = 1/r \\
\Gamma^1{}_{00} &= \tfrac{1}{2}N'\, e^{N-L} & \Gamma^2{}_{33} &= -\sin\theta \cos\theta \\
\Gamma^1{}_{11} &= \tfrac{1}{2}L' & \Gamma^3{}_{13} &= \Gamma^3{}_{31} = 1/r \\
\Gamma^1{}_{22} &= -r\, e^{-L} & \Gamma^3{}_{23} &= \Gamma^3{}_{32} = \cot\theta \\
\Gamma^1{}_{33} &= -r \sin^2\theta\, e^{-L}
\end{aligned} \qquad [33]$$

All other Christoffel symbols are zero.

The geodesic equation (5.22) can be expressed in terms of $\Gamma^\mu{}_{\alpha\beta}$. Since

$$\frac{dg_{\mu\nu}}{ds} = g_{\mu\nu,\alpha} \frac{dx^\alpha}{ds} \qquad [34]$$

Eq. (5.22) can be written as

$$g_{\mu\nu} \frac{d^2 x^\nu}{ds^2} + g_{\mu\nu,\alpha} \frac{dx^\alpha}{ds} \frac{dx^\nu}{ds} - \frac{1}{2} g_{\alpha\beta,\mu} \frac{dx^\alpha}{ds} \frac{dx^\beta}{ds} = 0 \qquad [35]$$

Multiplying this by $g^{\sigma\mu}$, we obtain

$$\frac{d^2 x^\sigma}{ds^2} + \Gamma^\sigma{}_{\alpha\beta} \frac{dx^\alpha}{ds} \frac{dx^\beta}{ds} = 0 \qquad [36]$$

EXERCISE 10. Multiply (35) by $g^{\sigma\mu}$ and show that (36) results.

The importance of the objects $\Gamma^\mu{}_{\alpha\beta}$ lies in that they can be used for the construction of tensors by differentiation of lower-rank tensors. Using Eqs. (16) and (30), one can show that the combination

$$\frac{\partial A^\mu}{\partial x^\beta} + \Gamma^\mu{}_{\alpha\beta} A^\alpha \qquad\qquad [37]$$

is a second-rank tensor if A^μ is a vector. It can easily be checked that the contributions from the last terms on the right sides of Eqs. (16) and (30) cancel against each other. Although neither $A^\mu{}_{,\beta}$ nor $\Gamma^\mu{}_{\alpha\beta}$ is a tensor, the combination (37) does then give a tensor.

The tensor (37) is called the *covariant derivative* of A^μ. It is usual to indicate a covariant derivative by a semicolon:

$$A^\mu{}_{;\beta} \equiv \frac{\partial A^\mu}{\partial x^\beta} + \Gamma^\mu{}_{\alpha\beta} A^\alpha = A^\mu{}_{,\beta} + \Gamma^\mu{}_{\alpha\beta} A^\alpha \qquad\qquad [38]$$

The word "covariant" as used in this context simply serves to distinguish $A^\mu{}_{;\beta}$ from the ordinary derivative $A^\alpha{}_{,\beta}$. Since we have already been using "covariant" to describe lower indices on a tensor, it might be better to call (38) the *invariant derivative;* but "covariant" is in general use, and what meaning is intended must be decided from the context. For example, (38) is sometimes said to be the covariant derivative of a contravariant vector.

The covariant derivative of a covariant vector is

$$B_{\mu;\beta} = \frac{\partial B_\mu}{\partial x^\beta} - \Gamma^\alpha{}_{\mu\beta} B_\alpha = B_{\mu,\beta} - \Gamma^\alpha{}_{\mu\beta} B_\alpha \qquad\qquad [39]$$

EXERCISE 11. Show that this is a second-rank tensor.

Note that an index with semicolon can be raised in the usual way,

$$A^{\mu;\beta} = g^{\beta\nu} A^\mu{}_{;\nu} \qquad\qquad [40]$$

For a tensor of rank higher than one, the covariant derivative differs from the ordinary derivative by as many extra terms as there are indices. This is a direct consequence of the fact that in the transformation laws (12) and (13), each of the tensor indices behaves as a vector index. For example,

$$B^{\mu\nu}{}_{;\beta} = \frac{\partial B^{\mu\nu}}{\partial x^\beta} + \Gamma^\mu{}_{\alpha\beta} B^{\alpha\nu} + \Gamma^\nu{}_{\alpha\beta} B^{\mu\alpha} \qquad\qquad [41]$$

and

$$A^{\mu\nu}{}_{\sigma;\beta} = \frac{\partial A^{\mu\nu}{}_{\sigma}}{\partial x^{\beta}} + \Gamma^{\mu}{}_{\alpha\beta}A^{\alpha\nu}{}_{\sigma} + \Gamma^{\nu}{}_{\alpha\beta}A^{\mu\alpha}{}_{\sigma} - \Gamma^{\alpha}{}_{\sigma\beta}A^{\mu\nu}{}_{\alpha} \qquad [42]$$

are the covariant derivatives of the tensors $B^{\mu\nu}$ and $A^{\mu\nu}{}_{\sigma}$, respectively.

By convention, the covariant derivative of a scalar field is simply taken to be the ordinary derivative: $\phi_{;\alpha} = \phi_{,\alpha}$.

As a consequence of (41), the covariant derivative of the second-rank tensor $A^{\mu}B^{\nu}$ can be written

$$(A^{\mu}B^{\nu})_{;\alpha} = A^{\mu}{}_{;\alpha}B^{\nu} + A^{\mu}B^{\nu}{}_{;\alpha} \qquad [43]$$

This shows that covariant differentiation obeys the usual rule for differentiating a product.

EXERCISE 12. Check (43).

The metric tensor $g_{\mu\nu}$ and its inverse $g^{\mu\nu}$ have zero covariant derivative,

$$g_{\mu\nu;\alpha} = 0 \qquad [44]$$

$$g^{\mu\nu}{}_{;\alpha} = 0 \qquad [45]$$

EXERCISE 13. Show this.

It follows that the operation of raising or lowering of indices commutes with covariant differentiation. For example,

$$(g^{\mu\nu}A_{\nu})_{;\alpha} = g^{\mu\nu}A_{\nu;\alpha} \qquad [46]$$

EXERCISE 14. Suppose we define objects $\Gamma_{\mu,\alpha\beta}$ by*

$$\Gamma_{\mu,\alpha\beta} = g_{\mu\nu}\Gamma^{\nu}{}_{\alpha\beta} = \tfrac{1}{2}(g_{\mu\alpha,\beta} + g_{\beta\mu,\alpha} - g_{\alpha\beta,\mu}) \qquad [47]$$

Show that the derivative $g_{\alpha\beta,\mu}$ can be written

$$g_{\alpha\beta,\mu} = \Gamma_{\alpha,\beta\mu} + \Gamma_{\beta,\alpha\mu} \qquad [48]$$

Hence, all the derivatives $g_{\alpha\beta,\mu}$ are zero if and only if all the Christoffel symbols are zero.

6.3 GEODESIC COORDINATES

It is always possible to find coordinates such that at *one* given point P the metric is that of flat space. Suppose that in some given coordinate

---

* The comma on Γ does *not* imply differentiation in this case.

system the metric is $g_{\mu\nu}(x)$. Introduce new coordinates x'^μ by a linear transformation

$$x'^\mu = b^\mu{}_\nu x^\nu \tag{49}$$

where the quantities $b^\mu{}_\nu$ are constants. The values of these constants can be chosen in such a way that $g'^{\mu\nu} = \eta^{\mu\nu}$. Since

$$g'^{\mu\nu} = \frac{\partial x'^\mu}{\partial x^\alpha} \frac{\partial x'^\nu}{\partial x^\beta} g^{\alpha\beta} \tag{50}$$

the metric in the new coordinates will equal $\eta^{\mu\nu}$ at the point P if

$$\eta^{\mu\nu} = b^\mu{}_\alpha b^\nu{}_\beta g^{\alpha\beta}(P) \tag{51}$$

This is a set of 10 equations for the 16 unknown constants $b^\mu{}_\nu$; there always exists a solution. Of course, the new metric will equal $\eta^{\mu\nu}$ only at the point P; everywhere else we will be left with whatever comes out of Eq. (50). Incidentally, our argument shows that if $g^{\mu\nu}$ is a constant throughout all of spacetime, then it can be transformed into $\eta^{\mu\nu}$ everywhere.

It is even possible to do better: at the given point P the first derivatives of $g^{\mu\nu}$ can be transformed away by a choice of coordinates. The transformation that accomplishes this is

$$x'^\mu = x^\mu - x^\mu(P) + \tfrac{1}{2}\Gamma^\mu{}_{\alpha\beta}(P)[x^\alpha - x^\alpha(P)][x^\beta - x^\beta(P)] \tag{52}$$

where $x^\mu(P)$ are the (old) coordinates of the point P. Eq. (52) gives

$$\left.\frac{\partial x'^\mu}{\partial x^\nu}\right|_{x=x(P)} = \delta_\nu{}^\mu \quad \text{and} \quad \left.\frac{\partial^2 x'^\mu}{\partial x^\sigma \partial x^\nu}\right|_{x=x(P)} = \Gamma^\mu{}_{\sigma\nu}(P) \tag{53}$$

which, when inserted into Eq. (30), gives $\Gamma'^\mu{}_{\alpha\beta}(P) = 0$ and hence $g'_{\mu\nu,\beta}(P) = 0$.

EXERCISE 15. Carry out the necessary calculations.

Note that (52) takes P as the origin of the new coordinates ($x'^\mu(P) = 0$). After performing this transformation we can perform a further transformation of the type (49) to change $g'^{\mu\nu}$ into $\eta^{\mu\nu}$.

The coordinates x'^μ which, at a given point, make the Christoffel symbols and the first derivatives of the metric zero and assign to the metric its flat spacetime values, are called *geodesic coordinates* (at the point). The name is due to the following: since $g'_{\alpha\beta,\mu} = 0$, the equation of motion of a particle (36) reduces to

$$\frac{d^2x'^{\mu}}{d\tau^2} = 0 \qquad\qquad [54]$$

that is, a particle placed at the point P will move with constant velocity or remain at rest as seen from the geodesic coordinates. This means that the coordinate frame is *instantaneously* in free fall with the same acceleration as the particle. It must be emphasized that the coordinates are geodesic only for one instant. The derivatives of the metric are zero only at one *spacetime* point P, i.e., at one place and at *one time;* if we want geodesic coordinates at a later instant of time, it will be necessary to carry out the transformation (52) for this later instant.

Geometrically the coordinates we have introduced correspond to replacing the curved space by a flat space tangent to the former at the point P. This can, of course, only be done over an infinitesimal region. In fact, as we will see later, the *second* derivatives of $g^{\mu\nu}$ cannot all be made zero by choice of coordinates.

It is also possible to introduce coordinates ("Fermi coordinates") such that the first derivatives of $g^{\mu\nu}$ are zero along every point on a given geodesic curve. This means that the origin of the coordinate frame is in free fall with the same motion as a particle on the given geodesic. The transformation to Fermi coordinates is a bit complicated and we will not give the explicit mathematical formulas (the procedure for the construction of such coordinates is outlined in Section 7.5).

Note that we can regard Eq. (54) as a consequence of the geodesic equation and, conversely, we can regard the geodesic equation as a consequence of Eq. (54). The latter is obvious since the transformation from general coordinates to geodesic coordinates can just as well be carried out in reverse; it then leads us from Eq. (54) to Eq. (36). This can be used as the basis for a derivation of the geodesic motion of a freely falling particle. We need only observe that a freely falling particle, next to a freely falling geometrodynamic clock, will not have any acceleration relative to the clock. Hence, in the coordinate frame moving with the clock, the equation of motion of the particle must be Eq. (54); when transformed to general coordinates this gives us the geodesic equation (36). This derivation is an alternative to that given in Section 5.4. The latter derivation was based on an appeal to the extremum condition for the proper time and it had the advantage that the (rather messy) transformation properties (see Eq. (29)) of the Christoffel symbols did not enter into it.

6.4 PARALLEL TRANSPORT

We can gain a better understanding of the meaning of the covariant derivative if we introduce the concept of *parallel transport* of a vector. Suppose we have set up geodesic coordinates at a point P and we transport a constant vector a'^{ν} parallel to itself for some distance $\delta x'^{\mu}$, starting at the origin $x'^{\mu} = 0$ (this origin coincides with P). Since the coordinates x'^{μ} are simply Cartesian coordinates, the components of a'^{ν} do not change during the parallel transport. If the vector is a'^{ν} before and $a'^{\nu} + \delta a'^{\nu}$ after parallel transport, then

$$\delta a'^{\nu} = 0 \tag{55}$$

However, the components a^{ν} of the vector as seen from the curvilinear coordinates *do change*. The components before and after the transport are, respectively,

$$a^{\alpha} = \frac{\partial x^{\alpha}}{\partial x'^{\nu}}\bigg|_{x'=0} a'^{\nu} \tag{56}$$

and

$$a^{\alpha} + \delta a^{\alpha} = \frac{\partial x^{\alpha}}{\partial x'^{\nu}}\bigg|_{x'=\delta x'}(a'^{\nu} + \delta a'^{\nu}) \tag{57}$$

To evaluate these expressions, we need $\partial x^{\alpha}/\partial x'^{\nu}$. The easiest way to obtain this derivative is to recognize that for small values of x'^{α} we can approximate Eq. (52) by

$$x^{\mu} \simeq x^{\mu}(P) + x'^{\mu} - \tfrac{1}{2}\Gamma^{\mu}_{\alpha\beta}(P)x'^{\alpha}x'^{\beta}$$

which gives

$$\frac{\partial x^{\alpha}}{\partial x'^{\nu}} \simeq \delta_{\nu}{}^{\alpha} - \Gamma^{\alpha}_{\nu\beta}x'^{\beta}$$

Substitution of this into Eqs. (56) and (57) results in

$$a^{\alpha} = a'^{\alpha} \tag{58}$$

and

$$a^{\alpha} + \delta a^{\alpha} = (a'^{\alpha} + \delta a'^{\alpha}) - \Gamma^{\alpha}_{\beta\nu}(a'^{\nu} + \delta a'^{\nu})\delta x'^{\beta}$$
$$= a'^{\alpha} - \Gamma^{\alpha}_{\beta\nu}a'^{\nu}\delta x'^{\beta} \tag{59}$$

The difference between the expressions (58) and (59) is

$$\delta a^{\alpha} = -\Gamma^{\alpha}_{\beta\nu}a'^{\nu}\delta x'^{\beta} \tag{60}$$

Since $\delta x = \delta x'$ at the point P, we can say that the change in a vector when parallel transported a distance δx^{β} is

$$\delta a^{\alpha} = -\Gamma^{\alpha}_{\beta\nu}a^{\nu}\delta x^{\beta} \tag{61}$$

For covariant vectors, the change introduced by parallel transport can be obtained by a similar calculation:

$$\delta a_\alpha = \Gamma^\nu_{\alpha\beta} a_\nu \delta x^\beta \qquad [62]$$

It is an immediate consequence of these formulas that if two vectors a^α and b_α are parallel transported together, then their scalar product remains unchanged. In fact,

$$\begin{aligned}
\delta(a^\alpha b_\alpha) &= a^\alpha \delta b_\alpha + b_\alpha \delta a^\alpha \\
&= a^\alpha \Gamma^\nu_{\alpha\beta} b_\nu \delta x^\beta - b_\alpha \Gamma^\alpha_{\beta\nu} a^\nu \delta x^\beta = 0
\end{aligned} \qquad [63]$$

As a corollary of this result it follows that the invariant "length" $a^\alpha a_\alpha$ of a vector remains constant as the vector is parallel transported.

In terms of the four-velocity

$$u^\alpha = \frac{dx^\alpha}{ds} \qquad [64]$$

the geodesic equation (36) can be written

$$du^\nu + \Gamma^\nu_{\alpha\beta} u^\alpha dx^\beta = 0 \qquad [65]$$

Comparison with Eq. (61) shows that u^ν is parallel transported along the geodesic curve. Consequently if we have some arbitrary vector a_α which is also parallel transported along the geodesic, then

$$\delta(u^\alpha a_\alpha) = 0 \qquad [66]$$

In two- or three-dimensional Riemannian space with a positive definite metric, Eq. (66) has a simple geometric interpretation: the vector u^α is tangent to the geodesic curve and hence a constant dot product of u^α and a^α means that as a^α is parallel transported along the geodesic, the angle between a^α and the geodesic remains constant. In our four-dimensional spacetime with a metric that is not positive definite, the scalar product cannot be expressed in terms of an angle between vectors and the geometric interpretation is not possible.

That parallel transport should produce changes in the components of a vector should not cause surprise. It even happens in flat spacetime when one uses polar coordinates. If a constant vector is parallel transported along some curve, its components with respect to rectangular coordinates do not change, but its components with respect to polar coordinates obviously do change (see Fig. 6.2).

Another, not quite so trivial, example is parallel transport on the surface of a sphere. Although we are mainly interested in the four-dimensional spacetime, the definition (17) for the metric tensor applies to a Riemannian space of any number of dimensions; we only have to

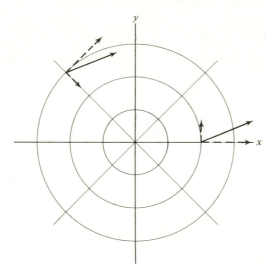

Fig. 6.2 *Parallel transport of a vector in flat spacetime. The vector does not change, but its components (broken lines) with respect to polar coordinates do change.*

keep in mind that in an *n*-dimensional space, the indices μ and ν take the values 1, 2, 3, . . . , *n*. Thus, the surface of a sphere of radius *b* is a two-dimensional curved space. If we use coordinates $x^1 = \theta$ and $x^2 = \phi$ (where ϕ and θ are the usual polar angles) on this surface, then distances along the sphere are given by

$$dl^2 = b^2 d\theta^2 + b^2 \sin^2 \theta d\phi^2 \qquad\qquad [67]$$

and $g^{(2)}_{\mu\nu}$ is a 2 × 2 tensor,

$$g^{(2)}_{\mu\nu} = \begin{pmatrix} b^2 & 0 \\ 0 & b^2 \sin^2 \theta \end{pmatrix} \qquad \mu,\nu = 1,2 \qquad\qquad [68]$$

Note that in this example it is more relevant to consider distances rather than intervals; we have inserted a superscript[2] to prevent any confusion with the metric of spacetime.

The geodesics on the surface of the sphere are great circles.

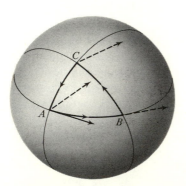

Fig. 6.3 *Parallel transport of a vector along a triangular path on a sphere.*

Fig. 6.3 shows the result of parallel transport of a vector along a closed triangular path formed by three intersecting great circles. According to Eq. (66), as the vector is moved along a great circle, the angle between the vector and the displacement must remain constant. Note that after parallel transport around the closed path *ABC,* the vector deviates from the original vector. As we will see in the next section, such a deviation is a characteristic of a curved space.

The connection between our concept of parallel transport and the covariant derivative is the following. Suppose we try to define the derivative of a vector field A^α in the usual way as

$$\lim_{dx \to 0} \frac{A^\alpha(x + dx) - A^\alpha(x)}{dx^\beta} \qquad [69]$$

The trouble with this expression is that the numerator is *not* a vector; although the difference between two vectors evaluated at the *same point* is a vector, the difference between two vectors evaluated at different points is *not a vector*. To get a numerator that is a vector, let us parallel transport $A^\alpha(x)$ from x to $x + dx$ before carrying out the subtraction. We then have

$$\frac{A^\alpha(x + dx) - A^\alpha(x) - \delta A^\alpha}{dx^\beta} = \frac{A^\alpha(x + dx) - A^\alpha(x)}{dx^\beta}$$
$$+ \Gamma^\alpha{}_{\mu\nu} A^\nu \frac{\delta x^\mu}{dx^\beta} \to A^\alpha{}_{,\beta} + \Gamma^\alpha{}_{\beta\nu} A^\nu \quad [70]$$

This is exactly our earlier expression for the covariant derivative.

Note that in local geodesic coordinates (in which $\Gamma^\alpha{}_{\beta\nu} = 0$) the covariant derivative coincides with the ordinary derivative.

6.5   THE RIEMANN CURVATURE TENSOR

Suppose we have a vector field A_α and we take two covariant derivatives of this — say with respect to μ and ν. We then obtain

$$A_{\beta;\mu;\nu} \qquad [71]$$

or

$$A_{\beta;\nu;\mu} \qquad [72]$$

depending on the order in which we carry out the differentiation. In general, these derivatives are *not* equal — that is, *covariant derivatives do not commute.*

EXERCISE 16. Show that in the exceptional case of covariant differentiation of a scalar field, the derivatives commute:

$$\phi_{;\mu;\nu} = \phi_{;\nu;\mu} \qquad\qquad [73]$$

(Hint: The first covariant derivative $\phi_{;\nu}$ is simply $\phi_{,\nu}$.)

Let us calculate the difference between (71) and (72). Since $A_{\beta;\mu}$ is a second-rank tensor, we have

$$A_{\beta;\mu;\nu} = A_{\beta;\mu,\nu} - \Gamma^{\sigma}{}_{\beta\nu}A_{\sigma;\mu} - \Gamma^{\sigma}{}_{\mu\nu}A_{\beta;\sigma}$$

and since A_β is a first-rank tensor,

$$\begin{aligned}
A_{\beta;\mu;\nu} = {} & (A_{\beta,\mu} - \Gamma^{\alpha}{}_{\beta\mu}A_\alpha)_{,\nu} \\
& - \Gamma^{\sigma}{}_{\beta\nu}(A_{\sigma,\mu} - \Gamma^{\alpha}{}_{\sigma\mu}A_\alpha) - \Gamma^{\sigma}{}_{\mu\nu}(A_{\beta,\sigma} - \Gamma^{\alpha}{}_{\beta\sigma}A_\alpha) \\
= {} & [A_{\beta,\mu,\nu} - \Gamma^{\alpha}{}_{\beta\mu}A_{\alpha,\nu} - \Gamma^{\sigma}{}_{\beta\nu}A_{\sigma,\mu} - \Gamma^{\sigma}{}_{\mu\nu}(A_{\beta,\sigma} - \Gamma^{\alpha}{}_{\beta\sigma}A_\alpha)] \\
& - \Gamma^{\alpha}{}_{\beta\mu,\nu}A_\alpha + \Gamma^{\sigma}{}_{\beta\nu}\Gamma^{\alpha}{}_{\sigma\mu}A_\alpha \qquad [74]
\end{aligned}$$

The expression for $A_{\beta;\nu;\mu}$ can be obtained from (74) by simply exchanging μ and ν. Obviously the quantity contained in the brackets in Eq. (74) is *symmetric* in μ,ν and hence will not contribute to $A_{\beta;\mu;\nu} - A_{\beta;\nu;\mu}$. The remaining terms give

$$\begin{aligned}
A_{\beta;\mu;\nu} - A_{\beta;\nu;\mu} = {} & (-\Gamma^{\alpha}{}_{\beta\mu,\nu} + \Gamma^{\alpha}{}_{\beta\nu,\mu})A_\alpha \\
& + (\Gamma^{\sigma}{}_{\beta\nu}\Gamma^{\alpha}{}_{\sigma\mu} - \Gamma^{\sigma}{}_{\beta\mu}\Gamma^{\alpha}{}_{\sigma\nu})A_\alpha \qquad [75]
\end{aligned}$$

We can rewrite this as

$$A_{\beta;\mu;\nu} - A_{\beta;\nu;\mu} = R^{\alpha}{}_{\beta\mu\nu}A_\alpha \qquad\qquad [76]$$

where

$$R^{\alpha}{}_{\beta\mu\nu} = -\Gamma^{\alpha}{}_{\beta\mu,\nu} + \Gamma^{\alpha}{}_{\beta\nu,\mu} + \Gamma^{\sigma}{}_{\beta\nu}\Gamma^{\alpha}{}_{\sigma\mu} - \Gamma^{\sigma}{}_{\beta\mu}\Gamma^{\alpha}{}_{\sigma\nu} \qquad [77]$$

The left side of (76) is a tensor; it is the difference between two tensors. Hence the right side must also be a tensor. Since A_α is arbitrary, the $4 \times 4 \times 4 \times 4$ component object $R^{\alpha}{}_{\beta\mu\nu}$ must be a tensor of rank four.* It is called the *Riemann curvature tensor*.

We can lower the index α to obtain a purely covariant tensor,

$$R_{\alpha\beta\mu\nu} = g_{\alpha\sigma}R^{\sigma}{}_{\beta\mu\nu} \qquad\qquad [78]$$

This tensor satisfies several important identities

$$R_{\alpha\beta\mu\nu} = -R_{\beta\alpha\mu\nu} \qquad\qquad [79]$$

---

* Of course, you can show explicitly that $R^{\alpha}{}_{\beta\mu\nu}$ is a tensor by using (77) and the transformation law of the Γ's. Try it; you won't like it.

$$R_{\alpha\beta\mu\nu} = -R_{\alpha\beta\nu\mu} \tag{80}$$

$$R_{\alpha\beta\mu\nu} = R_{\mu\nu\alpha\beta} \tag{81}$$

$$R_{\alpha\beta\mu\nu} + R_{\alpha\mu\nu\beta} + R_{\alpha\nu\beta\mu} = 0 \tag{82}$$

EXERCISE 17. Derive (79)–(82).

Although the tensor $R_{\alpha\beta\mu\nu}$ has 256 components, only 20 of these components are independent. Taking into account the identities (79) and (80), we see that $R_{\alpha\beta\mu\nu}$ is an antisymmetric tensor of second rank in $\alpha\beta$ and in $\mu\nu$. Since a second-rank antisymmetric tensor has 6 independent components, we are left with no more than 6×6 components for $R_{\alpha\beta\mu\nu}$. We can write these 6×6 components in the form of a 6×6 matrix R_{AB} where A stands for the *pair* $\alpha\beta$ and B for the *pair* $\mu\nu$; the values of A and B may be taken as 01, 02, 03, 12, 13, 23. The identity (81) says that R_{AB} is a symmetric matrix; a 6×6 symmetric matrix has 21 independent components. Finally, the identity (82) imposes *one* constraint* $R_{0123} - R_{0231} + R_{0312} = 0$ among these 21 components. Hence we are left with only 20 independent components for the Riemann tensor.

The Riemann tensor can be shown to be the only tensor that can be formed by taking linear combinations of the second derivatives of the metric. To prove this we can use a simple counting argument. We begin by taking geodesic coordinates x'^{μ} with origin at some point P and perform a coordinate transformation to a new set of coordinates

$$x''^{\mu} = x'^{\mu} + \frac{1}{6}a^{\mu}{}_{\alpha\beta\gamma}x'^{\alpha}x'^{\beta}x'^{\gamma} \tag{83}$$

where $a^{\mu}{}_{\alpha\beta\gamma}$ is a constant matrix. At the point $x'^{\mu} = 0$, this gives

$$\frac{\partial x''^{\mu}}{\partial x'^{\nu}} = \delta_{\nu}{}^{\mu} \qquad \frac{\partial^2 x''^{\mu}}{\partial x'^{\alpha}\partial x'^{\beta}} = 0 \qquad \frac{\partial^3 x''^{\mu}}{\partial x'^{\alpha}\partial x'^{\beta}\partial x'^{\gamma}} = a^{\mu}{}_{\alpha\beta\gamma}$$

Since the transformation law for $g_{\mu\nu}$ and $g_{\mu\nu,\alpha}$ only involves the first and second derivatives of x''^{μ}, it follows that $g'_{\mu\nu} = g''_{\mu\nu}$ and $g'_{\mu\nu,\alpha} = g''_{\mu\nu,\alpha}$. The transformation law for $g_{\mu\nu,\alpha,\beta}$, obtained from Eq. (20) by differentiating twice, involves the third derivative of x''^{μ} and hence $g'_{\mu\nu,\alpha,\beta}$ and $g''_{\mu\nu,\alpha,\beta}$ are different. In fact, the values of $a^{\mu}{}_{\alpha\beta\gamma}$ are at our disposal and we can adjust the values of the second derivatives $g''_{\mu\nu,\alpha,\beta}$ to some extent. There are 100 derivatives $g''_{\mu\nu,\alpha,\beta}$. Since there are 80 independent coefficients $a^{\mu}{}_{\alpha\beta\gamma}$, we can impose 80 conditions on these 100 derivatives.

---

* It is easy to see that unless all the indices in (82) are different, (82) coincides with one of (79)–(81).

EXERCISE 18. Show that only 80 of the coefficients $a^\mu{}_{\alpha\beta\gamma}$ are independent. (Hint: $x'^\alpha x'^\beta x'^\gamma$ in Eq. (83) is symmetric under exchanges of α, β, and γ; hence $a^\mu{}_{\alpha\beta\gamma}$ must also have this symmetry.)

We therefore conclude that there exist only 20 independent linear combinations of second derivatives which are unaffected by the choice of $a^\mu{}_{\alpha\beta\gamma}$. If we now want to construct *tensors* out of the second derivatives, we must use these 20 linear combinations exclusively; the reason for this is that the tensor transformation law only involves $\partial x''^\mu / \partial x'^\nu = \delta_\nu{}^\mu$ and therefore $a^\mu{}_{\alpha\beta\gamma}$ may not enter into it. What tensors can be constructed with the available 20 linear combinations? There is one and only one: the tensor $R''^\alpha{}_{\beta\mu\nu}$. This tensor has exactly 20 independent components and it therefore exhausts all the available linear combinations of second derivatives. The fact that the Riemann tensor is the only tensor which contains the second derivatives of the metric linearly will lead us to the conclusion (see Section 7.3) that this tensor must appear in the exact differential equations for the gravitational field.

Besides the purely algebraic identities (79)–(82), the Riemann tensor also satisfies some differential identities. The covariant derivative $R^\alpha{}_{\beta\mu\nu;\sigma}$ is a tensor of fifth rank. The following identity holds:

$$R^\alpha{}_{\beta\mu\nu;\sigma} + R^\alpha{}_{\beta\nu\sigma;\mu} + R^\alpha{}_{\beta\sigma\mu;\nu} = 0 \qquad [84]$$

To derive Eq. (84) we can use a trick. If we introduce geodesic coordinates, then $\Gamma^\mu{}_{\alpha\beta} = 0$ and

$$R^\alpha{}_{\beta\mu\nu} = -\Gamma^\alpha{}_{\beta\mu,\nu} + \Gamma^\alpha{}_{\beta\nu,\mu} \qquad [85]$$

Therefore (84) reduces to

$$-\Gamma^\alpha{}_{\beta\mu,\nu,\sigma} + \Gamma^\alpha{}_{\beta\nu,\mu,\sigma} - \Gamma^\alpha{}_{\beta\nu,\sigma,\mu} + \Gamma^\alpha{}_{\beta\sigma,\nu,\mu} - \Gamma^\alpha{}_{\beta\sigma,\mu,\nu} + \Gamma^\alpha{}_{\beta\mu,\sigma,\nu} = 0$$

which is a trivial identity. We have therefore established the equation (84) in geodesic coordinates. Since this equation is a *tensor* equation, its validity in one coordinate system implies its validity in all.

By contraction on the first and last index of $R^\alpha{}_{\beta\mu\nu}$ we obtain a second-rank tensor $R_{\beta\mu}$:

$$R_{\beta\mu} \equiv R^\alpha{}_{\beta\mu\alpha} \qquad [86]$$

This tensor is called the *Ricci tensor;* it is symmetric in $\beta\mu$:

$$R_{\beta\mu} = R_{\mu\beta} \qquad [87]$$

EXERCISE 19. Show this.

By a further contraction, we obtain a scalar

$$R \equiv R^{\beta}{}_{\beta} = R^{\alpha\beta}{}_{\beta\alpha} \qquad [88]$$

This one-component object is called the *curvature scalar.*
  The following differential equation holds identically:

$$(R^{\mu}{}_{\nu} - \tfrac{1}{2}\delta^{\mu}{}_{\nu}R)_{;\mu} = 0 \qquad [89]$$

The identities (84) and (89) are known as *Bianchi identities.*

EXERCISE 20. Derive Eq. (89). (Hint: Use the same method as for obtaining (84). You can also derive (89) directly from (84).)

EXERCISE 21. Show that the metric (32) and the Christoffel symbols (33) lead to the following components for the Riemann tensor:

$$\begin{array}{ll}
R^{0}{}_{101} = -\tfrac{1}{2}N'' + \tfrac{1}{4}L'N' - \tfrac{1}{4}(N')^2 & R^{0}{}_{202} = -\tfrac{1}{2}re^{-L}N' \\
R^{0}{}_{303} = -\tfrac{1}{2}re^{-L}N' \sin^2\theta & R^{1}{}_{212} = \tfrac{1}{2}re^{-L}L' \\
R^{1}{}_{313} = \tfrac{1}{2}re^{-L}L' \sin^2\theta & R^{2}{}_{323} = (1 - e^{-L}) \sin^2\theta
\end{array} \qquad [90]$$

All components not related to the above by the identities (79)–(82) are zero.

EXERCISE 22. Show that the Ricci tensor obtained from Eq. (90) has the components

$$R_{00} = e^{N-L}\left[-\tfrac{1}{2}N'' + \tfrac{1}{4}L'N' - \tfrac{1}{4}(N')^2 - \frac{N'}{r}\right]$$

$$R_{11} = \tfrac{1}{2}N'' - \tfrac{1}{4}L'N' + \tfrac{1}{4}(N')^2 - \frac{L'}{r}$$

$$R_{22} = e^{-L}[1 + \tfrac{1}{2}r(N' - L')] - 1$$

$$R_{33} = \sin^2\theta\, e^{-L}[1 + \tfrac{1}{2}r(N' - L')] - \sin^2\theta \qquad [91]$$

All other components are zero. Show that the curvature scalar has the value

$$R = e^{-L}\left[-N'' + \tfrac{1}{2}L'N' - \tfrac{1}{2}(N')^2 + \frac{2}{r}(L' - N') - \frac{2}{r^2}\right] + \frac{2}{r^2} \qquad [92]$$

  To see what $R^{\alpha}{}_{\beta\mu\nu}$ has to do with the *curvature* of spacetime, let us examine what happens to a *constant* vector a^{α} if it is parallel transported from a point P to a point P' along two *different* paths. Fig. 6.4 shows the points P and P' connected by a path PP_1P' consisting of a displacement $d\xi^{\mu}$ followed by a displacement $d\zeta^{\mu}$; path PP_2P' involves the same displacements, but carried out in the opposite order. All the displacements are infinitesimal.

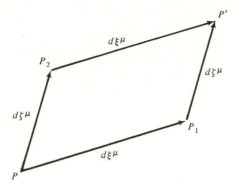

Fig. 6.4 *Spacetime points P,P′ connected by the two alternative
paths PP₁P′ and PP₂P′.*

The parallel transport of a^α from P to P_1 gives a vector

$$a^\alpha - \Gamma^\alpha_{\beta\mu}(P)a^\beta d\xi^\mu$$

and the further transport of this vector from P_1 to P' gives

$$a^\alpha - \Gamma^\alpha_{\beta\mu}(P)a^\beta d\xi^\mu - \Gamma^\alpha_{\sigma\nu}(P_1)[a^\sigma - \Gamma^\sigma_{\beta\mu}(P)a^\beta d\xi^\mu]d\zeta^\nu \qquad [93]$$

Since

$$\Gamma^\alpha_{\sigma\nu}(P_1) \simeq \Gamma^\alpha_{\sigma\nu}(P) + \Gamma^\alpha_{\sigma\nu,\mu}(P)d\xi^\mu \qquad [94]$$

we can write (93) as

$$a^\alpha - \Gamma^\alpha_{\beta\mu}a^\beta d\xi^\mu - \Gamma^\alpha_{\sigma\nu}a^\sigma d\zeta^\nu + \Gamma^\alpha_{\sigma\nu}\Gamma^\sigma_{\beta\mu}a^\beta d\xi^\mu d\zeta^\nu$$
$$- \Gamma^\alpha_{\sigma\nu,\mu}a^\sigma d\xi^\mu d\zeta^\nu \qquad [95]$$

where a term of third order in the (small) displacements has been omitted,
and where all Γ's are evaluated at the point P.

The parallel transport of a^α along PP_2P' gives a result similar to
(95) but with $d\xi$ and $d\zeta$ interchanged:

$$a^\alpha - \Gamma^\alpha_{\beta\mu}a^\beta d\zeta^\mu - \Gamma^\alpha_{\sigma\nu}a^\beta d\xi^\nu + \Gamma^\alpha_{\sigma\nu}\Gamma^\sigma_{\beta\mu}a^\beta d\zeta^\mu d\xi^\nu$$
$$- \Gamma^\alpha_{\sigma\nu,\mu}a^\sigma d\zeta^\mu d\xi^\nu \qquad [96]$$

If we exchange the (dummy) indices ν,μ in the last two terms of (95), we
find that the difference between the expressions (95) and (96) is

$$(\Gamma^\alpha_{\beta\nu,\mu} - \Gamma^\alpha_{\beta\mu,\nu} + \Gamma^\alpha_{\sigma\mu}\Gamma^\sigma_{\beta\nu} - \Gamma^\alpha_{\sigma\nu}\Gamma^\sigma_{\beta\mu})a^\beta d\xi^\nu d\zeta^\mu \qquad [97]$$

This shows that the result of parallel transport is *path dependent;* the
results obtained along two paths differ by a term involving the curvature
tensor,

$$\Delta a^{\alpha} = R^{\alpha}_{\ \beta\mu\nu} a^{\beta} d\xi^{\nu} d\zeta^{\mu} \qquad\qquad [98]$$

Since the change produced by parallel transport along PP_2P' is opposite to that produced along $P'P_2P$, it follows that Δa^{μ} may also be regarded as the change in the vector a^{μ} due to parallel transport around the closed loop $PP_1P'P_2P$.

EXERCISE 23. Show that the corresponding result for a covariant vector is

$$\Delta b_{\alpha} = R_{\alpha}^{\ \beta}_{\ \mu\nu} b_{\beta} d\xi^{\nu} d\zeta^{\mu} \qquad\qquad [99]$$

(Hint: Rather than go through the long calculation, use the condition $\Delta(b_{\alpha}a^{\alpha}) = 0$ for arbitrary a^{α}.)

Eqs. (98) and (99) show that successive parallel displacements do not commute. Since, as we have seen in the preceding section, the covariant derivatives can be defined in terms of parallel displacements, it is not at all surprising that successive covariant derivatives also fail to commute.

What does the tensor $R^{\alpha}_{\ \beta\mu\nu}$ have to do with curvature? In a flat space, the result of parallel transport cannot depend on the path taken. Hence any *path dependence is an indication of curvature*. The tensor $R^{\alpha}_{\ \beta\mu\nu}$ serves as a quantitative measure of this curvature; in a flat space one necessarily has $R^{\alpha}_{\ \beta\mu\nu} = 0$. It is possible to show the converse: if $R^{\alpha}_{\ \beta\mu\nu} = 0$ throughout the space, then the space is flat. In other words, if the curvature tensor is zero, then the metric can differ from $\eta_{\mu\nu}$ only by a coordinate transformation. A simple argument is the following: Choose a point P and introduce geodesic coordinate axes at that point. Then construct coordinates for all other points of the space by parallel transporting the axes from the point P to all other points. If $R^{\alpha}_{\ \beta\mu\nu} = 0$, the result is path independent and therefore uniquely defined. Since the axes have been set up by parallel transport, any vector transported with respect to the axes will suffer no change. Hence, $\Gamma^{\mu}_{\ \alpha\beta} = 0$ everywhere and therefore $g_{\mu\nu}$ is a constant. But we know that a constant metric can always be reduced to $\eta_{\mu\nu}$ by a coordinate transformation; this shows that the space is flat.

Given a metric tensor $g_{\mu\nu}$, we can decide unambiguously whether this describes flat spacetime in some complicated curvilinear coordinates or whether it describes curved spacetime by calculating the curvature tensor. We have seen that, in gravitational theory, $g_{\mu\nu}$ (or $h_{\mu\nu}$) plays a role similar to that of the vector potential of electrodynamics. The curvature tensor $R^{\alpha}_{\ \beta\mu\nu}$ plays a role analogous to that of the **E** and **B** fields of electrodynamics. A gravitational field is said to be present if, and only if, $R^{\alpha}_{\ \beta\mu\nu}$ has a nonzero component. As we will see in Section 7.5, the tensor $R^{\alpha}_{\ \beta\mu\nu}$ is the relativistic generalization of the tidal force tensor of Eq. (1.42).

TABLE 6.1 TENSOR ANALYSIS FORMULAS

Definition of contravariant vector	$A'^{\mu} = \dfrac{\partial x'^{\mu}}{\partial x^{\nu}} A^{\nu}$
Definition of covariant vector	$A'_{\mu} = \dfrac{\partial x^{\nu}}{\partial x'^{\mu}} A_{\nu}$
Spacetime interval	$ds^2 = g_{\mu\nu}dx^{\mu}dx^{\nu}$
Relation between covariant and contravariant metric tensor	$g_{\mu\alpha}g^{\alpha\nu} = \delta_{\mu}^{\ \nu}$
Christoffel symbol	$\Gamma^{\mu}_{\ \alpha\beta} = \tfrac{1}{2}g^{\mu\nu}(g_{\nu\alpha,\beta} + g_{\beta\nu,\alpha} - g_{\alpha\beta,\nu})$
Geodesic equation	$\dfrac{d^2x^{\mu}}{ds^2} + \Gamma^{\mu}_{\ \alpha\beta}\dfrac{dx^{\alpha}}{ds}\dfrac{dx^{\beta}}{ds} = 0$
Geodesic coordinates (at one spacetime point)	$g'_{\mu\nu} = \eta_{\mu\nu},\ g'_{\mu\nu,\alpha} = 0$
Covariant derivative of contravariant vector	$A^{\mu}_{\ ;\beta} = A^{\mu}_{\ ,\beta} + \Gamma^{\mu}_{\ \alpha\beta}A^{\alpha}$
Covariant derivative of covariant vector	$A_{\mu;\beta} = A_{\mu,\beta} - \Gamma^{\alpha}_{\ \mu\beta}A_{\alpha}$
Parallel transport	$\delta A^{\mu} = -\Gamma^{\mu}_{\ \alpha\beta}A^{\alpha}\delta x^{\beta}$ $\delta A_{\mu} = \Gamma^{\alpha}_{\ \mu\beta}A_{\alpha}\delta x^{\beta}$
Riemann tensor	$R^{\alpha}_{\ \beta\mu\nu} = -\Gamma^{\alpha}_{\ \beta\mu,\nu} + \Gamma^{\alpha}_{\ \beta\nu,\mu} + \Gamma^{\sigma}_{\ \beta\nu}\Gamma^{\alpha}_{\ \sigma\mu}$ $\quad - \Gamma^{\sigma}_{\ \beta\mu}\Gamma^{\alpha}_{\ \sigma\nu}$
Ricci tensor	$R_{\mu\nu} = R^{\alpha}_{\ \mu\nu\alpha}$
Commutation relation for derivatives	$A_{\beta;\mu;\nu} - A_{\beta;\nu;\mu} = R^{\alpha}_{\ \beta\mu\nu}A_{\alpha}$
Change in vector produced by parallel transport around a closed path $(0 \to d\xi^{\mu} \to d\xi^{\mu} + d\zeta^{\mu} \to d\zeta^{\mu} \to 0)$	$\Delta A^{\alpha} = R^{\alpha}_{\ \beta\mu\nu}A^{\beta}d\xi^{\nu}d\zeta^{\mu}$
Derivative of vector along a curve	$\dfrac{DA^{\mu}}{D\tau} = A^{\mu}_{\ ;\nu}\dfrac{dx^{\nu}}{d\tau} = \dfrac{dA^{\mu}}{d\tau} + \Gamma^{\mu}_{\ \alpha\beta}A^{\alpha}\dfrac{dx^{\beta}}{d\tau}$
Geodesic deviation	$\dfrac{D^2\xi^{\mu}}{D\tau^2} = -R^{\mu}_{\ \alpha\sigma\beta}\xi^{\sigma}\dfrac{dx^{\alpha}}{d\tau}\dfrac{dx^{\beta}}{d\tau}$

Table 6.1 summarizes the most important formulas of tensor analysis in a Riemannian geometry. The formulas are given in terms of a first-rank tensor (vector); the generalization to higher-rank tensors is obvious. The concepts of derivative along a curve and of geodesic deviation will be discussed in Section 7.5; the corresponding formulas have been included in Table 6.1 for the sake of completeness.

FURTHER READING

Introductions to differential calculus (i.e., tensor analysis in arbitrary coordinates) and Riemannian geometry are given in many books on relativity such as those listed in Chapter 2 and also Adler, Bazin, and Schiffer, *Introduction to General Relativity;* Weyl, *Space-Time-Matter;* and Eddington, *The Mathematical Theory of Relativity.*

A very neat and concise introduction is contained in Einstein's original paper "Die Grundlage der allgemeinen Relativitätstheorie," Ann. der Physik, **49,** 769 (1916) (translated in Lorentz, Einstein, Minkowski, and Weyl, *The Principle of Relativity*).

Misner, Thorne, and Wheeler, *Gravitation,* presents differential calculus both in the component formalism and in the abstract formalism of differential forms. The latter formalism leads to interesting graphical representations of vectors and tensors.

Eisenhart, *Riemannian Geometry,* is a classic reference on the geometry of curved spaces of an arbitrary number of dimensions. Sokolnikoff, *Tensor Analysis;* and Synge and Schild, *Tensor Calculus* are excellent mathematical treatises which include applications to physics.

REFERENCES

1. G. Ricci and T. Levi-Civita, Math. Ann. **54,** 125 (1901).
2. T. Levi-Civita, *The Absolute Differential Calculus* (Blackie and Son, London 1929).
3. H. Weyl, *Space-Time-Matter* (Dover, New York), p. 54.

PROBLEMS

1. Consider the *two-dimensional* space consisting of the surface of an ordinary sphere of radius a. If the familiar polar angles are used as coordinates in this space with

$$x^1 = \theta, \qquad x^2 = \phi$$

then

$$dl^2 = g_{\mu\nu}dx^\mu dx^\nu = a^2 d\theta^2 + a^2 \sin^2 \theta d\phi^2$$

and

$$g_{\mu\nu} = \begin{pmatrix} a^2 & 0 \\ 0 & a^2 \sin^2 \theta \end{pmatrix}$$

(a) Find $g^{\mu\nu}(\mu = 1,2; \nu = 1,2)$.
(b) Find all components of $\Gamma^\mu_{\alpha\beta}$ ($\mu = 1,2; \alpha = 1,2; \beta = 1,2$).
(c) Find R^1_{212} and R^2_{121}.
(d) Find R_{11} and R_{22}.
(e) Find R.

2. Consider the two-dimensional space consisting of the surface of a cylinder of radius a. Use cylindrical coordinates z, ϕ and define

$$x^1 = z, \qquad x^2 = \phi$$

Then

$$dl^2 = g_{\mu\nu}dx^\mu dx^\mu = dz^2 + a^2 d\phi^2$$

and $\quad g_{\mu\nu} = \begin{pmatrix} 1 & 0 \\ 0 & a^2 \end{pmatrix}$

Find items (a)–(e) as listed in Problem 1. Is the geometry of the cylindrical surface the same as that of a flat two-dimensional plane? Is the topology the same?

3. Suppose that we have a *three-dimensional* spacetime with an interval

$$ds^2 = g_{\mu\nu}dx^\mu dx^\nu = dt^2 - dz^2 - [a(t)]^2 d\phi^2$$

where $a(t)$ is some increasing function of time. The spatial part of this geometry may be regarded as the surface of an expanding cylinder with a radius $a(t)$ (see Problem 2). Find the nonzero components of the Riemann tensor $R^\alpha{}_{\beta\mu\nu}(\alpha,\beta,\mu,\nu = 0,1,2)$. Show that if, and only if, $\dot{a}(t) = $ constant, then the spacetime is flat.

4. The surface of a torus ("donut") is a two-dimensional curved surface. Suppose that the torus has a radius R measured from the center of symmetry to the middle line and a radius r measured from the middle line to the surface. Use the coordinates $x^1 \equiv \phi$ and $x^2 \equiv \theta$ to describe points on the surface; ϕ is the angle along the torus (measured around the axis of symmetry) and θ is the angle around its cross section (see Fig. 6.5). (a) Find

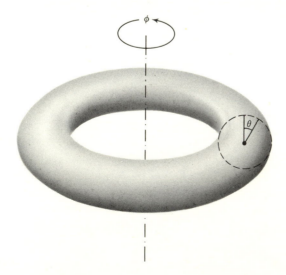

Fig. 6.5 *Coordinates on a torus.*

the metric tensor that describes the distances on the surface of the torus; find the Christoffel symbols. (b) Suppose that the vector $A^\mu = (0,1)$ is originally located at the point $x^\mu = (0,\pi/2)$. If this vector is parallel transported to $x^\mu = (\pi/2,\pi/2)$ along the curve $\theta = \text{constant} = \pi/2$, what will be the components of the transported vector? (c) Repeat the calculation for a vector $A^\mu = (1,0)$ that is parallel transported from $x^\mu = (0,0)$ to $x^\mu = (\pi/2,0)$ along the curve $\theta = \text{constant} = 0$.

5. Suppose that the metric of spacetime is

$$ds^2 = (x^0)^4(dx^0)^2 - 2e^{x^1}(dx^1)^2 - e^{-x^2}(dx^2)^2 - (dx^3)^2$$

Prove that all components of $R^\alpha_{\beta\mu\nu}$ are zero.

6. A curved spacetime has an interval

$$ds^2 = A(r)dt^2 - B(r)dr^2 - C(r)r^2d\theta^2 - C(r)r^2\sin^2\theta d\phi^2$$

where r, θ, ϕ are regarded as spherical coordinates, and where $A(r)$, $B(r)$, and $C(r)$ are given functions of r. Show that

(a) The circumference of the circle $r = r_1$ is

$$2\pi\sqrt{C(r_1)}r_1$$

(b) The area of the sphere $r = r_1$ is

$$4\pi r_1^2 C(r_1)$$

(c) The distance between the points $r = r_1$ and $r = r_2$ on a given radial line is

$$\int_{r_1}^{r_2}\sqrt{B(r)}dr$$

(d) The volume of the spherical shell $r_1 < r < r_2$ is

$$4\pi\int_{r_1}^{r_2}r^2C(r)\sqrt{B(r)}dr$$

7. Eq. (61) gives the change in a vector under parallel transport. Find the corresponding result for the parallel transport of a tensor of rank two.

8. Show that if the metric $g_{\mu\nu}$ is diagonal, then $\Gamma^\sigma_{\alpha\beta} = 0$ whenever σ, α, β are distinct.

9. In what follows we will use the notation g (or $|g_{\alpha\beta}|$) for the determinant of the matrix $g_{\alpha\beta}$. Show that the transformation law for this determinant is

$$g' = \left|\frac{\partial x^\alpha}{\partial x'^\beta}\right|^2 g$$

where $|\partial x^\alpha/\partial x'^\beta|$, called the *Jacobian*, is the determinant of the matrix $\partial x^\alpha/\partial x'^\beta$. (Hint: Take the determinant of the equation $g'_{\alpha\beta} = (\partial x^\mu/\partial x'^\alpha)(\partial x^\nu/\partial x'^\beta)g_{\mu\nu}$.)

Suppose that Φ is a scalar field. Show that the integral

$$\int \Phi \sqrt{-g}\,dx^0 dx^1 dx^2 dx^3$$

evaluated over a given (four-dimensional) region of spacetime, is a scalar.

10. Show that

$$\Gamma^\alpha{}_{\mu\alpha} = \frac{1}{\sqrt{-g}}\frac{\partial}{\partial x^\mu}\sqrt{-g}$$

where g is the determinant of the matrix $g_{\mu\nu}$. (Hint: Differentiation of the determinant gives $\partial g/\partial x^\mu = g g^{\alpha\beta}\,\partial g_{\alpha\beta}/\partial x^\mu$.)

11. Use the result of Problem 10 to show that the divergence of a vector field can be written as

$$A^\mu{}_{;\mu} = \frac{1}{\sqrt{-g}}\frac{\partial}{\partial x^\mu}(\sqrt{-g}\,A^\mu).$$

Hence, show that for a scalar field

$$\Phi^{\cdot\mu}{}_{;\mu} = \frac{1}{\sqrt{-g}}\frac{\partial}{\partial x^\mu}\left(\sqrt{-g}\,g^{\mu\nu}\frac{\partial}{\partial x^\nu}\Phi\right)$$

12. Use the result of Problem 10 to show that if $A^{\alpha\beta}$ is an antisymmetric tensor field, then

$$A^{\alpha\beta}{}_{;\beta} = \frac{1}{\sqrt{-g}}\frac{\partial}{\partial x^\beta}(\sqrt{-g}\,A^{\alpha\beta})$$

13. Find all the (nonzero) Christoffel symbols that correspond to the metric given by Eq. (21). Show that if Φ is a scalar field, then

$$\Phi_{,\mu}{}^{;\mu} = \frac{\partial^2\Phi}{\partial t^2} - \frac{1}{r}\frac{\partial^2}{\partial r^2}r\Phi - \frac{1}{r^2 \sin\theta}\frac{\partial}{\partial\theta}\sin\theta\frac{\partial\Phi}{\partial\theta} - \frac{1}{r^2 \sin^2\theta}\frac{\partial^2\Phi}{\partial\phi^2}$$

14. Show that if $B^{\mu\nu}$ is an antisymmetric tensor field, then

$$B_{\mu\nu;\alpha} + B_{\nu\alpha;\mu} + B_{\alpha\mu;\nu} = B_{\mu\nu,\alpha} + B_{\nu\alpha,\mu} + B_{\alpha\mu,\nu}$$

and

$$B^{\mu\nu}{}_{;\mu;\nu} = 0$$

15. Derive the following expression for the curvature scalar:

$$\sqrt{-g}\,R = \frac{\partial}{\partial x^\alpha}\,[g^{\mu\nu}\sqrt{-g}\,(-\Gamma^\alpha{}_{\mu\nu} + \delta_\mu{}^\alpha\Gamma^\beta{}_{\nu\beta})] - g^{\mu\nu}\sqrt{-g}\,(\Gamma^\beta{}_{\mu\alpha}\Gamma^\alpha{}_{\nu\beta}$$

$$- \Gamma^\alpha{}_{\mu\nu}\Gamma^\beta{}_{\alpha\beta})$$

where, as in Problems 9 and 10, g is the determinant of the matrix $g_{\mu\nu}$.

*16. Show that the set of all coordinate transformations forms a group ("the manifold mapping group"), and show that scalars, vectors, tensors, etc., form representations of this group.

* This problem requires some knowledge of group theory.

7. GEOMETRODYNAMICS

You boil it in sawdust: You salt it in glue:
You condense it with locusts and tape:
Still keeping one principal object in view—
To preserve its symmetrical shape.

Lewis Carroll, *The Hunting of the Snark*

In the preceding chapter our concern was with the kinematics of the geometry, i.e., the *description* of geometry and curvature. We now come to the dynamics of geometry, i.e., the *interaction* of geometry and matter. This interaction is the content of Einstein's dynamical equations for the gravitational field.

There are several routes that lead to Einstein's equations; they differ in their starting points. We will take as our starting point the equations of the linear approximation of Chapter 3. In addition, we will assume that the equations are of second differential order and possess *general invariance;* these assumptions are sufficient to determine the equations for the gravitational field completely.

That the linear equations imply the full nonlinear equations is a very remarkable feature of Einstein's theory. Given some complicated set of nonlinear equations, it is easy to derive the corresponding linear approximation, but in general if we only know the linear approximation, we cannot reconstruct the nonlinear equations. What permits us to perform this feat in gravitational theory is the requirement of general invariance. This requirement states that the equations should remain

250

unchanged under all transformations of the coordinates. Regarded naïvely, this seems to be a simple generalization of the requirement of Lorentz invariance, which states that the equations should remain unchanged under all Lorentz transformations. In his original work, Einstein sought to justify general invariance by appeal to a general principle of equivalence for reference frames in arbitrary (accelerated) motion. This was intended to be a generalization of the principle of special relativity which asserts the equivalence of reference frames in uniform motion. Thus, Einstein's theory was intended to be a theory of "general relativity." But in fact, the theory of special relativity is already as relativistic as can be. The principle of general invariance is not a relativity principle; rather, the principle of general invariance is a dynamical principle which imposes restrictions on the possible interactions of geometry and matter. It does this by insisting that the equations that describe these interactions have a high degree of symmetry.

Einstein also sought to justify general invariance through Mach's principle. The latter expresses the intriguing, if somewhat vague, notion that the inertia of bodies is somehow affected by the mass distribution of the universe. Although Einstein's theory does display some dependence of inertia on mass distribution, the fact is that Mach's principle only plays a quite marginal role in the theory.

Since Einstein's theory of "general relativity" is no more relativistic than special relativity, we prefer the name *geometrodynamics* for this theory. This name, coined by Wheeler, puts the emphasis where it belongs — on the dynamical geometry that acts on and reacts to matter.

7.1 GENERAL COVARIANCE AND INVARIANCE

We begin by introducing the important concept of *covariance* of an equation:*

An equation is said to be covariant under general coordinate transformations if the form of the equation is left unchanged by the transformations.

Obviously, any equation whose right side and left side are tensors of the same type is covariant. For example, the equation

$$A^\mu = B^\mu \tag{1}$$

where A^μ and B^μ are contravariant tensors of rank one, is covariant. A general coordinate transformation changes Eq. (1) into

---

* We have "covariant" vectors, "covariant" derivatives, and now "covariant" equations. The word "covariant" is unfortunately much overworked and which of the three meanings is intended must be guessed from context.

$$A'^{\mu} = B'^{\mu} \tag{2}$$

i.e., the *form* of the equation is unchanged.

An example of an equation that is *not* covariant is the following:

$$A^{\mu} = B_{\mu} \tag{3}$$

If we try to express this equation in terms of the transformed vectors A'^{μ} and B'_{μ}, it reads

$$\frac{\partial x^{\mu}}{\partial x'^{\alpha}} A'^{\alpha} = \frac{\partial x'^{\alpha}}{\partial x^{\mu}} B'_{\alpha} \tag{4}$$

This equation does not have the same form as (3). Of course, we can re-arrange Eq. (4) in several ways; but no matter what we do to it, we cannot get rid of the transformation coefficients $\partial x/\partial x'$ and $\partial x'/\partial x$. (However, note that if $B_{\mu} \equiv 0$, then Eq. (3) becomes

$$A^{\mu} = 0 \tag{5}$$

which is of course covariant.)

Note that although Eq. (3) is not covariant, Eq. (4) *is* covariant. To see this, let us perform a coordinate transformation to coordinates x''^{μ} and express A'^{μ} and B'^{μ} in terms of A''^{μ} and B''^{μ}:

$$\frac{\partial x^{\mu}}{\partial x'^{\alpha}} \left(\frac{\partial x'^{\alpha}}{\partial x''^{\beta}} A''^{\beta} \right) = \frac{\partial x'^{\alpha}}{\partial x^{\mu}} \left(\frac{\partial x''^{\beta}}{\partial x'^{\alpha}} B''_{\beta} \right) \tag{6}$$

By the chain rule for derivatives, this reduces to

$$\frac{\partial x^{\mu}}{\partial x''^{\beta}} A''^{\beta} = \frac{\partial x''^{\beta}}{\partial x^{\mu}} B''_{\beta} \tag{7}$$

which has exactly the same form as Eq. (4). This example shows that a noncovariant equation, such as Eq. (3), can be changed into an equivalent covariant equation. Since covariance is a statement about the *form* of an equation, it is not surprising that it depends on the form in which the equation is written.

We now lay down a commandment:

Principle of general covariance: All laws of physics shall be stated as equations covariant with respect to general coordinate trans-formations.

We will obey this commandment for the best of all reasons— it costs us nothing to do so. As we have seen, an equation that is not covariant can easily be transformed into an equivalent equation that is covariant. From a mathematical point of view, the covariance principle is therefore seen to be a triviality.

We can also recognize the trivial character of the covariance principle from a physical point of view. It must be possible to express physical laws in *any* coordinate system. This is obvious since ultimately these laws express nothing but *relationships* between spacetime co-incidences. The result of any experiment can always be reduced to a statement of the form ". . . when the ends of the two wires coming out of instrument A are made to *coincide* with the two terminals of instrument B, then the pointer on the face of instrument A will *coincide* with the third mark on the scale, etc.," which makes no reference to co-ordinates. Of course, coordinates are convenient crutches for describing coincidences—two events are coincident if their coordinates are the same. But the relationships between spacetime coincidences are independent of the coordinates used to describe them, and this independence finds its mathematical expression in the requirement that the laws have the same *form* in all coordinate systems.

That the principle of covariance imposes no restrictions on the content of the physical laws, but only on the form in which they are written, was recognized by Kretschmann[1] in a critical examination of Einstein's theory. The principle in no way implies that the laws must be "relativistic"; even Newton's equations can be written in generally covariant form. In reply to this criticism, Einstein wrote:

> . . . Even though it is true that one must be able to bring every empirical law into generally covariant form, yet the Principle has considerable heuristic force, which has proved itself in the problem of gravitation and which relies on the following. Of two theoretical systems which agree with experience, that one is to be preferred, which from the point of view of the absolute differential calculus is the simplest and most transparent. One need only try to bring Newtonian mechanics and gravitation into the form of . . . covariant equations (four-dimensional) and one will surely be convinced that the Principle rules out this theory practically if not theoretically![2]

We would like to restate Einstein's somewhat vague criterion for "simplest and most transparent" equations in a rigorous way. We can do this by appealing to the concept of *invariance* of the equations. Invariance goes beyond covariance in that it demands that not only the form but also the content of the equations be left unchanged by the coordinate transformation. Roughly, invariance is achieved by imposing the extra condition that all "constants" in the equation remain exactly the same. For example, covariance of the Maxwell equations under Lorentz transformations requires only that the equations have the same *appearance* in all Lorentz frames; it does not require that the velocity of light, which enters these equations as a parameter, be the same in all reference frames.

In this context, "constants" means not only those things which are true numerical constants, but anything that is independent of the state of matter. A better terminology is to call such things "absolute objects." Thus, an *absolute object* is any number or function, with one or more components, that is not dependent on the state of matter; that is, it is a quantity whose value is completely unaffected by any changes in the condition of the matter in the universe. For example, c and $\hbar$ are absolute objects; so are $\frac{1}{4}$, $\eta^{\mu\nu}$, $\partial x'^{\mu}/\partial x^{\nu}$ (where x' is some given function of x), etc. By contrast, *dynamical objects* are physical variables such as particle positions and momenta, field strengths, energy densities, etc., that do depend on the state of matter. Given any law of physics, we can then classify the quantities that appear in it into absolute objects and dynamical objects.[3,4]

We can now give a somewhat more precise definition of invariance:

An equation is said to be invariant *under a coordinate transformation if* (1) *it is covariant and* (2) *all absolute objects appearing in it are left unchanged.*

Note that this definition tells us that the equation (4) is not invariant. This equation is covariant, but it contains the absolute object $\partial x^{\mu}/\partial x'^{\alpha}$ and a coordinate transformation changes this object into $\partial x^{\mu}/\partial x''^{\alpha}$, in violation of item 2.

The requirement that an equation be invariant under some transformation imposes serious restrictions on the equation. Thus, invariance under Lorentz transformations ("principle of special relativity") is a *symmetry* requirement which puts tight limitations on acceptable equations. We have made use of these limitations in the construction of our linear gravitational theory. Note that the Lorentz symmetry requirement determines what possible absolute objects can appear in the equations: the admissible absolute objects must remain unchanged under Lorentz transformations. The only acceptable objects are therefore scalars (e.g., $c, \hbar, e, m$) and the tensor $\eta^{\mu\nu}$.* Physically, the Lorentz symmetry requirement of special relativity means that observers cannot distinguish between different inertial frames.

The gravitational theory of Einstein is based on the following postulate:

Principle of general invariance: All laws of physics must be invariant under general coordinate transformations.

This postulate demands that the laws have a higher symmetry, a symmetry that goes beyond Lorentz symmetry. One immediate consequence of the general symmetry requirement is that we must regard the metric of spacetime as dependent on the state of matter; the metric tensor must be a dynamical field, satisfying some field equation of its own. The reason is obvious: the general coordinate transformations *do not* leave $g_{\mu\nu}$ unchanged; hence the presence of $g_{\mu\nu}$ as an absolute object

---

* Tensors, such as $\eta_{\mu\nu}\eta_{\alpha\beta}$, constructed from $\eta_{\mu\nu}$ are, of course, also acceptable.

contradicts the requirement that all such objects keep their values unchanged under coordinate transformations. The only alternative is to suppose that $g_{\mu\nu}$ is *not* an absolute object, but rather a dynamical variable.

Since we have taken some pains to make it clear that covariance is trivial (and therefore unavoidable), but that invariance is nontrivial (and therefore avoidable), it will be well to give some arguments in favor of invariance.

One argument is based on our belief in the simplicity, or beauty, of nature. Mathematically, we may interpret a requirement for simplicity as a requirement for symmetry. We may say that the aim of theoretical physics is to maximize the symmetry of the (supposed) laws of nature subject to the constraints imposed by the available experimental evidence. A symmetry should be considered "broken" only if there are experiments that bear witness against the symmetry. Since general invariance is not in conflict with any present data, we feel strongly compelled to accept this symmetry.

Another argument, one that does not rely quite as much on faith in the kindness of nature, is the following: We have seen that the field equation (Eq. (3.48)) of the linear theory is invariant under the gauge transformation*

$$h^{\mu\nu}(x) \rightarrow h^{\mu\nu}(x) + \tfrac{1}{2}(\partial^\nu \Lambda^\mu + \partial^\mu \Lambda^\nu) \qquad [8]$$

Let us compare this gauge transformation with a coordinate transformation. Consider the infinitesimal coordinate transformation

$$x'^\mu = x^\mu + \xi^\mu \qquad [9]$$

where $\xi^\mu = \xi^\mu(x)$ is a function that is everywhere infinitesimal. Then

$$\frac{\partial x'^\mu}{\partial x^\alpha} = \delta_\alpha{}^\mu + \xi^\mu{}_{,\alpha} \qquad [10]$$

and

$$
\begin{aligned}
g'^{\mu\nu}(x') &= \frac{\partial x'^\mu}{\partial x^\alpha}\frac{\partial x'^\nu}{\partial x^\beta} g^{\alpha\beta}(x) \\
&= (\delta_\alpha{}^\mu + \xi^\mu{}_{,\alpha})(\delta_\beta{}^\nu + \xi^\nu{}_{,\beta})g^{\alpha\beta}(x) \\
&\simeq g^{\mu\nu}(x) + g^{\alpha\nu}(x)\,\xi^\mu{}_{,\alpha} + g^{\mu\beta}(x)\,\xi^\nu{}_{,\beta}
\end{aligned}
\qquad [11]
$$

In order to express everything as a function of x, we use a Taylor expansion

$$g'^{\mu\nu}(x') \simeq g'^{\mu\nu}(x) + g^{\mu\nu}{}_{,\alpha}(x)\,\xi^\alpha \qquad [12]$$

---

* In this equation indices are raised with the metric $\eta^{\mu\nu}$.

and therefore

$$g'^{\mu\nu}(x) = g^{\mu\nu}(x) - g^{\mu\nu}{}_{,\alpha}(x)\,\xi^\alpha + g^{\alpha\nu}(x)\,\xi^\mu{}_{,\alpha} + g^{\mu\beta}(x)\xi^\nu{}_{,\beta} \qquad [13]$$

If

$$g^{\mu\nu} = \eta^{\mu\nu} + \kappa h^{\mu\nu} \qquad [14]$$

where $h^{\mu\nu}$ is small, we obtain from (13), ignoring the very small terms of order ξh,

$$\kappa h'^{\mu\nu}(x) = \kappa h^{\mu\nu}(x) + \xi^{\mu,\nu} + \xi^{\nu,\mu} \qquad [15]$$

This is exactly the same as (8) provided we identify

$$\xi^\mu(x) = \frac{\kappa}{2}\Lambda^\mu(x) \qquad [16]$$

We therefore have shown that the gauge transformation of Section 3.2 (see Eq. (3.49)) is nothing but an *infinitesimal coordinate transformation:* the linear theory is invariant under infinitesimal coordinate transformations. This makes it quite natural to suppose that the general nonlinear theory is invariant under (finite) coordinate transformations. Note that we can regard general (finite) coordinate transformations as generalized gauge transformations. From this point of view, the symmetry expressed by the principle of general invariance would be described as a gauge symmetry.

As a final argument in favor of general invariance, we point out that this symmetry plays a crucial role in the general proof of the equality of inertial and gravitational mass of an arbitrary system. Here, as in Section 1.4, the gravitational mass is defined by the asymptotic gravitational field generated by the system and the inertial mass by the total energy of the system. We will not present the details, but it seems that without general invariance the theoretical proof of the equality $m_I = m_G$ becomes impossible.[21]

The physical interpretation of the principle of *special* relativity is that velocity is relative; no experiment can detect any intrinsic difference between reference frames in uniform motion with different velocities. It is tempting to give the principle of general invariance the interpretation that acceleration is also relative. Einstein found it hard to resist this temptation; he named his theory of gravitation the theory of *general relativity* because he thought that (locally) the phenomena observed in a gravitational field are indistinguishable from those observed in an accelerated system of reference and that

According to this conception one cannot speak of the *absolute acceleration* of a system of reference, just as in the ordinary theory of relativity
one cannot speak of the *absolute velocity* of a system.[5]

However, we have seen in Section 1.7 that the tidal effects allow
us to make an absolute distinction between the gravitational forces and
the pseudo-forces found in accelerated reference frames. It is therefore
false to speak of a general relativity of motion.

The principle of special relativity is a relativity principle, whereas
the principle of general invariance is not. The lack of relativity in Einstein's theory can also be understood in geometrical terms. In flat spacetime it is meaningless to speak of absolute positions or velocities because
the geometry has no identifiable features—it is exactly the same everywhere. On the other hand, in a Riemannian spacetime the features of
the geometry can be used to determine positions. Astronauts can describe
their location and velocity by reference to nearby bumps in the geometry.
For example, astronauts in orbit near the earth can take advantage of
the lack of uniformity of the earth's gravitational field to determine their
position. (Incidentally, the gravity gradiometer described in Section 1.7
was developed with some such intention. The ultimate nasty plan is to
install a gradiometer in a missile and use it to guide the missile to its
target by "feeling" the irregularities in the earth's gravitational field.[6])
Mathematically, the deviations of the geometry from uniformity are described by the Riemann tensor. Hence, we can say that the Riemann
tensor makes an absolute distinction, and destroys the relativity, between
different (small) reference frames in free fall. This geometrical point of
view is closely related to tidal forces because, as we will see in Section
7.5, the Riemann tensor is essentially the tidal force tensor.

Although the above argument makes it clear that Einstein's theory
is not a theory of "general relativity," we must be careful not to jump to
the conclusion that the theory is entirely lacking in any kind of relativity.
Obviously, the principle of general invariance includes the principle of
special relativity (Lorentz transformations are a special case of general
coordinate transformations). Hence Einstein's theory is as "relativistic"
as special relativity. In the context of the above example this means that
astronauts can use the features of the geometry to measure their velocity
relative to the earth, but they cannot measure any absolute velocity for
the earth because the earth, or, more precisely, the solar system, is
placed in a background geometry that is asymptotically flat.*

The principles of special relativity and general invariance are
different in kind because the symmetries involved in these principles
are different in kind. In the terminology of field theory: Lorentz in-

---

* This ignores the gravitational field produced by the galaxy and by the average mass distribution of the universe.

variance is an *algebraic* symmetry, general invariance is a *dynamic* symmetry.[7] The meaning of this is that Lorentz invariance tells us what algebraic combinations of components of (Lorentz) tensors can appear in the equations of physics; it places restrictions on *all* of the interactions that can occur in physics. General invariance tells us how the gravitational field can appear in the equations; it only places restrictions on the gravitational interactions and helps to determine the possible dynamical couplings between gravitation and matter.* The fundamental difference between these symmetries is related to the fact that Lorentz transformations perform the same operation at all points of spacetime (the Lorentz transformation coefficients are constants), while general coordinate transformations perform different operations at different points (the transformation coefficients are functions of spacetime).** The difference in the character of these symmetries is further illustrated by the fact that conservation laws for energy, momentum, and angular momentum may be derived from the Lorentz symmetry,[9] while no such conservation laws follow from general invariance. The proof of these conservation theorems is beyond the scope of our discussion; however, we may point out that these theorems are essentially generalizations of a well-known result of Lagrangian mechanics: if the Lagrangian has translational and rotational symmetry, then the momentum and angular momentum are conserved.

As an example of the help offered by the principle of general invariance in the search for the couplings between gravitation and matter, consider the problem of finding the equations that describe the effect of gravitation on electromagnetism. The Maxwell equations (see Eqs. (2.106), (2.107)) in flat spacetime are

$$F^{\mu\nu}{}_{,\mu} = 4\pi j^{\nu} \tag{17}$$

$$F^{\mu\nu,\alpha} + F^{\nu\alpha,\mu} + F^{\alpha\mu,\nu} = 0 \tag{18}$$

where, as always, the commas indicate *ordinary* derivatives. Obviously, these equations are not invariant under general coordinate transformations. The left sides are not tensors — we know that ordinary derivatives of tensors do not give tensors. In order to make the left sides into tensors we must replace the ordinary derivatives by covariant derivatives:

$$F^{\mu\nu}{}_{;\mu} = 4\pi j^{\nu} \tag{19}$$

---

* For the purposes of this statement we are concentrating on those aspects of general invariance that go beyond Lorentz invariance, i.e., we ignore that part of general invariance which overlaps with Lorentz invariance.

** In the terminology of field theory this is expressed by saying that the former are gauge transformations of the first kind, while the latter are gauge transformations of the second kind.[8]

$$F^{\mu\nu;\alpha} + F^{\nu\alpha;\mu} + F^{\alpha\mu;\nu} = 0 \qquad\qquad\qquad [20]$$

These equations satisfy the principle of general invariance. The gravitational field variables appear in these equations through the Christoffel symbols that are contained in the covariant derivatives. The gravitational field therefore affects the electromagnetic field; the light deflection given in Chapter 3 can be obtained by a study of the wave solutions of Eqs. (19) and (20).

Unfortunately, although the principle of general invariance places considerable restrictions on the possible interactions of gravitation and matter, it does not determine these interactions uniquely. Thus, Eq. (19) remains entirely consistent with general invariance if we add an extra term $(F^{\mu\nu}R)_{;\mu}$ to the left side.[*] To rule out such extra terms, it is customary to appeal to the *principle of minimal coupling*. This principle asserts that the equations of motion of matter in the presence of gravitation are to be obtained from those that hold in the absence of gravitation (equations of special relativity) by replacing $\eta^{\mu\nu}$ by $g^{\mu\nu}$ and ordinary derivatives by covariant derivatives; *no other changes are to be made*. Since in local geodesic coordinates the covariant derivative reduces to the ordinary derivative, we can also express this principle as follows: in local geodesic coordinates the equations of motion are those of special relativity.

This minimal coupling principle is *not a symmetry principle* and there is no good reason for it, except, of course, that it makes the life of physicists a bit simpler. In fact, this principle is not really a principle ("general law") because it cannot be applied to all the equations of physics. For example, if we apply the minimal coupling "principle" to the equation of motion of the spin of a rigid body, it would tell us that the equation of motion of the spin in local geodesic coordinates is $d\mathbf{S}/d\tau = 0$ since this is the equation that the spin satisfies in the absence of gravitation. But this is the wrong equation of motion for the spin. We know from Section 1.7 that tidal torques are present in the true equation (Eq. (1.53)) and that the motion of the spin depends on the Riemann tensor (the connection between the tidal force and the Riemann tensor will be further discussed in Section 7.4). Thus the "principle" of minimal coupling is an unreliable and dangerous thing. It may be used to determine the form that laws take in the presence of gravitation only as a last resort, when we are sure that these laws cannot be determined by other means. Furthermore, it is advisable to keep in mind the precedent set by nongravitational physics. It is well known that a minimal coupling principle can be formulated for electrodynamic interactions, and that this principle is *violated* in nature (by the existence of the anomalous magnetic moments of protons and neutrons). Hence, we had better be very skeptical of the analogous principle in geometrodynamics.

---

[*] This term satisfies the identity $(F^{\mu\nu}R)_{;\mu;\nu} = 0$, which is important for the consistency of Eq. (19).

7.2 MACH'S PRINCIPLE

The conclusion which we drew from the principle of general invariance
— that the metric $g_{\mu\nu}$ must be a dynamical field responsive to the state
of matter — agrees with certain ideas put forward by Mach in his critical
discussion of Newton's concept of absolute space.

In the view of Newton, acceleration is absolute. The inertial
pseudo-forces give an absolute measure of acceleration in space. In his
Principia Newton stated:

> The effects which distinguish absolute from relative motion are the
> forces of receding from the axis of circular motion. . . . If a vessel, hung
> by a long cord, is so often turned about that the cord is strongly twisted,
> then filled with water, and held at rest together with the water; there-
> upon, by the sudden action of another force, it is whirled about the
> contrary way, and while the cord is untwisting itself, the vessel con-
> tinues for some time in this motion; the surface of the water will at
> first be plain, as before the vessel began to move; but after that, the
> vessel, by gradually communicating its motion to the water, will make
> it begin sensibly to revolve, and recede by little and little from the
> middle, and ascend to the sides of the vessel, forming itself into a
> concave figure (as I have experienced), and the swifter the motion
> becomes, the higher will the water rise, till at last, performing its revo-
> lutions in the same times with the vessel, it becomes relatively at rest
> in it. This ascent of the water shows its endeavor to recede from the
> axis of its motion; and the true and absolute circular motion of the water,
> which is here directly contrary to the relative, becomes known, and may
> be measured by this endeavor. At first, when the relative motion of the
> water in the vessel was greatest, it produced no endeavor to recede
> from the axis; the water showed no tendency to the circumference, nor
> any ascent towards the sides of the vessel, but remained of a plain
> surface, and therefore its true circular motion had not yet begun. But
> afterwards, when the relative motion of the water had decreased, the
> ascent thereof towards the sides of the vessel proved its endeavor to
> recede from the axis; and this endeavor showed the real circular motion
> of the water continually increasing, till it had acquired its greatest
> quantity, when the water rested relatively in the vessel.[10]

In rebuttal Mach argued that the inertial forces should not be
regarded as indicating motion in absolute space, but rather as indicating
motion relative to the masses in the entire universe:

> The behavior of . . . bodies relative to [a given system of refer-
> ence] may be traced to their behavior relative to the distant celestial
> bodies. If we were to claim that we know more about the motion of
> bodies than the hypothetical behavior relative to the celestial bodies
> suggested by experience, we would render ourselves culpable of dis-
> honesty. When accordingly, we say that a body maintains its direction
> and speed in space, then this is nothing but an abbreviated reference
> to a consideration of the entire universe. . . .

> ... The experiment of Newton with the rotating vessel only teaches us that the relative rotation of the water against the walls of the vessel does not generate any appreciable centrifugal forces, but that the relative rotation against the mass of the earth and of the celestial bodies does generate such a force. Nobody can predict how the experiment would turn out quantitatively and qualitatively, if the walls of the vessel were to get thicker and more massive, ultimately a few miles thick. . . .[11]

In essence, Mach's view is that the inertial forces associated with the motion of a given mass are generated by all the other masses in the universe. This leads us to the following:

Mach's principle: The inertial properties of an object are determined by the energy-momentum throughout all space.[12]

On the basis of Mach's ideas we would expect that if a large mass is accelerated, it will tend to "drag" the inertial reference frame of nearby regions along and induce forces of an inertial type in nearby bodies. If the large mass has a linear acceleration, nearby bodies should experience a force in the same direction as this acceleration; if the large mass rotates, nearby bodies should experience centrifugal and Coriolis-type forces.[13] Of course these effects are not expected to be of large magnitude—a planet, a star, or even a galaxy has a mass that is very small compared to the total mass in the universe. But if we carry out our experiments very near to a massive body the effects may be detectable.

Einstein's theory of gravitation agrees with Mach's principle in that it predicts both of the above effects. For example, the rotation of the earth is predicted to drag the local inertial reference frames along. Near the surface of the earth, a free gyroscope with its axis perpendicular to the axis of rotation of the earth is calculated to precess at the rate of 0.05 seconds of arc per year (Lense-Thirring effect). An experiment is now being developed to measure this effect by means of a gyroscope carried by a satellite in orbit (see Section 8.3).

Naïvely, we might also argue that Mach's principle should lead to another effect: when bodies are placed near a large mass, their inertial mass should increase, with all the masses increasing in proportion.[13] However, from an operational point of view such a phenomenon makes no sense. Masses are defined relative to a standard of mass and therefore a proportional change in all masses is meaningless. A rigorous argument against any dependence of mass on space or time has been given in Section 5.5 (see Eqs. (5.26) and (5.27)).

Similar arguments[14] can be given to rule out the conjecture that a large mass might produce an anisotropy in the inertia of bodies moving in its vicinity. For example, one might conjecture that the large mass at the center of our galaxy increases the inertial mass for, say, motion towards or away from the center, and reduces it for motion in the transverse direction. But mass is a Lorentz *scalar,* and a dependence of mass on direction makes no sense at all. Hence the Hughes-Drever experiment,[15] which sought to test whether the motion of electrons in a lithium atom displays any anisotropy with respect to the galactic center, should be regarded as a test of the isotropy of electromagnetic forces rather than inertial forces. (The negative result of the experiment tells us that, on the freely falling earth, the dielectric properties of space are isotropic.)

We have stated that Einstein's theory displays certain specific "Machian" effects. In general terms we may say that in Einstein's theory the inertial properties of space are determined (locally) by the metric tensor through the geodesic equation. This led Einstein to formulate the Mach principle as a connection between metric and energy-momentum[2]: *The spacetime geometry is entirely determined by the energy-momentum tensor of matter.*

However, this formulation runs into a difficulty. The equations that relate the geometry to the energy-momentum are the Einstein equations (to be derived in the next section). These are *differential* equations for $g_{\mu\nu}$ and in order to determine their solution uniquely it is necessary to specify not only the source terms (energy-momentum tensor), but also the *boundary* conditions on $g_{\mu\nu}$. The need for boundary conditions can only be avoided in the case of a (spatially) closed universe, i.e., a universe which has no boundaries. Einstein regarded this as a strong argument in favor of a closed universe.[16]

Furthermore, Wheeler has pointed out that even in a closed universe it is necessary to specify more than just the energy-momentum tensor. In the general time-dependent case, the Einstein equations determine a unique solution only if the *initial values* for $g_{\mu\nu}$ and $\partial g_{\mu\nu}/\partial t$ are given.* Thus, in the context of Einstein's theory, a more precise formulation of Mach's principle is the following:[17] *The spacetime geometry is uniquely determined by the energy-momentum and by the initial conditions on the geometry.*

In a broad sense Mach's principle confirms us in the view that matter acts on the geometry. Since we have already found a field equation that, in the linear approximation, expresses the metric tensor in terms of the distribution of energy-momentum, Mach's principle does not really tell us anything new and plays only a secondary, supportive role in our development of gravitational theory.

7.3 EINSTEIN'S FIELD EQUATIONS

In our discussion of the linear theory (Chapter 3) we obtained an approximate equation for the gravitational field. Expressed in terms of the metric tensor $g_{\mu\nu} = \eta_{\mu\nu} + \kappa h_{\mu\nu}$, this equation takes the form (see Eq. (3.48)):

---

* t is some general time coordinate.

$$\partial_\lambda \partial^\lambda g_{\mu\nu} - \partial^\lambda \partial_\nu g_{\mu\lambda} - \partial^\lambda \partial_\mu g_{\nu\lambda} + \partial_\mu \partial_\nu g_\sigma{}^\sigma - \eta_{\mu\nu} \partial_\lambda \partial^\lambda g_\sigma{}^\sigma$$
$$+ \eta_{\mu\nu} \partial^\lambda \partial^\sigma g_{\lambda\sigma} = -\kappa^2 T_{\mu\nu} \quad [21]$$

In this equation all indices are raised with the Minkowski metric $\eta^{\mu\nu}$; for example, $g_\sigma{}^\sigma \equiv \eta^{\sigma\tau} g_{\sigma\tau}$.[*]

As we already remarked, the above equation suffers from several defects because it fails to take into account that the gravitational field gravitates. We will now derive the exact, nonlinear field equations. One possible derivation relies on a method of successive approximations to discover the exact energy-momentum tensor of the gravitational field;[18,19] the first step of this method was presented in Section 3.3. However, we will now use a different and much quicker method which relies heavily on general invariance.

To be precise, the exact, nonlinear field equation satisfied by the metric tensor $g_{\mu\nu}$ can be derived from the following assumptions:

1. the equation is invariant under general coordinate transformations
2. the equation reduces to Eq. (21) in the linear approximation
3. the equation is of second differential order, and is linear in second derivatives.

The first two assumptions require no further comment. The third assumption expresses a prejudice of physicists in favor of equations of the general form

$$f(\psi, \partial\psi)\partial^2\psi + g(\psi, \partial\psi) = 0 \qquad [22]$$

where f and g are functions of the field ψ and of the *first* derivative $\partial\psi$, but not of the second derivative $\partial^2\psi$. This form is suggested by our experience with the dynamical equations of Lagrangian mechanics. Note that the second derivative enters linearly in Eq. (22). Nonlinear equations of this special type are called *quasilinear*. Mathematical theorems on the uniqueness and existence of solutions, wave propagation, initial value problem, etc., can be proved for such quasilinear equations; very little is known about the solution of more general nonlinear equations.

To derive the gravitational field equation, we begin by asking what form the equation takes in local geodesic coordinates at some point x^μ. Since in these coordinates $g'_{\mu\nu} = \eta_{\mu\nu}$ and $g'_{\mu\nu,\alpha} = 0$, it is clear that the differential equation must be some expression involving *only* the second derivatives $g'_{\mu\nu,\alpha,\beta}$. By assumption 3, this implies that the equation is necessarily *linear*. But assumption 2 tells us that if the equation is linear it must have the form

---

[*] We adopt the general rule that in any equation written in the linear approximation indices are raised (or lowered) with $\eta^{\mu\nu}$; in any other equation in this and succeeding chapters, indices are raised with $g^{\mu\nu}$.

$$\partial'_\lambda \partial'^\lambda g'_{\mu\nu} - \partial'^\lambda \partial'_\nu g'_{\mu\lambda} - \partial'^\lambda \partial'_\mu g'_{\nu\lambda} + \partial'_\mu \partial'_\nu g'^\sigma_\sigma$$
$$- \eta_{\mu\nu} \partial'_\lambda \partial'^\lambda g'^\sigma_\sigma + \eta_{\mu\nu} \partial'^\lambda \partial'^\sigma g'_{\lambda\sigma} = -\kappa^2 T'_{\mu\nu} \quad [23]$$

Since (23) is the *exact* field equation in geodesic coordinates x'^μ, we can find the *exact* field equation in the original coordinates x^μ by a coordinate transformation from x'^μ to x^μ; it is of course at this stage that assumption 1 enters. Because the transformation $x'^\mu \to x^\mu$ is somewhat messy, it is more convenient to try to express the left side of Eq. (23) in terms of an object whose transformation law is known. In order to do this, we begin with the observation that in geodesic coordinates the Ricci tensor becomes

$$R'_{\mu\nu} = \tfrac{1}{2}(\partial'^\lambda \partial'_\lambda g'_{\mu\nu} - \partial'^\lambda \partial'_\nu g'_{\mu\lambda} - \partial'^\lambda \partial'_\mu g'_{\nu\lambda} + \partial'_\mu \partial'_\nu g'^\sigma_\sigma) \quad [24]$$

From this is follows that

$$R'_{\mu\nu} - \tfrac{1}{2} g'_{\mu\nu} R' = \tfrac{1}{2}(\partial'^\lambda \partial'_\lambda g'_{\mu\nu} - \partial'^\lambda \partial'_\nu g'_{\mu\lambda} - \partial'^\lambda \partial'_\mu g'_{\nu\lambda}$$
$$+ \partial'_\mu \partial'_\nu g'^\sigma_\sigma - \eta_{\mu\nu} \partial'_\lambda \partial'^\lambda g'^\sigma_\sigma + \eta_{\mu\nu} \partial'^\lambda \partial'^\sigma g'_{\lambda\sigma}) \quad [25]$$

In these equations indices are raised with the metric $\eta^{\mu\nu}$.

EXERCISE 1. Show that the general expression Eq. (6.77) reduces to Eq. (24) if $g'_{\mu\nu} = \eta_{\mu\nu}$ and $g'_{\mu\nu,\alpha} = 0$.

By comparing Eqs. (23) and (25), we see that our field equation in geodesic coordinates is simply

$$R'_{\mu\nu} - \tfrac{1}{2} g'_{\mu\nu} R' = -\tfrac{1}{2}\kappa^2 T'_{\mu\nu} \quad\quad\quad [26]$$

Since the transformation properties of $R'^{\mu\nu}$ are those of a tensor, the coordinate transformation from x'^μ to x^μ will give, with $\kappa^2 = 16\pi G$,

$$R_{\mu\nu} - \tfrac{1}{2} g_{\mu\nu} R = -8\pi G T_{\mu\nu} \quad\quad\quad [27]$$

This is *Einstein's field equation*. In cgs units, we must write

$$R_{\mu\nu} - \tfrac{1}{2} g_{\mu\nu} R = -\frac{8\pi G}{c^4} T_{\mu\nu} \quad\quad\quad [28]$$

That the exact nonlinear equations are implied by the linear equations (the converse is of course trivial) is a very remarkable feature of Einstein's theory. This very tight connection between the exact equations and the linear approximation would not exist were it not for the principle of general invariance.

The following chapters will deal with some solutions of Eq. (27). We will obtain the exact solution for the gravitational field surrounding a static, spherically symmetric mass (Schwarzschild solution) and that surrounding a rotating mass (Kerr solution); finally, we will apply (27) to cosmology and obtain solutions that describe the evolution of the large-scale geometry of the universe.

Unfortunately, the differential equation (27) is very complicated. The second derivatives of the metric field appear linearly, but the first derivatives enter quadratically. The dependence on $g_{\mu\nu}$ is even worse; the left side of Eq. (27) contains the inverse $g^{\mu\nu}$ (used to raise indices), and this is a horrible thing when expressed as a function of $g_{\mu\nu}$. No general procedure exists for solving the equations analytically; one guesses solutions as best one can.

Note that the quantity $T_{\mu\nu}$ appearing on the right side of Eq. (27) is the energy-momentum tensor of *matter*. The energy-momentum of the gravitational field is already included (implicitly) on the left side of this equation. Essentially, all the nonlinear terms on the left side arise from the energy-momentum tensor of the gravitational field.

In empty space, the field equation reduces to

$$R_{\mu\nu} = 0 \tag{29}$$

EXERCISE 2. Show this. (Hint: Contract μ,ν in Eq. (27).)

We can also put Eq. (27) into the form

$$R_\mu{}^\nu - \tfrac{1}{2}\delta_\mu{}^\nu R = -8\pi G T_\mu{}^\nu \tag{30}$$

by raising an index. Since the left side of Eq. (30) satisfies the Bianchi identity

$$(R_\mu{}^\nu - \tfrac{1}{2}\delta_\mu{}^\nu R)_{;\nu} = 0 \tag{31}$$

it follows that

$$T_\mu{}^\nu{}_{;\nu} = 0 \tag{32}$$

This expresses the exchange of energy momentum between matter and the gravitational field. Using the definition of covariant derivative, Eq. (32) becomes

$$\partial_\nu T_\mu{}^\nu + \Gamma^\nu{}_{\alpha\nu} T_\mu{}^\alpha - \Gamma^\alpha{}_{\mu\nu} T_\alpha{}^\nu = 0 \tag{33}$$

This equation is the generalization of our approximate equation (3.68).

EXERCISE 3. Show that (33) implies (3.68) if the linear approximation is used. [Hint: Show that in the linear approximation

$$\Gamma^\mu{}_{\alpha\beta} = \tfrac{1}{2}\eta^{\mu\nu}\kappa(h_{\nu\alpha,\beta} + h_{\beta\nu,\alpha} - h_{\alpha\beta,\nu}).]$$ [34]

The equation of motion of a particle can be obtained from (33) by methods similar to those of Section 3.3. Of course we already know the equation of motion of a particle — it is the geodesic equation. Nevertheless, the calculation based on Eq. (33) is important in principle because we must make sure that the equations of motion are not contradictory.

To obtain the equation of motion from (33), it is simplest to go to geodesic coordinates. Then Eq. (33) reduces to

$$\partial'_\nu T'_\mu{}^\nu = 0$$ [35]

Although this equation is strictly valid only at one point, in the case of a sufficiently small (pointlike) particle we can regard it as valid throughout the volume of the particle. If we apply the methods of Section 3.3, we find that the equation of motion is

$$\frac{dp'_\mu}{d\tau} = 0$$ [36]

that is, as seen in the local geodesic coordinates the energy and the momentum of the particle do not change. This agrees with the result (6.54) which we obtained from the geodesic equation. We can therefore conclude that the particle motion predicted by Eq. (33) is geodesic motion.[20]

Since in Einstein's theory all particles obey the geodesic equation of motion, it follows that the Galileo principle of equivalence is satisfied. It can also be shown that the Newton principle of equivalence is satisfied; that is, for any arbitrary system (not necessarily a particle), the gravitational mass equals the inertial mass.[21] The principle of general invariance plays an important role in this proof.

A final general remark about the field equations (27). Since the tensor $R_{\mu\nu}$ is symmetric, the number of equations is ten. However, not all these equations are independent; the Bianchi identity, consisting of a set of four equations, shows that we have not ten, but only *six* independent differential equations. Since the unknown functions are the *ten components* $g_{\mu\nu}(x)$, the Einstein equations do not determine the fields entirely. We have encountered this situation already in the linear case where the solutions suffered from an ambiguity due to gauge transformation; we eliminated the ambiguity by using the Hilbert gauge condition. In the present case, the ambiguity in the solution for $g_{\mu\nu}(x)$ arises from the ambiguity in the choice of coordinates. This comes as no surprise since the gauge transformations of the linear theory are in fact (infinitesimal) coordinate transformations. That the Einstein field equations determine the field $g_{\mu\nu}(x)$ only up to a general coordinate transformation is actually a good thing: it would be absurd that the field equations should determine not only the geometry, but also prescribe what coordinates we must use to describe the geometry.

In order to remove the ambiguity in the solution for $g_{\mu\nu}(x)$, one imposes extra "coordinate" conditions. For example, one can use the *harmonic* coordinate condition

$$g^{\mu\nu}\Gamma^{\alpha}_{\mu\nu} = 0 \tag{37}$$

This condition replaces the Hilbert condition of the linear theory.

EXERCISE 4. Show that in the linear approximation, (37) reduces to the Hilbert condition.

Other coordinate conditions are also possible. It is often convenient to impose a purely algebraic condition, rather than a differential equation, on the $g_{\mu\nu}$. One such possibility are *time-orthogonal* coordinates* with

$$g_{00} = 1, \qquad g_{0k} = 0 \tag{38}$$

In general, the condition (38) is not very useful because the coordinate frame in which this condition holds will usually not be rigid; relative to the rectangular coordinates at infinity, the coordinate frame will have to bend and twist in a horrible way. We will make our choice of coordinate condition whenever and however it is convenient.

7.4 ANOTHER APPROACH TO EINSTEIN'S EQUATIONS; THE COSMOLOGICAL TERM

The connection between Eq. (32) and geodesic motion can be used as a basis for an alternative derivation of Einstein's equations. The argument is the following: From the analysis of spacetime measurements of Section 5.5 we know that free particles must move along geodesics. Therefore, as seen in the local geodesic reference frame, their energy and momentum do not change. Consider a cloud of such particles (cloud of dust). The energy-momentum tensor of the cloud is

$$T^{\mu\nu} = \rho_0 u^{\mu} u^{\nu} \tag{39}$$

where ρ_0 is the proper mass density (see Eq. (2.76)). Since the momentum of each particle is constant, the energy momentum tensor (39) is conserved and satisfies (35) in the geodesic coordinates. Expressed in general coordinates, this conservation equation reads

$$T_{\mu}{}^{\nu}{}_{;\nu} = 0 \tag{40}$$

* Also called Gaussian coordinates.

We have shown that Eq. (40) may be regarded as a consequence of the geodesic equation. Although the derivation is only valid for a cloud of free particles, we may *postulate* that Eq. (40) should hold for any physical system. Such a postulate may be regarded as an application of a minimal coupling principle; essentially we are assuming that the coupling of gravitation to arbitrary forms of matter is as simple as the coupling of gravitation to (free) particles.

In the field equation, $T_{\mu\nu}$ should appear as source of the gravitational field. The field equation should therefore be of the form

$$G_{\mu\nu} = -8\pi G T_{\mu\nu} \qquad [41]$$

where $G_{\mu\nu}$ is some expression involving the metric and derivatives of the metric. According to Eq. (40), the object $G_{\mu\nu}$ must satisfy the identity

$$G_{\mu}{}^{\nu}{}_{;\nu} = 0 \qquad [42]$$

To proceed with this derivation of Einstein's field equations, we combine the condition (42) with the previous assumptions 1 and 3 (but we do *not* assume 2). General invariance demands that $G_{\mu\nu}$ be a tensor. We know from Section 6.5 that the only tensor which contains the second derivatives linearly is $R_{\mu\nu\alpha\beta}$. Hence the only tensors available for the construction of $G_{\mu\nu}$ are $R_{\mu\nu\alpha\beta}$ and $g_{\mu\nu}$. The most general second-rank tensor that can be built out of these, with $R_{\mu\nu\alpha\beta}$ entering linearly, is

$$G_{\mu\nu} = aR_{\mu\nu} + bg_{\mu\nu}R + \Lambda g_{\mu\nu} \qquad [43]$$

where a, b, and Λ are constants. Since $g_{\mu}{}^{\nu}{}_{;\nu} \equiv 0$, the condition (42) reduces to

$$(aR_{\mu}{}^{\nu} + b\delta_{\mu}{}^{\nu}R)_{;\nu} = 0 \qquad [44]$$

In view of the Bianchi identity (6.89), the condition (44) will be fulfilled if $b = -\frac{1}{2}a$.

We are now left with a field equation

$$a(R_{\mu\nu} - \tfrac{1}{2}g_{\mu\nu}R) + \Lambda g_{\mu\nu} = -8\pi G T_{\mu\nu} \qquad [45]$$

To determine a and Λ, we must take into account that in the linear, non-relativistic limit, Eq. (45) should agree with the equation for the Newtonian potential. This requires $a = 1$ and $\Lambda = 0$ and gives us, again, the Einstein equations.

The quantity $\Lambda g_{\mu\nu}$ appearing in Eq. (45) is called the *cosmological term*. Let us consider the possibility that Λ is very small, but not zero. If Λ is sufficiently small, then within the solar system Eq. (45) will agree very well with the equation for the Newtonian potential. However, even a small value of Λ could have drastic effects on the evolution of the universe (see Section 10.3).

To grasp the implications of a nonzero value of Λ, let us look at the linear approximation corresponding to Eq. (45) (with $a = 1$). Since Λ is certainly small, we can approximate $\Lambda g_{\mu\nu} \cong \Lambda \eta_{\mu\nu}$ and with the usual gauge condition (Eq. (3.50)) we obtain

$$\partial_\lambda \partial^\lambda (h_{\mu\nu} - \tfrac{1}{2}\eta_{\mu\nu}h) = -\kappa T_{\mu\nu} - \frac{2\Lambda}{\kappa}\,\eta_{\mu\nu}$$

This can also be written as

$$\partial_\lambda \partial^\lambda h_{\mu\nu} = -\kappa(T_{\mu\nu} - \tfrac{1}{2}\eta_{\mu\nu}T) + \frac{2\Lambda}{\kappa}\,\eta_{\mu\nu}$$

EXERCISE 5. Show this.

In the absence of matter ($T_{\mu\nu} = 0$), the equation for the static Newtonian potential $\Phi = \tfrac{1}{2}\kappa h_{00}$ (see Eq. (3.100)) is then

$$\nabla^2 \Phi = -\Lambda \tag{46}$$

If we compare this with Eq. (1.10), we recognize that the Λ-term in Einstein's equation corresponds to a uniform effective mass density

$$\rho_{\text{eff}} = -\frac{\Lambda}{4\pi G}$$

Thus if Λ is positive, the vacuum has an effective mass density which is negative, and the converse is also true.

If we arbitrarily set $\Phi = 0$ at the origin, then in spherical coordinates Eq. (46) has the solution

$$\Phi = -\frac{\Lambda}{6}\,r^2 \tag{47}$$

EXERCISE 6. Show this.

This potential indicates that between any two particles there acts an effective harmonic oscillator force of $\Lambda r/3$ per unit mass, attractive in the case $\Lambda < 0$, and repulsive in the case $\Lambda > 0$.

For large distances, the potential (47) and the force grow very large. But for small distances they are negligible. We can set an upper limit on Λ by noting that the Newtonian inverse square law seems to be valid for the motion of galaxies in multiple systems of galaxies. For example, Page has applied the inverse square law to determine the mass of binary systems of galaxies, i.e., pairs of galaxies in a tight orbit about each other.[23] The masses obtained for binary systems (via the virial theorem) are in reasonable agreement with independent mass determination of the components (via the rotation curve; see Section 10.6). This implies that the deviations from the inverse square law are not significant, that is, within regions of a size at least as large as these binary systems, the acceleration $\Lambda r/3$ is small compared to the gravitational acceleration GM/r^2. From the dimension ($r \sim 10^{23}$ cm), and the mass ($M \sim 10^{11} M_\odot$) of typical binary systems we can then set the limit

$$|\Lambda| < 10^{-33}/\text{sec}^2$$

If it could be established that the deviations from the inverse square law are also insignificant in groups or clusters of galaxies of large dimensions (such as, e.g., the M 81 group), then the above limit on Λ could be reduced by a factor of 100 or 1,000. Unfortunately, our ignorance of the dynamics of large clusters is very deep. There are very serious discrepancies between different methods of mass determination (see Section 10.6), and until these discrepancies are resolved we cannot use large clusters to set any firm limits on Λ.

The effects of the Λ-term will be most noticeable in the large-scale motion of the universe. Hence, cosmological data sets the best limits on Λ. Our universe is expanding (see Section 10.3). The attractive Newtonian gravitational force between the masses in the universe tends to slow down this expansion. The extra force associated with the cosmological term tends to slow down the expansion if $\Lambda < 0$ and speed it up if $\Lambda > 0$. From the available data on the expansion one can set the limit

$$|\Lambda| < 10^{-35}/\text{sec}^2$$

on the value of the cosmological constant (see Section 10.3).

Although the implications of the Λ-term are cosmological, the origin of this term probably is to be found in quantum theory rather than cosmology. As we have seen, this term can be interpreted as an energy density of the vacuum. The quantum theoretical vacuum is a very active place. It is crawling with electron-positron pairs, proton-antiproton pairs, photons, gravitons, etc., which are spontaneously created and destroyed by vacuum fluctuations. It is not surprising that the vacuum should have an energy density; the surprise is rather that the energy density is as small as our limit on the cosmological term indicates: only 10^{-29} g/cm^3 or 10^{-50} erg/cm^3. Quantum theoretical calculations usually give an infinite value for the energy density. This indicates a defect in our theories, but it also indicates that the possibility of a nonzero value of Λ should not be rejected peremptorily.

In the context of Chapter 3, we did not attempt to include a term $\Lambda\eta_{\mu\nu}$ in our field equations, because we wanted to work with weak fields in an asymptotically flat spacetime. The cosmological term violates this assumption. Obviously, at large distances the potential (47) becomes very large, and this contradicts both the weak field condition ($\Phi << c^2$) and the boundary condition at infinity ($\Phi \rightarrow 0$). However, a nonzero value of Λ does not entirely destroy the validity of the approximate equations of Chapter 3. If we are dealing with a reasonable central mass M, then there will exist a wide range of radii

$$\frac{c}{\sqrt{\Lambda}} >> r >> \frac{GM}{c^2}$$

in which the spacetime is very nearly flat; in this range the linear theory is a good approximation.

7.5 TIDAL FORCES IN GEOMETRODYNAMICS

We have defined the Riemann curvature tensor in terms of entirely geometrical concepts. But this tensor also has a simple physical interpretation: $R^{\alpha}{}_{\beta\mu\nu}$ is the tidal force field, i.e., it gives the relative acceleration between two particles in free fall. Before we can calculate this relative acceleration, we must introduce the concept of *differentiation along a curve*. What is involved here is the following. Suppose you have a curve, possibly, but not necessarily, a geodesic. Displacements along the curve may be indicated by giving the corresponding increment in the "length" s or the proper time τ. Suppose the curve is surrounded by a vector field A^{μ}. If we want to know the rate of change of A^{μ} along the curve, we must calculate

$$\frac{dA^{\mu}}{d\tau}$$

However, this is not a good measure of how much A^{μ} is "really" changing since part, or maybe all, of the contribution to $dA^{\mu}/d\tau$ may be due to the curvilinear coordinates used to define the components A^{α}. Also, $dA^{\mu}/d\tau$ is *not* a vector (for the same reason that $\partial A^{\mu}/\partial x^{\nu}$ is not a tensor). A better measure of the rate of change of A^{μ} along the curve is the *vector* quantity

$$\frac{DA^{\mu}}{D\tau} \equiv A^{\mu}{}_{;\nu}\frac{dx^{\nu}}{d\tau} = \frac{dA^{\mu}}{d\tau} + \Gamma^{\mu}{}_{\alpha\beta}A^{\alpha}\frac{dx^{\beta}}{d\tau} \qquad [48]$$

This is a vector because it is a product of the tensor $A^{\mu}{}_{;\nu}$ and the vector $dx^{\nu}/d\tau$. The quantity $DA^{\mu}/D\tau$ is called the *derivative along the curve*.

EXERCISE 7. $dA^\mu/d\tau$ is not a vector, but $dx^\nu/d\tau$ is a vector. Explain this. [Hint: x^ν is not a vector (see Eq. (6.1)), but dx^ν is a vector (see Eq. (6.3)).]

In terms of parallel transport, we can understand Eq. (48) as follows: $DA^\mu/D\tau$ is obtained from $[A^\mu(\tau + d\tau) - A^\mu(\tau)]/d\tau$ by subtracting from the latter expression that part which is entirely due to parallel transport of the vector $A^\mu(\tau)$ from the point τ to the point $\tau + d\tau$ on the curve (compare Eq. (6.70)), and by then taking the limit $d\tau \to 0$.

The expression (48) is valid for any curve. Let us now concentrate on geodesics and find the *second* derivative along a *geodesic* curve:

$$
\begin{aligned}
\frac{D^2 A^\mu}{D\tau^2} &= \frac{D}{D\tau}\left(\frac{D A^\mu}{D\tau}\right) = \frac{d}{d\tau}\left(\frac{D A^\mu}{D\tau}\right) + \Gamma^\mu{}_{\alpha\beta}\frac{D A^\alpha}{D\tau}\frac{dx^\beta}{d\tau} \\
&= \frac{d}{d\tau}\left(\frac{d A^\mu}{d\tau} + \Gamma^\mu{}_{\alpha\beta}A^\alpha\frac{dx^\beta}{d\tau}\right) + \Gamma^\mu{}_{\alpha\beta}\frac{D A^\alpha}{D\tau}\frac{dx^\beta}{d\tau} \\
&= \frac{d^2 A^\mu}{d\tau^2} + \Gamma^\mu{}_{\alpha\beta,\nu}\frac{dx^\nu}{d\tau}A^\alpha\frac{dx^\beta}{d\tau} + 2\Gamma^\mu{}_{\alpha\beta}\frac{d A^\alpha}{d\tau}\frac{dx^\beta}{d\tau} \\
&\quad - \Gamma^\mu{}_{\alpha\beta}A^\alpha\Gamma^\beta{}_{\kappa\lambda}\frac{dx^\kappa}{d\tau}\frac{dx^\lambda}{d\tau} + \Gamma^\mu{}_{\alpha\beta}\Gamma^\alpha{}_{\kappa\lambda}A^\kappa\frac{dx^\lambda}{d\tau}\frac{dx^\beta}{d\tau} \quad [49]
\end{aligned}
$$

The next to last term comes from evaluating $d^2x^\beta/d\tau^2$ with the geodesic equation.

We can now begin our calculation of the relative acceleration of two particles moving on neighboring geodesics by writing down the equation of motion of each. If, for a given value of τ (same τ for both), the coordinates of the particles are $x^\mu(\tau)$ and $x^\mu(\tau) + \xi^\mu(\tau)$, then

$$
\frac{d^2 x^\mu}{d\tau^2} + \Gamma^\mu{}_{\alpha\beta}(x)\frac{dx^\alpha}{d\tau}\frac{dx^\beta}{d\tau} = 0 \tag{50}
$$

$$
\frac{d^2(x^\mu + \xi^\mu)}{d\tau^2} + \Gamma^\mu{}_{\alpha\beta}(x + \xi)\left(\frac{dx^\alpha}{d\tau} + \frac{d\xi^\alpha}{d\tau}\right)\left(\frac{dx^\beta}{d\tau} + \frac{d\xi^\beta}{d\tau}\right) = 0 \tag{51}
$$

We will assume that ξ^μ and $d\xi^\mu/d\tau$ are infinitesimal; this means that the particles are near to each other and remain near for a fairly long time. If we approximate

$$
\Gamma^\mu{}_{\alpha\beta}(x + \xi) \simeq \Gamma^\mu{}_{\alpha\beta}(x) + \Gamma^\mu{}_{\alpha\beta,\sigma}\xi^\sigma \tag{52}
$$

and keep only first-order terms in ξ, the difference between (51) and (50) gives us

$$
\frac{d^2\xi^\mu}{d\tau^2} = -\Gamma^\mu{}_{\alpha\beta,\sigma}\xi^\sigma\frac{dx^\alpha}{d\tau}\frac{dx^\beta}{d\tau} - 2\Gamma^\mu{}_{\alpha\beta}\frac{d\xi^\alpha}{d\tau}\frac{dx^\beta}{d\tau} \tag{53}
$$

The second derivative of ξ^μ along the geodesic can then be calculated by substituting Eq. (53) into (49), with the following result:

$$\frac{D^2 \xi^\mu}{D\tau^2} = -R^\mu{}_{\alpha\sigma\beta}\xi^\sigma \frac{dx^\alpha}{d\tau}\frac{dx^\beta}{d\tau} \qquad [54]$$

EXERCISE 8. Carry out the calculation leading to (54).

Eq. (54) is called the equation of *geodesic deviation*. This equation gives the relativistic generalization of our Newtonian result for the tidal force.

To bring out the similarity between Eq. (54) and the corresponding Newtonian equation (Eq. (1.41)), we will transform the former equation into freely falling geodesic coordinates (Fermi coordinates). These coordinates are constructed by the following procedure: Suppose that P is a point on the geodesic curve determined by Eq. (50). To be concrete, we may think of P as representing the initial position of the particle which moves along this geodesic. For a start, take the point P as origin and introduce geodesic coordinates with $g'_{\mu\nu} = \eta_{\mu\nu}$ and $g'_{\mu\nu,\alpha} = 0$ at this point. Take the time axis in the direction of the geodesic and take as time coordinate the proper time along the geodesic. At the point P, select the three directions for the space axes so that these axes are perpendicular to each other and to the time axis (see Fig. 7.1). Now let the origin of the spatial coordinates fall freely along the geodesic so that the particle always remains at the origin. Let the directions of the axes at each point along the geodesic be determined by parallel transport of the initial directions from the point P. This construction defines space and time axes, and coordinates, for the region near the geodesic.

Since the directions of the axes of this freely falling coordinate frame have been defined by parallel transport, if any arbitrary vector is

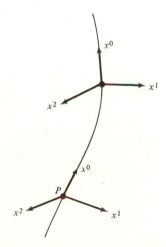

Fig. 7.1 *Free-fall coordinates near a geodesic.*

parallel transported along the geodesic, then its components with respect to the freely falling coordinate frame cannot change. It follows that $\Gamma'^{\mu}{}_{\alpha\beta} = 0$ everywhere along the geodesic; this means that $g'_{\mu\nu,\alpha} = 0$ and that $g'_{\mu\nu}$ must have the same value everywhere along the geodesic as at the point P, i.e., $g'_{\mu\nu} = \eta_{\mu\nu}$. The coordinates are therefore geodesic everywhere along the geodesic curve.

A concrete picture is helpful: the freely falling geodesic coordinates may be thought of as obtained by taking a small three-dimensional rectangular coordinate grid and a clock, and letting them fall along the geodesic. The grid must fall without rotation, i.e., it must be parallel transported (physically, the absence of rotation can be tested with a gyroscope; see Section 8.3).

Note that since $\Gamma'^{\mu}{}_{\alpha\beta} = 0$, the derivative $DA'^{\mu}/D\tau$ along the geodesic curve (Eq. (48)) reduces to the ordinary derivative $dA'^{\mu}/d\tau$, or dA'^{μ}/dt', in the freely falling geodesic coordinates.

Let us express Eq. (54) in these coordinates. Since the particle with the equation of motion (50) remains forever at rest at the origin of our coordinates, we have

$$\frac{dx'^{\alpha}}{d\tau} = (1,0,0,0) \tag{55}$$

Furthermore, we may assume that $\xi'^{0} = 0$; this simply means that the particle accelerations are compared at equal times. Then Eq. (54) reduces to

$$\frac{d^2\xi'^k}{dt'^2} = -R'^k{}_{0l0}\xi'^l \tag{56}$$

The tidal force in the freely falling geodesic reference frame is therefore

$$f^k = -mR'^k{}_{0l0}\xi'^l \tag{57}$$

where m is the mass of the particle and ξ'^l its displacement from the origin. Note that these equations are only valid if the displacement and velocity are small (strictly speaking: $\xi'^l \to 0$, $d\xi'^l/d\tau \to 0$).

To show that the geometrodynamic tidal force is in accord with the Newtonian expression (Eq. (1.40)), it remains to be demonstrated that the old (Eq. (1.39)) and new (Eq. (6.77)) definitions of $R^k{}_{0l0}$ agree in the case of weak static gravitational fields.

In the linear approximation, the two terms quadratic in Γ can be omitted from Eq. (6.77). Furthermore, in this approximation

$$\Gamma^{\alpha}{}_{\beta\mu} \simeq \frac{\kappa}{2}\,\eta^{\alpha\sigma}(h_{\sigma\beta,\mu} + h_{\mu\sigma,\beta} - h_{\beta\mu,\sigma}) \tag{58}$$

This results in

$$R^{\alpha}{}_{\beta\mu\nu} \simeq -\frac{\kappa}{2} \, \eta^{\alpha\sigma}(h_{\mu\sigma,\beta,\nu} - h_{\beta\mu,\sigma,\nu} - h_{\nu\sigma,\beta,\mu} + h_{\beta\nu,\sigma,\mu}) \qquad [59]$$

Therefore

$$R^{k}{}_{0l0} \simeq \frac{\kappa}{2} \, (h_{lk,0,0} - h_{0l,k,0} - h_{0k,0,l} + h_{00,k,l}) \qquad [60]$$

In the Newtonian limit all the terms containing time derivatives are to be omitted. Using Eq. (3.100),

$$\tfrac{1}{2}\kappa h_{00} = \Phi \qquad [61]$$

We then find

$$R^{k}{}_{0l0} = \frac{\partial^2\Phi}{\partial x^k \partial x^l} \qquad [62]$$

in agreement with Eq. (1.39).

The tidal force equation (57) can be used for measuring the components $R^{k}{}_{0l0}$ of the Riemann tensor. We have already discussed in Section 1.7 how such measurements of the tidal force can be carried out; we emphasize once more that these measurements can be performed *locally*.

What about the other components of the Riemann tensor? These components can be obtained by measuring the components $R^{k}{}_{0l0}$ in a sufficiently large number of other reference frames that have some velocity relative to the original reference frame. Lorentz transformations between these reference frames relate the components $R^{k}{}_{0l0}$ in one frame to the general components $R^{\alpha}{}_{\beta\mu\nu}$ in another and therefore can be used to find all the components. The Riemann tensor is therefore entirely determined by tidal force measurements.

To conclude this section, we will show that the tidal force (57) given by geometrodynamics satisfies the condition

$$\frac{\partial f^k}{\partial \xi'^k} = 0 \qquad [63]$$

This means that the tidal force field can be represented by field lines. Of course, we have already come across the condition (63) in Newton's theory (Eq. (1.43)), and in the linear tensor theory (Eq. (4.43)), but we will now give a general proof using the exact Einstein equations.

According to Eq. (57),

$$\frac{\partial f^k}{\partial \xi'^k} = - mR'^k{}_{0l0} \frac{\partial \xi'^l}{\partial \xi'^k} = - mR'^k{}_{0k0} \qquad [64]$$

Since $R^\alpha{}_{\beta\mu\nu}$ is antisymmetric in $\mu\nu$, we have $R'^0{}_{000} \equiv 0$ and Eq. (64) is equivalent to

$$\frac{\partial f^k}{\partial \xi'^k} = - mR'^\mu{}_{0\mu0} = mR'_{00} \qquad [65]$$

where $R'_{00} \equiv R'^\mu{}_{00\mu}$ is a component of the Ricci tensor. The Einstein field equation *in vacuo* tells us that $R'_{\mu\nu} = 0$, and hence Eq. (63) is valid as long as we remain outside of the mass distribution which generates the gravitational field.

FURTHER READING

The clear distinctions between covariance and invariance and between absolute and dynamical objects are due to Anderson and are rigorously spelt out by him in his book, *Principles of Relativity Physics,* and also in his article "Relativity Principles and the Role of Coordinates in Physics" (*in* Chiu and Hoffmann, eds., *Gravitation and Relativity*). (It should be noted that whereas the article uses the misleading term "relativity group," the book uses the much better terms "symmetry group" or "invariance group.")

Wigner, *Symmetries and Reflections,* gives a very nice qualitative description of the differences between algebraic (or geometric) symmetries and dynamical symmetries.

An interesting discussion of general invariance as a gauge symmetry, analogous to the gauge symmetry of electrodynamics, is contained in the article "Invariant Theoretical Interpretation of Interaction" by Utiyama (see ref. 8).

Mach, *Die Mechanik* (translated as *The Science of Mechanics,* see the reading list in Chapter 1), contains a profound criticism of Newton's concepts of space, time, and mass. This book exerted a strong influence on Einstein.

Mach's principle is discussed from somewhat different points of view in the following articles:

Wheeler, "Mach's Principle as Boundary Condition for Einstein's Equations" (see ref. 12);

Dicke, "The Many Faces of Mach" (*in* Chiu and Hoffmann, eds., *Gravitation and Relativity*);

Dicke, "Mach's Principle" (*in* Klauder, ed., *Magic Without Magic: John Archibald Wheeler*).

Elementary discussions are given in Sciama's books *The Unity of the Universe* and *The Physical Foundations of General Relativity;* the first is philosophical and the second speculates on a very literal interpretation of Mach's principle that lies outside of geometrodynamics.

Fock, *The Theory of Space, Time, and Gravitation,* strongly condemns Einstein for describing the theory of gravitation as "general relativity." Although Fock's criticisms are excessive, they do help to clear the air.

The term "geometrodynamics" which we use to replace "general relativity" was originally introduced by Wheeler in a much broader sense to describe a theory that attempts to interpret matter as an excited state of the geometry. According to this view "there is nothing in the world except empty curved space. Matter, charge, electromagnetism, and other fields are manifestations of the bending of space. *Physics is geometry.*" [22] The program of this aspiring project is outlined in the books *Einsteins Vision* and *Geometrodynamics* by Wheeler, and also in the article "Geometrodynamics" by Fletcher (*in* Witten, ed., *Gravitation: An Introduction to Current Research*).

A complete list of the various routes that lead to Einstein's equations is given in Misner, Thorne, and Wheeler, *Gravitation*.

The derivation of the equations of motion of test particles, with and without spin, from the Einstein field equations is given in the articles "Spinning Test Particles in General Relativity I" by Papapetrou and "Spinning Test Particles in General Relativity II" by Corinaldesi and Papapetrou (see ref. 20). A different and older method is that of Infeld and Schild in "The Motion of Test Particles in General Relativity" (see ref. 24).

A careful mathematical treatment of Fermi coordinates is given in Synge, *Relativity: The General Theory*. The physical interpretation of the curvature tensor and a method for measuring it are also described in this book. Other methods for the measurement of the curvature tensor have been described by Bertotti, "The Theory of Measurement in General Relativity" (*in* Møller, ed., *Evidence for Gravitational Theories*) and by Wigner, *Symmetries and Reflections*.

REFERENCES

1. E. Kretschmann, Ann. Physik **53**, 575 (1917).
2. A. Einstein, Ann. Physik **55**, 241 (1918). Quoted with the permission of the Estate of Albert Einstein.
3. A. Trautman, Usp. Fiz. Nauk. **89**, 3 (1966); translated in Sov. Phys.-Uspekhi **9**, 319 (1966).
4. J. L. Anderson, *Principles of Relativity Physics* (Academic Press, New York, 1967).
5. A. Einstein, Ann. Physik **35**, 898 (1911); a translation is given in H. A. Lorentz, A. Einstein, M. Minkowski, and H. Weyl, *The Principle of Relativity* (Dover, New York). Quoted with the permission of the Estate of Albert Einstein.
6. R. L. Forward, "Moving-Base Rotating Gravity Gradiometer Development" (Hughes Research Laboratories, Research Report No. 461).
7. S. Weinberg, *in* S. Deser, M. Grisaru, and H. Pendleton, eds., *Lectures on Elementary Particles and Quantum Field Theory* (MIT Press, Cambridge, 1970), vol. 1.
8. R. Utiyama, Phys. Rev. **101**, 1597 (1956).
9. S. S. Schweber, *An Introduction to Relativistic Quantum Field Theory* (Harper and Row, New York, 1961), p. 207.
10. I. Newton, *Principia* (University of California Press, Berkeley, 1962), vol. 1, pp. 10–11. Originally published by the University of California Press; reprinted by permission of The Regents of the University of California.

11. E. Mach, *Die Mechanik* (9th Ed., Brockhaus, Wiesbaden, 1933), pp. 226–27.
12. J. A. Wheeler, *in* H.-Y. Chiu and W. F. Hoffman, eds., *Gravitation and Relativity* (Benjamin, New York, 1964), p. 303.
13. A. Einstein, *The Meaning of Relativity* (5th ed., Princeton University Press, Princeton, 1955), p. 100.
14. R. H. Dicke, Phys. Rev. Lett. **7**, 359 (1961).
15. V. W. Hughes, *in* H.-Y. Chiu and W. F. Hoffman, eds., *Gravitation and Relativity* (Benjamin, New York, 1964), p. 106.
16. See ref. 13, p. 107.
17. See ref. 12, p. 333.
18. S. N. Gupta, Rev. Mod. Phys. **29**, 334 (1957).
19. S. Deser, Gen. Rel. and Grav. **1**, 9 (1970).
20. For a more rigorous derivation of the equation of motion, see A. Papapetrou, Proc. Roy. Soc. London **A209**, 248 (1951); E. Corinaldesi and A. Papapetrou, Proc. Roy. Soc. London **A209**, 259 (1951).
21. H. C. Ohanian, J. Math. Phys. **14**, 1892 (1973).
22. J. A. Wheeler, *Geometrodynamics* (Academic Press, New York, 1962), p. 225.
23. T. L. Page, Ap. J. **116**, 63 (1952); Ap. J. **132**, 910 (1960).
24. L. Infeld and A. Schild, Rev. Mod. Phys. **21**, 408 (1949).

PROBLEMS

1. Show that none of the following equations is covariant (A^μ, B^μ are vectors, $A^{\mu\nu}$ is a tensor):

$$A_\mu = \frac{1}{B^\mu}$$
$$g_{\mu\nu}A^\alpha = \Gamma^\alpha{}_{\mu\nu}$$
$$A^{\mu\nu}{}_{,\sigma} = 0$$
$$A^0 = B^1$$

2. Show that Newton's equation of motion, $(\partial/\partial x^0)p^k = F^k$, can be put into the generally covariant form

$$\frac{\partial x'^\alpha}{\partial x^0} \frac{\partial}{\partial x'^\alpha} \left(\frac{\partial x^k}{\partial x'^\mu} p'^\mu \right) = \frac{\partial x^k}{\partial x'^\mu} F'^\mu$$

[Hint: Define $p'^\mu \equiv (\partial x'^\mu/\partial x^l)p^l$ and $F'^\mu \equiv (\partial x'^\mu/\partial x^l)F^l$.]

3. Show that Einstein's equation can also be written as

$$R_{\mu\nu} = -\frac{8\pi G}{c^4} (T_{\mu\nu} - \tfrac{1}{2}g_{\mu\nu} T)$$

4. Evaluate $R^k{}_{0l0}$ for the gravitational wave of the type given by Eq. (4.25). Then use Eq. (7.57) to calculate the tidal force and check that your result agrees with Eqs. (4.41) and (4.42).

5. According to Mach's ideas, a light ray passing near a rotating body should suffer an extra deflection because it tends to be carried along by the rotation. That part of the wavefront which is nearer the star tends to be carried along more than the part which is farther away — this results in an extra deflection of the wavefront. Are the results of Problem 3.11 in agreement with Mach's ideas?

6. The total mass and radius of some multiple systems of galaxies are listed in the following table. For a galaxy at the outer edge of each of these systems, calculate the ratio of the Newtonian gravitational force to the "harmonic oscillator force" associated with the cosmological term (see Eq. (47)). Assume that $\Lambda = 10^{-35}/\text{sec}^2$. Is the cosmological force important?

Multiple System	Number of Galaxies	M	r
Typical binary	2	$10^{11}\ M_\odot$	10^{23} cm
M 81 (inner subgroup)	5	$4 \times 10^{11}\ M_\odot$	10^{24} cm
Coma cluster	1,000	$3 \times 10^{15}\ M_\odot$	10^{25} cm

7. Show that in a region in which the energy-momentum tensor differs from zero, the divergence of the tidal force (in geodesic coordinates) is

$$\frac{\partial f^k}{\partial \xi'^k} = -8\pi G(T'_{00} - \tfrac{1}{2}T')m$$

Check that this agrees with Exercise 1.13.

8. THE SCHWARZSCHILD SOLUTION

> When they anticipate the time, let
> them be put to death without mercy;
> when their reckoning is behind the
> time, let them be put to death without
> mercy.
>
> The Marquis of Yin, pronouncing sentence
> on Hsi and Ho, ministers of the Board of
> Astronomy, for miscalculation of an eclipse.
> *The Shu Ching,* 2150 B.C.

For the discussion of the deflection, the retardation, and the redshift of light in the gravitational field of the sun, the linear approximation of Chapter 3 was quite adequate. The same may be said for the gyroscope precession effect, which is another experimental test of gravitational theory that is now in preparation. However, the perihelion precession of planetary orbits in the field of the sun cannot be treated by the linear approximation. We need an exact, or at least more exact, solution of the gravitational field equations. This is the Schwarzschild solution. The importance of this solution goes much beyond its application in the solar system. It is believed that the gravitational collapse of any nonrotating, electrically neutral star will necessarily lead to the Schwarzschild geometry as its final result; even if the initial geometry is asymmetric, any deviations from the symmetric Schwarzschild field will ultimately be radiated away in the form of gravitational waves.

In what follows we will solve the Einstein equations only in the vacuum surrounding the mass distribution. The solution in the interior of a mass distribution can be carried out analytically only in some exceptional cases, such as the case of an ideal incompressible fluid or the case of a fluid without pressure (cloud of dust); these special cases are of little interest in the real world. The solution for the case of matter with a realistic equation of state can only be carried out numerically.

8.1 THE SPHERICALLY SYMMETRIC, STATIC FIELD

We want to find the gravitational field surrounding a spherically symmetric mass distribution at rest. It is obvious that this gravitational field should then also have spherical symmetry. Furthermore, we will require that the field be *static*. The latter requirement goes beyond time independence. To be precise, a field is said to be *static* if it is both time independent and time symmetric, i.e., unchanged by time reversal. If the field is merely time independent, then it is said to be *stationary*.[1] An example from electrodynamics will make the distinction clear: Consider the magnetic field of a loop of wire carrying a steady current. This field is stationary, i.e., time independent. However, the field is not static because under time reversal the current, and hence also the magnetic field, reverses direction.

For a field with spherical symmetry it is natural to use polar coordinates t, r, θ, ϕ. The angular coordinates θ, ϕ are unambiguous, the measurement of these coordinates only depends on our ability to divide a circumference concentric with the origin into equal parts; we can do this even though we do not know the metric. The "radial" coordinate r is ambiguous because we do not yet know its precise relation to the measurement of distance. For a start we will treat r simply as a parameter that identifies different spherical surfaces concentric with the origin. The "time" coordinate t suffers from similar ambiguities. (We will of course insist that as $r \to \infty$ and the space becomes flat, increments in r and t should equal the increments in the true distance and time, respectively).

The static character and the spherical symmetry of the field impose some restrictions on the form of the spacetime interval. To get at these restrictions it is convenient to introduce rectangular coordinates t, x, y, z related to the polar coordinates t, r, θ, ϕ in the usual way. The demand for a static solution forbids terms of the type $dxdt, dydt, dzdt$ in the expression for spacetime interval because these terms change sign when we perform the time reversal $t \to -t$. The possible remaining terms are dt^2 and products formed with dx, dy, dz (taken two at a time).

To meet the demand for spherical symmetry, the interval must be a function of only those combinations of x, y, z, dx, dy, dz that are left invariant by (spatial) rotations. The only such combinations are

$$(x^2 + y^2 + z^2)^{1/2}$$
$$dx^2 + dy^2 + dz^2 \qquad\qquad [1]$$
$$xdx + ydy + zdz$$

or functions of these (the quantities (1) are the only three-dimensional scalar products that can be formed out of the three-dimensional vectors *x, y, z* and *dx, dy, dz*). If we reintroduce the polar coordinates, then the quantities (1) may be written

$$r$$
$$dr^2 + r^2 d\theta^2 + r^2 \sin^2 \theta d\phi^2 \qquad [2]$$
$$rdr$$

and the interval must be of the form

$$ds^2 = A(r)dt^2 - B(r)(dr^2 + r^2 d\theta^2 + r^2 \sin^2 \theta d\phi^2) - C(r)dr^2 \qquad [3]$$

Here $A(r)$, $B(r)$, and $C(r)$ are some unknown functions of r. These functions are to be determined by solving Einstein's differential equations.

It is possible to simplify (3) further by introducing a new radial coordinate:

$$r' = r\sqrt{B(r)}, \qquad \theta' = \theta, \qquad \phi' = \phi, \qquad t' = t \qquad [4]$$

This has the advantage that the terms involving angular displacements reduce to $r'^2 d\theta'^2 + r'^2 \sin^2 \theta' d\phi'^2$. Furthermore, all terms involving dr'^2 may be combined into one and we obtain

$$ds^2 = A'(r')dt'^2 - B'(r')dr'^2 - r'^2 d\theta'^2 - r'^2 \sin^2 \theta' d\phi'^2 \qquad [5]$$

Here A' and B' are some new functions which, if desired, can be expressed in terms of A, B, C.

It is convenient to omit the primes in Eq. (5) and to write the unknown functions A', B' as exponentials:

$$ds^2 = e^{N(r)}dt^2 - e^{L(r)}dr^2 - r^2 d\theta^2 - r^2 \sin^2 \theta d\phi^2 \qquad [6]$$

With $x^0 \equiv t$, $x^1 \equiv r$, $x^2 \equiv \theta$, $x^3 \equiv \phi$, the metric tensor corresponding to (6) is

$$g_{\mu\nu} = \begin{pmatrix} e^N & 0 & 0 & 0 \\ 0 & -e^L & 0 & 0 \\ 0 & 0 & -r^2 & 0 \\ 0 & 0 & 0 & -r^2 \sin^2 \theta \end{pmatrix} \qquad [7]$$

The unknown functions are now $N(r)$ and $L(r)$; we will use Einstein's equations to find them.

The Christoffel symbols for a metric tensor of the form (7) have been given in Chapter 6 (see Eq. (6.33)). We repeat them here for convenience (the primes indicate derivatives; $N' = \partial N/\partial r$, $L' = \partial L/\partial r$):

$$\Gamma^0_{\ 01} = \Gamma^0_{\ 10} = \tfrac{1}{2}N' \qquad \Gamma^2_{\ 12} = \Gamma^2_{\ 21} = 1/r$$
$$\Gamma^1_{\ 00} = \tfrac{1}{2}N'e^{N-L} \qquad \Gamma^2_{\ 33} = -\sin\theta\cos\theta$$
$$\Gamma^1_{\ 11} = \tfrac{1}{2}L' \qquad \Gamma^3_{\ 13} = \Gamma^3_{\ 31} = 1/r \qquad [8]$$
$$\Gamma^1_{\ 22} = -r\,e^{-L} \qquad \Gamma^3_{\ 23} = \Gamma^3_{\ 32} = \cot\theta$$
$$\Gamma^1_{\ 33} = -r\sin^2\theta e^{-L}$$

All other Christoffel symbols are zero.

The Ricci tensor and the curvature invariant may be calculated from (8) (see Eq. (6.91)); only the diagonal components of $R_\mu^{\ \nu}$ are non-zero. The Einstein equations *in vacuo*,

$$R_\mu^{\ \nu} - \tfrac{1}{2}\delta_\mu^{\ \nu}R = 0 \qquad [9]$$

become

$$R_0^{\ 0} - \tfrac{1}{2}R = -e^{-L}\left(\frac{L'}{r} - \frac{1}{r^2}\right) - \frac{1}{r^2} = 0 \qquad [10]$$

$$R_1^{\ 1} - \tfrac{1}{2}R = e^{-L}\left(\frac{N'}{r} + \frac{1}{r^2}\right) - \frac{1}{r^2} = 0 \qquad [11]$$

$$R_2^{\ 2} - \tfrac{1}{2}R = e^{-L}\left(\frac{N''}{2} - \frac{L'N'}{4} + \frac{N'^2}{4} + \frac{N'-L'}{2r}\right) = 0 \qquad [12]$$

$$R_3^{\ 3} - \tfrac{1}{2}R = e^{-L}\left(\frac{N''}{2} - \frac{L'N'}{4} + \frac{N'^2}{4} + \frac{N'-L'}{2r}\right) = 0 \qquad [13]$$

These equations can be integrated very easily. Consider Eq. (10); it can be written

$$e^{-L}(-rL' + 1) = 1 \qquad [14]$$

This has the unique solution

$$e^L = \frac{1}{1 - C/r} \qquad [15]$$

where C is a constant.

EXERCISE 1. Check that Eq. (14) implies (15). (Hint: The substitution $f(r) = e^{-L}$ simplifies the differential equation (14)).

By subtraction of Eq. (10) from (11) we see that

$$L' = -N' \qquad [16]$$

from which

$$L = -N + \text{constant} \qquad [17]$$

But at large distances ($r \to \infty$), the metric tensor (7) must reduce to that of flat spacetime in spherical coordinates. Hence both L and N must tend to zero in this limit and the constant $= 0$,

$$L = -N \qquad [18]$$

Therefore the solution for $N(r)$ is

$$e^{N(r)} = 1 - C/r \qquad [19]$$

As a final step, we must check that (15) and (19) also satisfy the differential equations (12) and (13) which have not been used so far.

EXERCISE 2. Check this.

According to Eqs. (15) and (19) the interval for the spherically symmetric field has the form

$$ds^2 = (1 - C/r)dt^2 - \frac{dr^2}{1 - C/r} - r^2 d\theta^2 - r^2 \sin^2 \theta d\phi^2 \qquad [20]$$

To find the value of C, we want to compare this expression with the result obtained from the linear theory (Eq. (5.29)),

$$ds^2 \simeq \left(1 - \frac{2GM}{r}\right)dt^2 - \left(1 + \frac{2GM}{r}\right)(dx^2 + dy^2 + dz^2) \qquad [21]$$

where M is the central mass.

We cannot compare (20) and (21) directly because although both of these equations are written in polar coordinates, the radial variables are defined in a somewhat different way: the variable r of Eq. (20) is such that the measured circumference of a circle concentric with the origin is $2\pi r$, while the variable r of Eq. (21) is such that the circumference is approximately $2\pi r(1 + GM/r)$.

EXERCISE 3. Show this.

Let us change Eq. (20) by introducing a new coordinate r':

$$r' = \tfrac{1}{2} \sqrt{r^2 - Cr} + \tfrac{1}{2}r - \tfrac{1}{4}C \qquad [22]$$

$$r = r'\left(1 + \frac{C}{4r'}\right)^2 \qquad [23]$$

This transformation gives

$$ds^2 = \left(\frac{1 - C/4r'}{1 + C/4r'}\right)^2 dt^2 - (1 + C/4r')^4 \, (dr'^2 + r'^2 d\theta^2$$

$$+ r'^2 \sin^2 \theta d\phi^2) \quad [24]$$

EXERCISE 4. Derive Eq. (24).

The coordinates used in Eq. (24) are called *isotropic*. In the weak field limit ($r' \to \infty$),

$$ds^2 \simeq \left(1 - \frac{C}{r'}\right) dt^2 - \left(1 + \frac{C}{r'}\right)(dr'^2 + r'^2 d\theta^2 + r'^2 \sin^2 \theta d\phi^2) \quad [25]$$

Obviously this has the same form as (21). By comparison we see that

$$C = 2GM$$

With this, Eq. (20) becomes

$$ds^2 = (1 - 2GM/r)dt^2 - \frac{dr^2}{(1 - 2GM/r)} - r^2 d\theta^2 - r^2 \sin^2 \theta d\phi^2$$

$$[26]$$

This is called the *Schwarzschild solution*.

It must be kept in mind that the mass M is the total mass of the system; the mass-energy contributed by the gravitational fields is to be included in M. It is clear that only if M is the total energy of the system will the principle of equivalence, i.e., $M_I = M_G$, be satisfied. We will not here give the proof that the gravitational mass of the system which produces the field (26) is in fact equal to the inertial mass of the system. Such a proof can be given, but requires some knowledge of the properties of the exact energy-momentum tensors of matter and of gravitation.[2]

An additional remark: the above calculations were, of course, based on the assumption that the cosmological constant is zero. If we want to drop this assumption, then we must replace Eq. (9) by

$$R_\mu{}^\nu - \tfrac{1}{2}\delta_\mu{}^\nu R = -\Lambda \delta_\mu{}^\nu$$

This equation leads to a Schwarzschild solution

$$ds^2 = \left(1 - \frac{2GM}{r} - \frac{\Lambda r^2}{3}\right) dt^2 - \frac{dr^2}{1 - \dfrac{2GM}{r} - \dfrac{\Lambda r^2}{3}} - r^2 d\theta^2$$

$$- r^2 \sin^2 \theta d\phi^2 \quad [27]$$

EXERCISE 5. Prove this.

Note that this metric does not become asymptotically flat as $r \to \infty$. However, since Λ is very small (see Section 7.4), there is a range of radii

$$1/\sqrt{\Lambda} \gg r \gg GM$$

in which the metric is nearly flat. For values of r below this range, the effect of the mass M dominates; for values of r above this range, the effect of the cosmological term dominates. (At very large values of r we must also take into account the large-scale curvature of the universe; see Chapter 10).

   We know from Section 3.5 that in the Newtonian approximation the quantity $(g_{00} - 1)/2$ is to be identified with the Newtonian potential (see Eq. (3.100)):

$$\Phi = (g_{00} - 1)/2$$

According to (27) this gives

$$\Phi = -\frac{GM}{r} - \frac{\Lambda r^2}{6}$$

The second term on the right side of this equation represents a cosmological correction to the Newtonian potential; this term agrees with Eq. (7.47)).

   For the motion of the planets, the cosmological correction is completely insignificant. We will, therefore, assume that $\Lambda = 0$ until further notice.

8.2   THE MOTION OF PLANETS; PERIHELION PRECESSION

The equation of motion of a particle in a gravitational field (see Eq. (6.36)) is

$$\frac{d^2 x^\mu}{d\tau^2} + \Gamma^\mu{}_{\alpha\beta} \frac{dx^\alpha}{d\tau} \frac{dx^\beta}{d\tau} = 0 \qquad [28]$$

The following quantity is a constant of the motion:

$$g_{\mu\nu}\frac{dx^\mu}{d\tau}\frac{dx^\nu}{d\tau} = \text{constant} = 1 \qquad [29]$$

This equation is obviously true by definition of $d\tau$. But we can also use the equation of motion to show explicitly that the left side of (29) is a constant. Thus, (29) can be regarded as a first integral of the equations of motion.

EXERCISE 6. Show that $g_{\mu\nu}(dx^\mu/d\tau)(dx^\nu/d\tau) = $ constant from the equation of motion. (Hint: Write (28) as

$$\frac{d}{d\tau}\left(g_{\mu\nu}\frac{dx^\nu}{d\tau}\right) - \tfrac{1}{2}g_{\alpha\beta,\mu}\frac{dx^\alpha}{d\tau}\frac{dx^\beta}{d\tau} = 0 \qquad [30]$$

and multiply by $dx^\mu/d\tau$.)

For the special case of a planet moving in the Schwarzschild field of the sun, the equations of motion are

$$\ddot{r} + \tfrac{1}{2}N'e^{N-L}(\dot{t})^2 + \tfrac{1}{2}L'(\dot{r})^2 - re^{-L}(\dot{\theta})^2 - r\sin^2\theta e^{-L}(\dot{\phi})^2 = 0 \qquad [31]$$

$$\ddot{\theta} + \frac{2}{r}\dot{r}\dot{\theta} - \cos\theta\sin\theta(\dot{\phi})^2 = 0 \qquad [32]$$

$$\ddot{\phi} + \frac{2}{r}\dot{r}\dot{\phi} + 2\cot\theta\,\dot{\phi}\dot{\theta} = 0 \qquad [33]$$

$$\ddot{t} + N'\dot{r}\dot{t} = 0 \qquad [34]$$

where dots indicate derivatives with respect to τ, $\dot{t} \equiv dt/d\tau$, $\dot{r} = dr/d\tau$, etc.

EXERCISE 7. Derive Eqs. (31)–(34).

We will assume that the orbit is in the $\theta = \pi/2$ plane. Eq. (32) shows that if $\theta = \pi/2$ and $\dot{\theta} = 0$ initially, then $\ddot{\theta} = 0$ and the orbit remains in this plane.

Eqs. (33) and (34) can now be integrated directly, giving

$$\dot{\phi} = A/r^2 \qquad [35]$$

$$\dot{t} = Be^{-N} \qquad [36]$$

where A, B are constants.

Eq. (31) is somewhat harder to solve, but we can bypass it if we use the first integral of the motion given by (29):

$$e^N(\dot{t})^2 - e^L(\dot{r})^2 - r^2(\dot{\theta})^2 - r^2 \sin^2 \theta(\dot{\phi})^2 = 1 \qquad [37]$$

If we insert $\dot{\phi}$ and $\dot{t}$ from (35) and (36), $\theta = \pi/2$, and $e^{-L} = e^N = 1 - 2GM/r$, we obtain

$$\frac{B^2}{1 - 2GM/r} - \frac{(\dot{r})^2}{1 - 2GM/r} - \frac{A^2}{r^2} = 1 \qquad [38]$$

It is convenient to introduce the variable

$$u = \frac{1}{r} \qquad [39]$$

which is also often used in nonrelativistic celestial mechanics. Then

$$\dot{r} = \frac{dr}{d\phi}\dot{\phi} = -\frac{1}{u^2}\frac{du}{d\phi}\frac{A}{r^2} = -\frac{du}{d\phi}A \qquad [40]$$

and Eq. (38) reduces to

$$B^2 - A^2\left(\frac{du}{d\phi}\right)^2 - A^2u^2(1 - 2GMu) = (1 - 2GMu) \qquad [41]$$

By differentiating this with respect to ϕ we obtain a second-order differential equation for the orbit:

$$\frac{d^2u}{d\phi^2} + u - \frac{GM}{A^2} - 3GMu^2 = 0 \qquad [42]$$

This differs from the corresponding orbital equation of Newtonian theory,

$$\frac{d^2u}{d\phi^2} + u - \frac{GM}{A^2} = 0 \qquad [43]$$

by the term $3GMu^2$.

The relativistic correction to planetary motion is extremely small. This can be seen by comparing the second and fourth terms on the left side of Eq. (42). These terms differ by a factor GMu, or, in cgs units,

$$\frac{GM}{rc^2} \qquad [44]$$

Even for the planet Mercury, this is a very small number; with $M = M_\odot \simeq 2.0 \times 10^{33}$ g, $r = 5.5 \times 10^{12}$ cm, Eq. (44) gives

$$\frac{GM}{rc^2} \sim 3 \times 10^{-8} \qquad\qquad [45]$$

Since $3GMu^2$ is small compared to the other terms, it will be sufficient to use a method of successive approximations. We first write down the solution of the Newtonian equation,

$$u = \frac{GM}{A^2}\,[1 + \epsilon \cos(\phi - \phi_0)] \qquad\qquad [46]$$

where ϵ and ϕ_0 are constants. The equation (46) is that of an ellipse with an eccentricity ϵ and a perihelion located at $\phi = \phi_0$.

EXERCISE 8. Check that (46) solves Eq. (43).

If we now replace the small term $3GMu^2$ in Eq. (42) by its Newtonian approximation as given by (46), we obtain

$$\frac{d^2u}{d\phi^2} + u - \frac{GM}{A^2} - \frac{3(GM)^3}{A^4} - \frac{6\epsilon(GM)^3}{A^4} \cos(\phi - \phi_0)$$

$$- 3\epsilon^2 \frac{(GM)^3}{A^4} \cos^2(\phi - \phi_0) = 0 \quad [47]$$

For nearly circular orbits, ϵ is small and we can neglect the term $\propto \epsilon^2$. The term $3(GM)^3/A^4$ can also be ignored; it is equivalent to a change in the constant GM/A^2 and produces no interesting observable effects. We are then left with

$$\frac{d^2u}{d\phi^2} + u - \frac{GM}{A^2} - \frac{6\epsilon(GM)^3}{A^4} \cos(\phi - \phi_0) = 0 \qquad [48]$$

The solution of this differential equation is

$$u = \frac{GM}{A^2}\,[1 + \epsilon \cos(\phi - \phi_0)] + \frac{3\epsilon(GM)^3}{A^4}\,\phi \sin(\phi - \phi_0) \qquad [49]$$

which can also be written

$$u \simeq \frac{GM}{A^2}\left[1 + \epsilon \cos\left(\phi - \phi_0 - \frac{3(GM)^2}{A^2}\,\phi\right)\right] \qquad [50]$$

EXERCISE 9. Check that Eq. (48) has the solution (49). Check that Eq. (50) reduces to (49) if the quantity $(GM)^2\phi/A^2$ is small.

Eq. (50) represents an orbit that is a precessing ellipse. The argument of the cosine changes by 2π when ϕ changes by

$$\Delta\phi = 2\pi \left[1 - \frac{3(GM)^2}{A^2} \right]^{-1} \simeq 2\pi \left[1 + \frac{3(GM)^2}{A^2} \right] \qquad [51]$$

This shows that the angular distance between one perihelion and the next is *larger than* 2π by

$$6\pi \frac{(GM)^2}{A^2} \qquad [52]$$

This quantity gives the angular precession of the perihelion per revolution. Note that the perihelion precesses in the direction of motion, the *perihelion advances* ($\Delta\phi > 2\pi$; see Fig. 8.1).

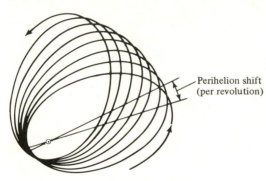

Perihelion shift
(per revolution)

Fig. 8.1 *Planetary orbit with perihelion precession. The angular distance between one perihelion and the next is $\Delta\phi > 2\pi$.*

For a nearly circular orbit Eq. (46) gives

$$\frac{GM}{A^2} \simeq \frac{1}{r} \qquad [53]$$

where r is the radius of the orbit. Then the angular advance of the perihelion per revolution is, in cgs units,

$$6\pi \frac{GM}{rc^2} \qquad [54]$$

For the case of Mercury this gives $\sim 0.10''$ per revolution or $\sim 40''$ per century.

In our calculation we have treated the eccentricity as small. For an orbit of appreciable eccentricity,* a careful calculation shows that the factor $1/r$ in Eq. (54) must be replaced by $\frac{1}{2}(1/r_{max} + 1/r_{min})$. Taking this into account gives a predicted precession of 43.03″ per century.

In an investigation of the motion of Mercury, Clemence[3] took into account all the observations of the planet carried out between 1765 and 1940. The value of the perihelion precession deduced from these ob-

---

* Mercury has $\epsilon = 0.206$ and $r_{max}/r_{min} = 1.52$.

servations is 43.11″ per century. Table 8.1[4] gives the observed and predicted precessions for the innermost planets.

TABLE 8.1 PERIHELION PRECESSION OF PLANETS

	Observed Perihelion Advance	Predicted Perihelion Advance
Mercury	$(43.11 \pm 0.45)''$ per century	$(43.03)''$ per century
Venus	8.4 ± 4.8	8.6
Earth	5.0 ± 1.2	3.8

The entries listed as "observed" perihelion advance have been corrected for the purely Newtonian perturbations of the orbits. Thus, the perturbation of Mercury's orbit by other planets amounts to ~500″ per century. Furthermore, the equinoctial precession of the celestial coordinates used in planetary observations gives an apparent advance of ~5,000″ per century. All these known corrections have been subtracted out in Table 8.1. No correction has been made for a possible quadrupole moment of the sun. The agreement between observed and predicted values is as good as can be expected.

The results of Table 8.1 are based on optical observations extending over very many years. Some new results have recently been obtained by means of interplanetary radar observations. Shapiro et al.[5] succeeded in a determination of the perihelion advance of Mercury after only five years of observation with the Haystack (Massachusetts) and Arecibo (Puerto Rico) radio telescopes. In addition to the perihelion advance, 19 other parameters, such as the orbital parameters of the three innermost planets, the mass of Mercury, the mean equatorial radii of Mercury and Venus, etc., had to be determined from the radar data because they were not known with sufficient precision. The uncertainty in the observed value of the perihelion advance is only ~2%; the observed and predicted values agree to within this uncertainty.

The asteroid Icarus, discovered in 1949, has also been used to test the perihelion advance formula.[6] As of now, this is not a very sensitive test because the value determined by observation has an uncertainty of ~20%. The large uncertainty arises from inaccuracies in the positions of the stars used as reference points in the photographic observations of Icarus.

The precession of the orbit can be quite large in the case of close binary star systems. For a system consisting of two white dwarfs or two neutron stars of mass $1 M_{\odot}$ separated by a distance of 10^{11} cm, Eq. (54) gives a periastron advance of 3×10^{-5} radians per revolution.* In one

---

* Although Eq. (54) was derived for the motion of a small mass in the field of a much larger mass, it will give an order of magnitude estimate even if the two masses are comparable.

year the system goes through 10^3 revolutions, and hence the periastron advance comes to $\sim 2^0$/year.

Of course such a close binary system cannot be resolved optically, and we cannot observe the position of the orbit directly. The best we can do is to measure the orbital velocities by means of Doppler shifts and deduce the perihelion precession from gradual drifts in these Doppler shifts. For example, if the orbit has an appreciable eccentricity, then as the orbit precesses relative to our line of sight the maximum and minimum components of the velocity along the line of sight will gradually change.

In order to detect the periastron advance, we need a binary system in which Doppler shifts can be measured with extreme precision. The recently discovered pulsar PSR 1913+16 appears to meet our requirements.[7] Pulsars are rotating neutron stars which emit radio pulses with a very well defined period. The pulsar PSR 1913+16 emits one pulse every 0.059030 ± 0.000001 seconds. The pulse frequency as measured at the earth displays a Doppler shift which indicates a periodic change in the velocity of the pulsar; the velocity along the line of sight oscillates between 80 km/sec and -320 km/sec with a period of 7.752 hr. This implies that the pulsar is in an orbit around some companion, probably a neutron star or a black hole. The dimension of the orbit is $\sim 10^{11}$ cm, the eccentricity is $\epsilon = 0.61$, and the predicted periastron advance is $4°$/year. Further precise measurements of the arrival times of the pulses should allow us to determine not only the periastron advance, but also the gravitational redshift of the pulses.

8.3 GEODETIC PRECESSION IN THE SCHWARZSCHILD FIELD

Given a direction (unit vector) at some point in spacetime, we can transfer this direction to another point by two alternative procedures. Mathematically, we can parallel transport the unit vector. Physically, we can align the axis of spin of a (small) torque-free gyroscope with the initial vector, and then use the spin as indicator of direction while we bodily transport the gyroscope. If the spacetime curve along which the transport is made is a geodesic curve, and if the gyroscope is *spherical,* then these two alternative operations give the same result. We have seen in Section 6.4 that a parallel transported vector has constant components as seen in a reference frame in free fall along the geodesic. In this reference frame, the spin of a gyroscope also remains constant. In fact, Eq. (6.54) is the equation of motion of a free (except for gravitational forces) particle in this reference frame. The equation of motion of a solid body can be deduced from Eq. (6.54) and is given by Eq. (1.53):

$$\frac{dS^n}{dt} = \epsilon^{knl} R'^k{}_{oso} \left(-I^{ls} + \frac{1}{3} \, \delta_l^s I^{rr} \right) \tag{55}$$

This vanishes identically if, and only if, the gyroscope has the moment of inertia tensor of a sphere ($I^{ls} \propto \delta_i^s$).

A warning: the agreement between parallel transport and gyroscope transport only holds along geodesics. For example, it is well known that if a gyroscope in *flat* spacetime is moved along a nongeodesic curve (accelerated gyroscope), then it will change its direction relative to the fixed stars (Thomas precession) and therefore deviate from the direction given by parallel transport of a vector (in flat spacetime, parallel transport of a vector maintains its direction relative to the fixed stars, i.e., the components remain constant in an inertial rectangular coordinate system).

We know that in curved spacetime, parallel transport does have drastic effects on the direction of a vector and we therefore expect similar effects on a gyroscope. Let us calculate what happens to a spherical gyroscope that is in free fall near the earth. To be specific, let us assume that the gyroscope is in a circular orbit of radius r_0. In the local reference frame that falls freely with the gyroscope the spin does not change, but in any other reference frame it will change. To calculate the rate of change, take a unit vector **n** whose direction coincides with the spin axis in the freely falling geodesic frame. With respect to an arbitrary reference frame the rate of change of this unit vector is that given by Eq. (7.48):

$$\frac{Dn^\mu}{D\tau} = \frac{dn^\mu}{d\tau} + \Gamma^\mu_{\alpha\beta} n^\alpha \frac{dx^\beta}{d\tau} = 0 \qquad [56]$$

i.e.,

$$\frac{dn^\mu}{d\tau} = -\Gamma^\mu_{\alpha\beta} n^\alpha \frac{dx^\beta}{d\tau} \qquad [57]$$

We will evaluate this in a reference frame x'^μ which has the same velocity for one instant as the freely falling frame, but does not rotate. By absence of rotation we mean that the reference frame does not rotate with respect to the fixed stars or, equivalently, that it does not rotate with respect to the r, θ, ϕ-coordinates of the Schwarzschild solution.

Of course, we could also directly evaluate (57) in the Schwarzschild coordinates. But nonrotating coordinates moving with the gyroscope are more relevant than nonrotating coordinates centered on the earth because in practice one will have to place the gyroscope in a satellite and measure the change in the spin, relative to star directions, by means of devices installed *in* the satellite.

In the comoving reference frame x'^μ, we have $dx'^\beta/d\tau = (1,0,0,0)$ and hence the rate of change of the space part of the vector n^μ is

$$\frac{dn'^k}{d\tau} = -\Gamma'^k{}_{l0} n'^l \qquad\qquad [58]$$

The vector n'^k gives a direction in space and hence has no time part in the comoving frame ($n^0 = 0$).*

Fig. 8.2 shows the gyroscope and the unit vector n^k at one instant. The instantaneous velocity in the (nonrotating) reference frame x^μ centered on the earth has a value

$$v \simeq \sqrt{GM/r} \qquad\qquad [59]$$

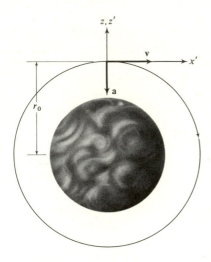

Fig. 8.2 *The comoving reference frame* x' y' z'.

and is along the x-axis; the acceleration has a value

$$a \simeq -GM/r^2 \qquad\qquad [60]$$

and is along the z-axis.

In order to find the coordinate transformation from the reference frame x^μ to the comoving frame x'^μ, let us begin by ignoring the acceleration of the latter. Then we have a simple Lorentz transformation with velocity v along the x-axis. Since v is not very large for an earth satellite, we can ignore terms of order v^2 and write the approximate equations

$$x^0 = x'^0 + vx'^1 \qquad\qquad [61]$$

$$x^1 = x'^1 + vx'^0 \qquad\qquad [62]$$

$$x^2 = x'^2 \qquad\qquad [63]$$

* This is another advantage of the x'^μ-coordinates. In x^μ-coordinates, $n^0 \neq 0$ and the length of the space part of the vector is variable.

$$x^3 = x'^3 + r_0 \qquad\qquad [64]$$

Note that the origin of x'^μ-coordinates is at the radial distance r_0 (Eq. (64)). The x'^μ-coordinates in Eqs. (61)–(64) define an instantaneous co-moving frame, i.e., a frame in which the gyroscope is at rest at one instant. Since we want to calculate a time derivative (see Eq. (58)), we must also define the comoving frame at the next instant. Obviously, in order to avoid spurious rotation of n'^k, the comoving frame at the next instant must have the same orientation as the comoving frame at the initial instant. But it is known from special relativity* that even if successive instantaneous co-moving frames are not rotated relative to one another, they are nevertheless rotated relative to the reference frame x^μ. In the later reference frame, the angle between two instantaneous comoving frames separated by a (small) time interval x'^0 is $\frac{1}{2}vax'^0$, counterclockwise in Fig. 8.2. This is called the *Thomas precession*; it arises from the Lorentz transformation of angles. Since we want to evaluate the precession of the gyroscope in a frame that does not rotate relative to the frame x^μ, we must subtract the Thomas precession of the instantaneous comoving frame. This modifies Eqs. (61)–(64) as follows:

$$x^1 = x'^1 + vx'^0 - \tfrac{1}{2}vax'^0x'^3 \qquad\qquad [65]$$

$$x^3 = x'^3 + r_0 + \tfrac{1}{2}vax'^0x'^1 \qquad\qquad [66]$$

The extra terms in Eqs. (65) and (66) represent an ordinary spatial rotation by an angle $\frac{1}{2}vax'^0$ (to first order in this angle).

We can now use Eqs. (61), (63), (65), and (66) to calculate $g'_{\mu\nu}$ according to the transformation equation $g'_{\mu\nu} = (\partial x^\alpha/\partial x'^\mu)(\partial x^\beta/\partial x'^\nu)\,g_{\alpha\beta}$ where $g_{\alpha\beta}$ is the approximate Schwarzschild metric (21); for the earth, $GM/rc^2 << 1$ and hence this linear approximation to $g_{\alpha\beta}$ suffices. The result is $g'_{\mu\nu} =$

$$
\begin{pmatrix}
1 - 2GM/r - v^2 & -4vGM/r + \tfrac{1}{2}vax'^3 & 0 & -\tfrac{1}{2}vax'^1 \\
4vGM/r + \tfrac{1}{2}vax'^3 & -(1 + 2GM/r) + v^2 & 0 & 0 \\
0 & 0 & -(1 + 2GM/r) & 0 \\
-\tfrac{1}{2}vax'^1 & 0 & 0 & -(1 + 2GM/r)
\end{pmatrix}
$$
$$[70]$$

In this expression terms that involve aGM/r and a^2 have been neglected. According to Eq. (60), such terms are proportional to $(GM/r)^2$; this is a very small quantity. On the other hand, terms of order $va \propto (GM/r)^{3/2}$ have been kept; these terms are somewhat larger.

In Eq. (58) we need all the Christoffel symbols of the type $\Gamma'^k{}_{10}$. These must be calculated from Eq. (70). One finds

$$\Gamma'^1{}_{30} = -\Gamma'^3{}_{10} = -2v\frac{GM}{r^2} - \tfrac{1}{2}va \qquad\qquad [71]$$

*See, for instance, Ref. 1, p. 56.

All other $\Gamma'^k{}_{l0}$ symbols are zero. With Eq. (60),

$$\Gamma''_{30} = -\Gamma'^3{}_{10} = -\frac{3}{2} v \frac{GM}{r^2}$$

The rate of change of n'^k is then

$$\frac{dn'^1}{d\tau} = \frac{3}{2} v \frac{GM}{r^2} n'^3$$

$$\frac{dn'^2}{d\tau} = 0$$

$$\frac{dn'^3}{d\tau} = -\frac{3}{2} v \frac{GM}{r^2} n'^1 \qquad [72]$$

In three-dimensional vector notation this can also be written as

$$\frac{d\mathbf{n}}{d\tau} = \left(\frac{3}{2} \frac{GM}{r^3} \mathbf{r} \times \mathbf{v} \right) \times \mathbf{n} \qquad [73]$$

From Eq. (73) we see that $\mathbf{n}$ precesses about the axis of the orbit with an angular velocity

$$\mathbf{\Omega}_G = \frac{3}{2} \frac{GM}{r^3} \mathbf{r} \times \mathbf{v} \qquad [74]$$

The spin of the gyroscope precesses at the same rate; this is the *geodetic precession* effect.

Although we have derived Eq. (74) for a circular orbit, it may be shown that the result is valid in general. Note that the rate of change of the spin can be written

$$\frac{d\mathbf{S}}{d\tau} = \frac{3}{2} \frac{GM}{mr^3} \mathbf{L} \times \mathbf{S} \qquad [75]$$

where $\mathbf{L} = m\mathbf{r} \times \mathbf{v}$ is the *orbital* angular momentum of the gyroscope. In Newtonian mechanics, we would interpret the right side of Eq. (75) as a torque due to spin-orbit coupling with a potential (in cgs units)

$$U = \frac{3}{2} \frac{GM}{mc^2 r^3} \mathbf{L} \cdot \mathbf{S} \qquad [76]$$

EXERCISE 11. Show that the "torque" on the right side of Eq. (75) is correctly given by $-\partial U/\partial \theta$ (where θ is the angle between $\mathbf{L}$ and $\mathbf{S}$).

A spinning particle with the potential (76) will not only suffer a spin precession, but will also suffer translational accelerations caused by a spin-orbit force. This suggests that a spinning body in a gravitational field will *deviate from geodesic motion*. The conjecture is confirmed by a careful derivation of the translational equation of motion for spinning bodies in geometrodynamics.[8] Although the deviation from geodesic motion is quite small and practically undetectable, it is of some theoretical importance in that it shows that *Galileo's principle of equivalence fails for spinning particles*. For example, electrons, protons, neutrons, etc., do not move exactly on geodesics in a gravitational field.

EXERCISE 12. What is the ratio of $\partial U / \partial r$ to GMm/r^2 for an electron at the surface of the earth?

For a gyroscope in orbit 500 miles above the earth, with the spin in the plane of the orbit (see Fig. 8.3), the geodetic precession (74) amounts to

$$\frac{3}{2} \frac{GM}{c^2 r^2} \sqrt{\frac{GM}{r}} \sim 6.9''/\text{year} \qquad [77]$$

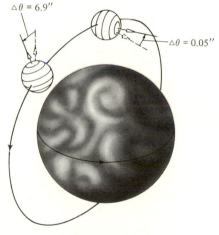

Fig. 8.3 *Two gyroscopes in a 500-mile polar orbit. The first gyroscope, with its spin in the plane of the orbit, suffers (mainly) geodetic precession. The second gyroscope, with its spin perpendicular to the plane of the orbit and also perpendicular to the axis of the earth, suffers exclusively Lense-Thirring precession. The angles give the amount of precession per year. (After Everitt, Fairbank, and Hamilton,* in *Klauder, ed.,* Magic without Magic: John Archibald Wheeler)

An experiment to measure this small effect is now being developed by Fairbank et al. (Stanford University).[9] A spherical gyroscope rotor of quartz ~ 4 cm in diameter placed in a satellite will be used. In order to avoid any extra precession produced by the torque (55), the rotor must be a very precise and homogenous sphere; deviations from spherical shape larger than 10^{-6} cm are unacceptable. One of the advantages of carrying out this experiment in an orbiting satellite is that the electrical suspension of the rotor will operate with next to no friction; in a laboratory experiment on the surface of the earth, the frictional torques on the rotor would be intolerably large. Incidentally, the direction of the instantaneous spin axis will be determined by taking advantage of a superconducting effect. The surface of the quartz sphere will be plated with a thin layer of superconducting material (the entire assembly will be kept at liquid helium temperature). Such a rotating superconductor is endowed with a magnetic moment (London moment) parallel to the spin; a change in the direction of the magnetic moment is easily detected and serves as an indicator of the change of spin.

The experiment is also expected to measure another precession effect caused by the rotation of the earth. The gravitational field of the earth is not exactly a Schwarzschild field; rather, it is the field of a rotating mass (see Eq. (3.168)). If one takes into account the extra terms in the metric associated with the rotation of the earth, one finds an extra precession

$$\mathbf{\Omega}_{LT} = \frac{G}{r^3}\left[\frac{3\mathbf{r}}{r^2}(\mathbf{S}_e \cdot \mathbf{r}) - \mathbf{S}_e\right] \qquad [78]$$

where $\mathbf{S}_e$ is the spin angular momentum of the earth. This is the Lense-Thirring effect.[10]

EXERCISE 13. Repeat the calculations (69)–(72), taking into account the extra off-diagonal terms in the metric (3.168).

In Newtonian terms, the precession (78) may be thought of as due to a spin-spin coupling between the spins of the gyroscope and the earth. For a gyroscope in a polar orbit at 500 miles, the *average* precession rate (average over one orbit) that results from Eq. (78) is $\sim 0.05''/$ year, in the same direction as the earth's rotation.

The Lense-Thirring effect is a manifestation of Mach's principle. It shows that the inertial reference frame of the gyroscope is, to some extent, dragged along by the rotation of the earth. An apparently paradoxical feature of Eq. (78) is that for a gyroscope located over the earth's equator, the drag induces a precession *opposite* to $\mathbf{S}_e$; for $\mathbf{S}_e \cdot \mathbf{r} = 0$, we obtain $\mathbf{\Omega}_{LT} \propto -\mathbf{S}_e$. This has been explained by Schiff[11] in terms of the decrease of the drag effect with distance: for a gyroscope over the equator, the side of the rotor nearest the earth is dragged with the earth more than the side farthest from the earth; hence the induced rotation is opposite to the rotation of the earth. In terms of a very simple picture: we have two adjacent wheels with their rims touching and coupled (weakly) by "friction"; the wheels will then obviously rotate in opposite directions. For a gyroscope over the pole, the precession is of course in the direction of $\mathbf{S}_e$.

As a final remark: the calculations (69)–(72) did not use the exact Schwarzschild solution and could have been carried out in Chapter 3 in the context of the linear theory. But at that stage we did not have the necessary mathematical tools.

8.4 THE BIRKHOFF THEOREM

Our derivation of the Schwarzschild solution started from the assumption that the metric is spherically symmetric and static. The latter assumption

is not really necessary because it is implicitly contained in the former—
it can be shown that any spherically symmetric vacuum solution of
Einstein's equations must be static and must agree with the Schwarzschild solution.* This is *Birkhoff's theorem.* As a consequence of this
theorem, the field produced by a spherically symmetric mass distribution is always the static Schwarzschild field, regardless of whether the
mass is static, collapsing, expanding, or pulsating. (This statement is
based on the implicit assumption that the mass distribution has no (net)
electric charge; if it does, then there will exist electric fields in the
surrounding space and the vacuum condition $T^{\mu\nu} = 0$ will be violated.)

The proof of the theorem proceeds as follows: We have seen that
if the metric is spherically symmetric and static, then it must have the
form given by Eq. (5),

$$ds^2 = A(r)dt^2 - B(r)dr^2 - r^2 d\theta^2 - r^2 \sin^2 \theta d\phi^2 \qquad [79]$$

where we have omitted the primes. If we do not wish to make any assumptions about the static character of the metric, then we will have to
let A and B depend on time. Furthermore, we will have to include an
extra term proportional to *drdt;* this term is not static, but it is spherically
symmetric. Thus, the most general spherically symmetric expression for
the interval is

$$ds^2 = A(r,t)dt^2 - B(r,t)dr^2 - r^2 d\theta^2 - r^2 \sin^2 \theta d\phi^2$$
$$ - 2F(r,t)drdt \quad [80]$$

The term involving *drdt* can be eliminated by a change in the time
coordinate. We introduce a new time coordinate t' such that

$$dt' = (Adt - Fdr)Q \qquad [81]$$

where Q is a function of r and t which is to be chosen so as to make the
right side of Eq. (81) into a perfect differential.

EXERCISE 13. Show that this requires that Q satisfy the differential equation

$$\frac{\partial}{\partial t}(FQ) = \frac{\partial}{\partial r}(AQ)$$

From Eq. (81) we obtain

$$Adt^2 - 2Fdrdt = \frac{1}{Q^2 A} dt'^2 - \frac{F^2}{A} dr^2 \qquad [82]$$

---

* Of course it is possible to change the *form* of any solution by a coordinate transformation.
Therefore, two solutions will be said to agree if they differ by no more than a coordinate
transformation; such solutions are physically identical.

and hence

$$ds^2 = \frac{1}{Q^2 A} dt'^2 - \left(B + \frac{F^2}{A}\right) dr^2 - r^2 d\theta^2 - r^2 \sin^2 \theta d\phi^2 \qquad [83]$$

It is convenient to drop the prime on t and to write the functions as exponentials:

$$ds^2 = e^{N(r,t)} dt^2 - e^{L(r,t)} dr^2 - r^2 d\theta^2 - r^2 \sin^2 \theta d\phi^2 \qquad [84]$$

Note that this differs from Eq. (6) only in that N and L are now functions of both r and t.

The calculation of the curvature tensor for (84) is only slightly more complicated than the corresponding calculation for (6). The resulting Einstein equations are as follows (the primes and dots indicate derivatives with respect to r and t, respectively):

$$R_0{}^0 - \tfrac{1}{2} R = -e^{-L}\left(\frac{L'}{r} - \frac{1}{r^2}\right) - \frac{1}{r^2} = 0 \qquad [85]$$

$$R_1{}^1 - \tfrac{1}{2} R = e^{-L}\left(\frac{N'}{r} + \frac{1}{r^2}\right) - \frac{1}{r^2} = 0 \qquad [86]$$

$$R_2{}^2 - \tfrac{1}{2} R = e^{-L}\left(\frac{N''}{2} - \frac{L'N'}{4} + \frac{N'^2}{4} + \frac{N' - L'}{2r}\right)$$
$$- e^{-N}\left(\frac{\ddot{L}}{2} + \frac{\dot{L}^2}{4} - \frac{\dot{L}\dot{N}}{4}\right) = 0 \quad [87]$$

$$R_3{}^3 - \tfrac{1}{2} R = e^{-L}\left(\frac{N''}{2} - \frac{L'N'}{4} + \frac{N'^2}{4} + \frac{N' - L'}{2r}\right)$$
$$- e^{-N}\left(\frac{\ddot{L}}{2} + \frac{\dot{L}^2}{4} - \frac{\dot{L}\dot{N}}{4}\right) = 0 \quad [88]$$

$$R_0{}^1 = e^{-L}\frac{\dot{L}}{r} = 0 \qquad [89]$$

$$R_1{}^0 = -e^{-N}\frac{\dot{L}}{r} = 0 \qquad [90]$$

Eqs. (89) and (90) imply that $\dot{L} = 0$, so that L is time independent. If this is substituted into Eqs. (85)–(88), these equations become identical to Eqs. (10)–(13). Hence the solution of the equations can be carried out in exactly the same manner as before and we obtain (see Eqs. (14)–(16)).

$$L' = -N' \qquad\qquad [91]$$

The solution of this differential equation is

$$L = -N + h(t) \qquad\qquad [92]$$

where $h(t)$ is an arbitrary function of time.
With L as given by Eq. (15), and

$$e^N = (1 - C/r)e^{-h(t)} \qquad\qquad [93]$$

we therefore have

$$ds^2 = (1 - C/r)e^{-h(t)}dt^2 - \frac{dr^2}{1 - C/r} - r^2 d\theta^2 - r^2 \sin^2 \theta d\phi^2 \qquad [94]$$

This differs from the Schwarzschild solution only by the factor $e^{-h(t)}$ in the first term. This time-dependent factor can be eliminated by a further transformation of the time coordinate. If we use a new coordinate such that

$$dt' = e^{-h(t)/2}dt$$

then the expression (104) becomes identical to (20). The spherically symmetric solution therefore agrees with the Schwarzschild solution to within a coordinate transformation.

It is a corollary of Birkhoff's theorem that a spherically symmetric mass distribution produces no gravitational field inside an empty spherical cavity centered on the mass distribution. This result is of course well known in the Newtonian case (see Exercise 1.4). In the geometro-dynamic case, the Birkhoff theorem guarantees that the solution inside the cavity must be of the form given by Eq. (20). Since an empty cavity cannot contain any singularities, we must take $C = 0$ and hence the spacetime is flat inside the cavity.

FURTHER READING

The general properties of the motion of test particles in a Schwarzschild geometry are described in Rees, Ruffini, and Wheeler, *Black Holes, Gravitational Waves, and Cosmology*, and in Misner, Thorne, and Wheeler, *Gravitation*.

The "interior" Schwarzschild solution (i.e., the solution for the metric inside a spherically symmetric ball of incompressible fluid) is given in Møller, *The Theory of Relativity;* Tolman, *Relativity, Thermodynamics and Cosmology*, and other references. This solution is of little astrophysical relevance.

The theory of the geodetic and Lense-Thirring precessions of a gyroscope is well explained in the article "Motion of a Gyroscope According to Einstein's Theory of Gravitation" by Schiff (see ref. 11). Details on the proposed experimental test will be found in the article "From Quantized Flux in Superconductors to Experiments on Gravitation and Time-Reversal Invariance" by Everitt, Fairbank, and Hamilton (see ref. 9).

REFERENCES

1. C. Møller, *The Theory of Relativity* (Oxford University Press, Oxford, 1952), p. 250.
2. H. C. Ohanian, J. Math. Phys. **14**, 1892 (1973).
3. G. M. Clemence, Rev. Mod. Phys. **19**, 361 (1947).
4. R. L. Duncombe, Astron. J. **61**, 174 (1956).
5. I. I. Shapiro, G. H. Pettengill, M. E. Ash, R. P. Ingalls, D. B. Campbell, and R. B. Dyce, Phys. Rev. Lett. **28**, 1594 (1972).
6. I. I. Shapiro, M. E. Ash, and W. B. Smith, Phys. Rev. Lett. **20**, 1517 (1968); I. I. Shapiro, W. B. Smith, M. E. Ash, and S. Herrick, Astron. J. **76**, 588 (1971).
7. R. A. Hulse and J. H. Taylor, Ap. J. **195**, L51 (1975).
8. A. Papapetrou, Proc. Roy. Soc. London **A209**, 248 (1951); E. Corinaldesi and A. Papapetrou, Proc. Roy. Soc. London **A209**, 259 (1951).
9. C. W. F. Everitt, W. M. Fairbank, and W. O. Hamilton, *in* J. R. Klauder, ed., *Magic Without Magic: John Archibald Wheeler* (Freeman, San Francisco, 1972).
10. J. Lense and H. Thirring, Phys. Zeits. **19**, 156 (1918).
11. L. I. Schiff, Proc. Natl. Acad. Sci. (U.S.) **46**, 871 (1960).

PROBLEMS

1. Consider two concentric coplanar circles, $r =$ constant, in the Schwarzschild geometry. Suppose that the measured lengths of their circumferences are L_1 and L_2. What is the radial coordinate difference Δr between these circles? What is the measured radial distance between them?

   Take two circles around the sun with $L_1 = 2\pi R_\odot$ and $L_2 = 4\pi R_\odot$. By how many centimeters does the measured radial distance between them differ from the result expected in a flat space?

2. Find a formula for the actual measured volume in the spherical shell $r_1 < r < r_2$ in a space with the Schwarzschild metric (26). If the shell is taken around the sun with $r_1 = R_\odot$ and $r_2 = 2R_\odot$, by how many cubic centimeters does the volume differ from the result expected in a flat space? What fractional deviation does this represent?

3. Derive the following *exact* time dilation formula for clocks at rest in the Schwarzschild field:

$$\frac{d\tau_2}{d\tau_1} = \frac{\sqrt{1 - 2GM/r_2}}{\sqrt{1 - 2GM/r_1}}$$

Compare this with the approximate formula given by Eq. (5.20). What is the fractional difference between the time dilations predicted by the exact and approximate formulas if the first clock is at the surface of the sun and the second at infinity? What if the first clock is at the surface of a neutron star?

4. An observer at $r = r_1$ in a Schwarzschild field sends a light signal in the radial direction towards $r = r_2$ (where $r_2 < r_1$). (a) What is the coordinate velocity dr/dt of the signal? (b) Suppose the signal is reflected at $r = r_2$ and returns to r_1. How long, as measured by t-time, does the signal take to return? (c) How long does the signal take to return according to the clock of the observer?

5. Estimate the periastron precession, in degrees per year, for each of the close binary systems listed in Table 4.2.

6. The eccentricity of the orbit of the pulsar PSR 1113+16 is $\epsilon = 0.61$. The predicted periastron advance is 4° per year. Carefully draw the orbit with this eccentricity at $t = 0$ and at $t = 1$ year, superposing the second drawing on the first. How many orbital revolutions intervene between the two orbits that you have drawn? (See Section 8.2 for needed data).

   Assuming that the mass of the pulsar is 0.3 $M_\odot$, that of the companion 1.5 $M_\odot$, and that the semi-major axis of the orbital ellipse is 1.7×10^{11} cm, find the time dilation factor for the pulsar pulses at periastron and at apastron. Find the fractional difference. Take into account both the gravitational potential and the velocity, but ignore the (longitudinal) Doppler shift.

7. The geodesic equation for a light signal is

$$\frac{d^2x^\mu}{d\sigma^2} + \Gamma^\mu{}_{\alpha\beta}\frac{dx^\alpha}{d\sigma}\frac{dx^\beta}{d\sigma} = 0$$

where σ is a parameter defined along the worldline (for a massive particle we would take $\sigma = \tau =$ (proper time), but for a photon the parameter has no physical interpretation). The following is a constant of the motion

$$g_{\mu\nu}\frac{dx^\mu}{d\sigma}\frac{dx^\nu}{d\sigma} = 0$$

Derive the following equation for the orbit of a light ray in the plane $\theta = \pi/2$ in the Schwarzschild field:

$$\frac{d^2u}{d\phi^2} + u - 3GMu^2 = 0$$

where $u = 1/r$. Show that this equation has the solution

$$u = (3GM)^{-1} = \text{constant}$$

This means that a light ray can be in a circular orbit of radius $3GM/c^2$ around a black hole. The spherical surface of this radius is called the *photosphere*.
   Show that the circular orbit is unstable, that is, show that if u is slightly larger (smaller) than $(3GM)^{-1}$, then the "acceleration" $d^2u/d\phi^2$ is positive (negative).

8. A given gravitational field has a spacetime interval (in isotropic coordinates)

$$ds^2 = \left(1 - \frac{2GM}{r'} + \frac{aG^2M^2}{r'^2}\right)dt^2$$

$$- \left(1 + \frac{2GM}{r'} + \frac{bG^2M^2}{r'^2}\right)(dr'^2 + r'^2 \, d\theta^2 + r'^2 \sin^2\theta d\phi^2)$$

where a and b are numerical constants. Making suitable approximations, find the perihelion precession for a planet that moves in this gravitational field. Show that if $a = 2$, then your result agrees with Eq. (54). Show that this is true regardless of the value of b. This proves that in order to obtain the correct perihelion precession, it is sufficient to know the terms of order G^2M^2 in g_{00} and the terms of order GM in the other components of the metric. Transform the approximate (second-order) solution of Problem 3.8 into isotropic coordinates and show that it gives the correct perihelion precession.

9. Show that Eq. (38) can be written as $(\dot{r})^2 + \left(1 - \frac{2GM}{r}\right)\left(1 + \frac{A^2}{r^2}\right) = B^2$ so that

$$`V`(r) = \left(1 - \frac{2GM}{r}\right)\left(1 + \frac{A^2}{r^2}\right)$$

can be regarded as an effective potential for the radial part of the motion. Plot $`V`$ as a function of $r/2GM$ for the case $A/GM = 4$.

A stable circular orbit must correspond to a minimum in the effective potential. Show that the minimum in $`V`$ occurs at

$$r_0 = 6GM/[1 - \sqrt{1 - 12(GM/A)^2}]$$

Show that r_0 is always larger than $6GM$. Hence, conclude that there exists no stable circular orbit with a radius less than $6GM$.

10. Show that for a particle in an orbit in a Schwarzschild field

$$\frac{d\phi}{dt} = \frac{A}{r^2B}\left(1 - \frac{2GM}{r}\right)$$

and show that for a *circular* orbit

$$B^2 = \left(1 - \frac{2GM}{r}\right)\left(1 + \frac{A^2}{r^2}\right)$$

Use these results to prove that a particle in a stable circular orbit of radius r_0 (see Problem 9 for the value of r_0) satisfies Kepler's law

$$\left(\frac{d\phi}{dt}\right)^2 r_0^3 = GM = \text{constant}$$

11. Suppose that the rotor of a gyroscope has the shape of an ellipsoid of revolution with semiaxes a and $b = a(1 + \delta)$ (where $\delta << 1$). If the gyroscope is in orbit around the earth at a height of 500 miles, *estimate* the magnitude of the tidal torques that act on it. For what value of δ will the tidal torque be comparable to the "geodetic torque" given by Eq. (75)?

12. For a gyroscope in a satellite in a circular orbit around the earth at a height of 500 miles, evaluate the average of Ω_{LT} (see Eq. (78)) over one revolution. Do this both for a polar orbit, and for an equatorial orbit. Express your answer in seconds of arc per year. (The moment of inertia of the earth is $I = 0.331MR^2$ where M is the mass and R the equatorial radius.)

13. Find a potential with spin-spin coupling such that the torque produced by this potential gives the precession Ω_{LT} of Eq. (78). Compare the translational force that your potential exerts on an electron at the surface of the earth with the ordinary gravitational force. How do these forces compare at the surface of a neutron star with mass $\sim M_\odot$, radius ~ 10 km, and a rotation period of 0.033 sec?

14. Suppose that initially we have a spherical mass m_1 of radius R_1 surrounded by a Schwarzschild field. Next, a uniform, thin, spherical shell of radius R_2 is placed concentrically around the mass m_1. Given that the total mass of the configuration (spherical mass plus shell) is m_2, find the solution of the Einstein equations in the space $r > R_2$ outside the shell and in the space $R_1 < r < R_2$ between the shell and the spherical mass; use Schwarzschild coordinates so that the solution has the form of Eq. (20). Explain why g_{00} and g_{11} have discontinuities at $r = R_2$. (Hint: What does Eq. (10) say about L' if the right side of this equation is not zero, but instead has a sharp peak (a delta function) representing the energy density T_0^0 of the thin shell?)

　　Show that the discontinuity in g_{00} can be eliminated by changing the time coordinate in the region $R_1 < r < R_2$ by a constant factor.

9. BLACK HOLES AND GRAVITATIONAL COLLAPSE

Abandon all hope ye who enter here.

Dante Alighieri, *The Inferno*

The dimensionless quantity GM/rc^2 may be regarded as a measure of the strength of the gravitational field. This quantity enters into the formulas for light deflection, light retardation, redshift, perihelion precession, etc. The small magnitude of the relativistic gravitational effects in the solar system is related to the small magnitude of this quantity; even at the solar surface, GM/rc^2 is only 2×10^{-6}. Large relativistic effects are found in the gravitational field of a very compact mass, when $GM/rc^2 \lesssim 1$. For example, at a radius $r = 3GM/c^2$ the deflection of light in the Schwarzschild field becomes so large that a light signal will move in a closed circular orbit around the central mass.*

The relativistic effects become spectacular when $r = 2GM/c^2$. The corresponding gravitational fields are so strong that nothing can escape from their grip. Light signals, particles, and even spaceships with arbitrarily powerful engines, are inexorably pulled inwards.

The existence of a "no-escape" radius for light is suggested by the following very naïve calculation based on Newton's gravitational theory:

---

* Unfortunately, the orbit is unstable (see Problem 8.7).

At a radius *r*, the escape velocity for a particle projected upwards is $v = \sqrt{2GM/r}$. Since this expression is independent of the mass of the particle, we may hope that it can be applied to the case of light. For $v = c$, we then obtain a "no-escape" radius of $r = 2GM/c^2$ which, by dumb luck, happens to agree with the result of a careful relativistic calculation. If light cannot escape, nothing can because even spaceships with arbitrarily powerful engines always have velocities below that of light. Although this naïve calculation gives the correct value for the "no-escape" radius, it gives the wrong picture of what happens. Light or particles emitted at or inside the "no-escape" radius do not first rise, then stop, and then fall back. Rather, they fall immediately and never even begin to move in the outward direction.

A region of spacetime into which signals can enter, but from which no signal can ever emerge, is called a black hole.

9.1 SINGULARITIES AND PSEUDOSINGULARITIES

The Schwarzschild solution

$$ds^2 = (1 - 2GM/r)dt^2 - \frac{dr^2}{(1 - 2GM/r)} - r^2 d\theta^2 - r^2 \sin^2 \theta d\phi^2 \quad [1]$$

develops a "singularity" as $r \rightarrow 2GM$. At this point,

$$g_{00} = 1 - 2GM/r \rightarrow 0 \quad\quad [2]$$

and

$$g_{11} = \frac{-1}{1 - 2GM/r} \rightarrow -\infty \quad\quad [3]$$

The quantity (in cgs units)

$$r_S = \frac{2GM}{c^2}$$

is called the *Schwarzschild radius*, or the *gravitational radius* of the mass *M*. For a mass $M = M_\odot = 2.0 \times 10^{33}$ g,

$$r_S = 3.0 \text{ km} \quad\quad [4]$$

If the mass has an actual radius larger than r_S, then (2) and (3) need not concern us since the solution (1) is only applicable in the region exterior to the mass. If an object of $M \sim M_\odot$ is to have a radius smaller than r_S, it must have a density of $\sim 10^{16}$ g/cm³, even larger than that of a neutron star. In fact, as we will see, an object that has collapsed to a radius smaller than r_S is unable to come to equilibrium and will continue to collapse; the gravitational forces are so strong that nothing can resist them.

Note that the critical density at which an object has a size smaller than its Schwarzschild radius r_S decreases with mass. For example, suppose that a "galaxy" consists of 10^{11} stars, each of a mass $\sim M_\odot$, uniformly distributed over a spherical volume. In this case the critical density is only 10^{-6} g/cm³; at this density the typical distance between adjacent stars in the galaxy will be about the same as the earth-sun distance. For now, let us suppose that there is no mass density anywhere (except perhaps at $r = 0$). The Schwarzschild solution is an exact vacuum solution of Einstein's equations and we would like to discuss it as such.

The surface $r = r_S$ is a surface of *infinite redshift* (relative to $r = \infty$). A clock placed (at rest) near $r = r_S$ shows a proper time

$$d\tau = \sqrt{(1 - 2GM/r)} \, dt \qquad [5]$$

which approaches zero as $r \to r_S$, i.e., the clock runs infinitely slow compared to a clock at infinity. This means that if an astronaut near $r = r_S$ sends out light pulses, or signal rockets, with a time interval of 1 sec (as shown by his clock) between one pulse and the next, then an observer at large distance will receive these pulses with a time interval much larger than 1 sec (as shown by *his* clock) between successive pulses.

Eq. (5) gives an infinite time dilation for a clock at *rest* at $r = r_S$. Actually, a clock cannot remain at rest on this surface. The vanishing of $d\tau$ is characteristic of the worldline of a light signal and hence only a light signal (aimed in the outward direction) can remain at rest at $r = r_S$. (The infinite redshift surface is sometimes called the *static limit* because material particles cannot remain at rest on it.)

If r is in the range $r_S > r > 0$, then the Schwarzschild solution is free of singularities. However, in this region

$$g_{00} = (1 - r_S/r) < 0$$

$$g_{11} = \frac{-1}{1 - r_S/r} > 0 \qquad [6]$$

The signs of g_{00} and g_{11} have become the opposite of what is normal. In the region $r < r_S$, t is a *spacelike* coordinate, and r is a *timelike* coordinate. Since the metric is a function of r, and since r now measures the progress of time, it follows that when $r < r_S$, the metric is actually *time dependent*. We should be a bit more careful: the static character of the metric in the region $r > r_S$ is a consequence of our particular choice of coordinates; if we were to use new coordinates x'^μ which, with respect to t-, r-, θ-, ϕ-

coordinates, deform in time, then the metric as seen in these new coordinates would of course be time dependent. However, the assertion that the metric in the *interior* of r_S is not static holds true in all possible coordinates, the reason being that no matter how the new coordinates x'^μ are defined in terms of the old t-, r-, θ-, ϕ-coordinates, an advance of time ($ds^2 > 0$) is impossible unless $dr^2 > 0$; hence advance of time necessarily entails a change in r, and hence a change in the metric.

It is very important to recognize that the Schwarzschild "singularity" at $r = r_S$ is *not a physical singularity*. The "singularity" in Eqs. (2) and (3) is spurious; it arises from the choice of coordinates and can be eliminated by a change of coordinates. An astronaut falling through the horizon will not feel anything unusual. In his immediate vicinity physics will go on in the way it always does in a freely falling spaceship. Of course there will be tidal forces, and these tidal forces will grow stronger and stronger as the astronaut falls deeper and deeper into the hole. However, the tidal forces remain finite at $r = r_S$; only at $r = 0$ do the tidal forces become infinite and indicate the presence of a real physical singularity.

Mathematically, the absence of any local singularity at $r = r_S$ can be seen from the evaluation of the Riemann curvature tensor; it turns out that, even though in Schwarzschild coordinates g_{00} and g_{11} misbehave, all components of $R^\alpha{}_{\beta\mu\nu}$ are finite at $r = r_S$ when calculated in local geodesic coordinates.

For example, consider the component $R^0{}_{101}$. This component is given by (see Eq. (6.90)):

$$R^0{}_{101} = -\tfrac{1}{2}N'' + \tfrac{1}{4}L'N' - \tfrac{1}{4}(N')^2$$

$$= \frac{r_S}{r}\frac{1}{1 - r_S/r} \tag{7}$$

At $r = r_S$, this function is singular. However, the singularity is spurious, just as that in Eq. (3). To recognize this, let $x^\mu = (t_0, r_0, \pi/2, \phi_0)$ be a given point in the equatorial plane and introduce geodesic coordinates x'^μ at this point by means of the transformation

$$
\begin{aligned}
x'^0 &= (t - t_0)\sqrt{1 - r_S/r_0} + \ldots \\
x'^1 &= (r - r_0)/\sqrt{1 - r_S/r_0} + \ldots \\
x'^2 &= r_0(\theta - \pi/2) + \ldots \\
x'^3 &= r_0(\phi - \phi_0) + \ldots
\end{aligned}
\tag{8}
$$

The triple dots in these equations stand for quadratic terms, such as those appearing in Eq. (6.52). These quadratic terms are, of course, crucial to achieve the geodesic coordinate condition $\Gamma'^\alpha{}_{\mu\nu} = 0$; but we are now only interested in the transformation law for the Riemann tensor and this law does not depend on the quadratic terms. The transformation coefficients at $x'^\mu = 0$ are

$$\frac{\partial x^0}{\partial x'^0} = \frac{1}{\sqrt{1 - r_S/r_0}} \qquad \frac{\partial x^1}{\partial x'^1} = \sqrt{1 - r_S/r_0}$$

$$\frac{\partial x^2}{\partial x'^2} = \frac{1}{r_0} \qquad \frac{\partial x^3}{\partial x'^3} = \frac{1}{r_0} \qquad\qquad [9]$$

all other coefficients are zero.

EXERCISE 1. Show that this coordinate transformation changes the Schwarzschild metric (at the point $(t_0, r_0, \pi/2, \phi_0)$) into $g'_{\mu\nu} = \eta_{\mu\nu}$, and therefore eliminates the singularity in the metric.

Note that the coordinate transformations given by (11) are singular if r_0 coincides with r_S. The elimination of the "singularity" of the Schwarzschild solution hinges on the use of a singular coordinate transformation. We will of course adopt the view that the coordinates that go bad at $r = r_S$ are the Schwarzschild coordinates $x^\mu = (t, r, \theta, \phi)$; the geodesic coordinates x'^μ can be given the physical interpretation of belonging to a reference frame in free fall (see Section 7.5) and they must therefore necessarily be good coordinates.

For the transformed Riemann tensor we find

$$R'^0_{\ 101} = \frac{\partial x'^0}{\partial x^\alpha} \frac{\partial x^\beta}{\partial x'^1} \frac{\partial x^\mu}{\partial x'^0} \frac{\partial x^\nu}{\partial x'^1} R^\alpha_{\ \beta\mu\nu}$$

$$= \frac{\partial x'^0}{\partial x^0} \frac{\partial x^1}{\partial x'^1} \frac{\partial x^0}{\partial x'^0} \frac{\partial x^1}{\partial x'^1} R^0_{\ 101}$$

$$= (1 - r_S/r_0) \frac{r_S}{r_0^3} \frac{1}{1 - r_S/r_0}$$

$$= \frac{r_S}{r_0^3} \qquad\qquad [10]$$

Now let r_0 approach r_S; the function given by Eq. (10) remains finite. This shows that in the x'^μ coordinates the $R'^0_{\ 101}$ component of the Riemann tensor is free of singularities. It is easy to check that all other components of the Riemann tensor are also free of singularities. *The tidal force therefore remains finite at $r = r_S$.*

Note that at $r_0 = 0$, the expression (10) diverges. Hence the tidal forces diverge at the "origin" and we have a true singularity. Since near this point r is a timelike coordinate and t a spacelike coordinate, this singularity happens at a given instant of time ($r = 0$) in all of space (at all values of t). Thus the time-dependent, dynamic geometry in the interior region evolves into a singularity and comes to an end. The geometry on the outside of course remains static forever.

In some respects the Schwarzschild pseudosingularity at $r = r_S$ is similar to a pseudosingularity found in a rotating coordinate system in *flat* spacetime. Consider the coordinate transformation from $x^\mu = (t,x,y,z)$ to $x'^\mu = (t,\rho,\phi,z)$ defined by

$$t = t \qquad\qquad [11]$$

$$x = \rho \cos (\phi + \omega t) \qquad\qquad [12]$$

$$y = \rho \sin (\phi + \omega t) \qquad\qquad [13]$$

$$z = z \qquad\qquad [14]$$

The coordinates (t,ρ,ϕ,z) are cylindrical coordinates rotating with angular velocity ω. The metric $g_{\mu\nu} = \eta_{\mu\nu}$ of flat spacetime is transformed into

$$g'_{\mu\nu} = \begin{pmatrix} 1 - \rho^2\omega^2 & 0 & -\rho^2\omega & 0 \\ 0 & -1 & 0 & 0 \\ -\rho^2\omega & 0 & -\rho^2 & 0 \\ 0 & 0 & 0 & -1 \end{pmatrix}$$

with

$$ds^2 = (1 - \rho^2\omega^2)dt^2 - d\rho^2 - \rho^2 d\phi^2 - dz^2 - 2\omega\rho^2 d\phi dt \qquad\qquad [15]$$

EXERCISE 2. Carry out the transformation from $g_{\mu\nu}$ to $g'_{\mu\nu}$.

Obviously the surface

$$\rho = \frac{1}{\omega} \qquad\qquad [16]$$

is a surface of infinite redshift. Although the metric (15) misbehaves at this point, this is entirely the fault of the x'^μ-coordinates; the flat spacetime clearly has no singularity at $\rho = 1/\omega$. The trouble is that a point at rest in x'^μ-coordinates at $\rho = 1/\omega$ is moving at the speed of light with respect to the x^μ-coordinates; the time dilation, and hence the redshift, are infinite. This example shows very clearly that the presence of surfaces of infinite redshift depends on the choice of coordinates.

9.2 THE SCHWARZSCHILD HORIZON

Although the surface $r = r_S$ has no unusual properties of a local kind, it does have some very unusual properties of a global kind. As we will see from a careful analysis of the spacetime geometry, the surface $r = r_S$ is an *event horizon*. By this is meant that from the region $r < r_S$ it is impossible to send signals of any kind to the region $r > r_S$. The surface $r = r_S$ acts as a "one-way membrane": signals can be sent in, but not out. We say that this is a global (or nonlocal) property because in order to test it we must send signals and check what happens to them in the long run.

A region of spacetime enclosed in such an event horizon or "one-way membrane" is called a *black hole* (see Fig. 9.1).

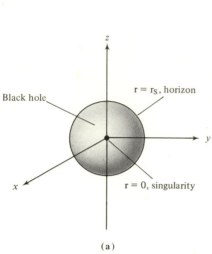

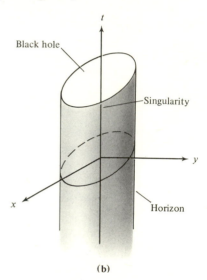

Fig. 9.1(a)   *The Schwarzschild horizon in x-, y-, z-coordinates. The horizon is a sphere of radius r_S.*

Fig. 9.1(b)   *The Schwarzschild horizon in x-, y-, t-coordinates. The horizon is a cylinder of radius r_S. Note that these diagrams are only intended to display coordinates and not true distances.*

The example discussed in the preceding section (see Eq. (15)) shows that the presence of surfaces of infinite redshift depends on the choice of coordinates. On the other hand, the presence or absence of horizons does not depend on the choice of coordinates. This does not mean that at a horizon spacetime possesses a noticeable local singularity. But it does mean that all observers, using whatever coordinates they like, agree on the existence and location of the surfaces across which two-way communication is impossible. Horizons refer to global rather than local properties of the space, but that does not make horizons any less real.

Although the surface of infinite redshift and the horizon coincide in the case of the Schwarzschild solution, there exist other solutions of Einstein's equations for which these surfaces are separated (for example, see the Kerr solution of Section 9.4). If the redshift, for static observers, is infinite on some given surface, this does not by itself necessarily prevent communication across that surface. It could happen that for a sender located *inside* the infinite redshift surface, the redshift, relative to an observer at large distance, is again reduced to a finite value. Also, it should be kept in mind that the total redshift of a signal depends not only on the location of the sender in the gravitational field, but also on the velocity of the sender. For a sender in motion, in some suitable direction, the Doppler shift may compensate for part, or all, of the gravitational redshift. Thus, an "infinite redshift surface" is only defined relative to a special class of observers, with specified motions. On the other hand, a horizon is an absolute feature of spacetime, completely independent of the state of motion of the observers.

To see how signals are cut off, consider a light signal propagating in the radial direction. The velocity of this light signal, with respect to (t,r) coordinates, follows from Eq. (1) by taking $ds^2 = 0$:

$$\frac{dr}{dt} = \pm(1 - r_S/r) \qquad [17]$$

As $r \to r_S$, this coordinate velocity tends to zero. Fig. 9.2 shows the (forward) light cones obtained from Eq. (17) at different values of r. In the exterior of the black hole ($r > r_S$), the axis of the light cones is parallel to the t-axis. In the interior of the black hole ($r < r_S$) the axis of the light cones is parallel to the r-axis. The strange orientation of the light cones in the interior region is a simple consequence of reversal of the character of the coordinates in this region: r is a timelike coordinate and t is a space-like coordinate. Note that in this region the quantity dt/dr gives what we would normally call the "velocity," i.e., the ratio of spacelike increment to timelike increment.

The existence of an horizon at $r = r_S$ is obvious from Fig. 9.2. Signals must necessarily travel in a spacetime direction that lies within the light cone. Since the light cones in the black hole region are oriented towards $r = 0$, any signal in this region is unavoidably pulled towards decreasing values of r and can never leave the black hole.

Although signals cannot emerge from the black hole, they can enter it freely. The curve in Fig. 9.2 is the worldline of a light signal that travels inwards. This curve is obtained by integrating Eq. (17). The signal follows the worldline AB to $t = \infty$, and then it follows the worldline BC to $r = 0$. As measured in (t,r) coordinates, the signal velocity tends to zero as $r \to r_S$ and the signal takes an infinite t-time to reach $r = r_S$. Hence from the point of view of an observer at infinity, whose clocks give t-time, the signal never reaches the horizon.

To make this concrete, suppose that the black hole is surrounded by some dust which scatters a small part of the light signal so that the position of the signal becomes visible to an observer at infinity. Then this observer will see the lighted spot asymptotically approach the Schwarz-schild radius without ever reaching it. This is, in part, a manifestation of

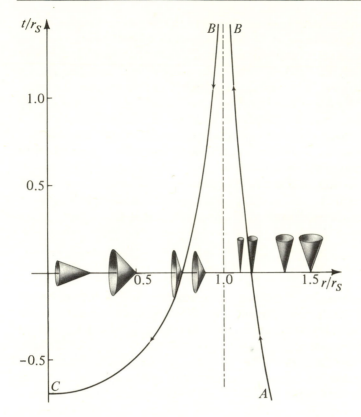

Fig. 9.2 *The forward light cones near and inside the black hole.
As* r → ∞, *the light cone assumes its usual shape, i.e.,*
$dr/dt = \pm 1$. *The curve* **ABBC** *is the worldline of an in-
going light signal.*

the gravitational time dilation effect. From the point of view of an astronaut in free fall in the vicinity of the light signal at $r = r_S$, no such slowing
down takes place; the signal always has the speed of light relative to
him; the signal (and the astronaut!) both cross into the black hole in a
finite proper time.

If the observer at infinity keeps in touch with the astronaut by
radio and television, he finds that the motion and metabolic rate of this
astronaut are slowed down in much the same way as the motion and
vibration rate of the light signal. The astronaut appears to go into "slow
motion" and his movement freezes asymptotically. The last syllable he
sends through his radio before crossing into the black hole is drawn out
to infinite length when received by the outside observer; the words he
speaks after that remain inside the black hole.

Not only will all signals of an astronaut who has entered the black
hole remain trapped, but the astronaut himself is trapped. His worldline
intersects the singularity at $r = 0$ and when he comes near that place, the
infinite tidal forces pull him to bits and pieces; since *all* worldlines within

the astronaut's future light cone terminate on $r = 0$, the collision with the singularity cannot be avoided; not even a spaceship with the most powerful rocket engine can resist the pull of gravity in a black hole. (In fact, the use of rocket engines inside a black hole is not recommended; it can never delay the collision with the singularity, only speed it up! (see Problem 9.4).)

Incidentally, we can also conclude that if a (spherical) star is compressed, by some astrophysical process, to a radius smaller than its Schwarzschild radius, then gravitational collapse necessarily ensues. We need only consider a particle on the surface of the star and remember that according to the Birkhoff theorem the exterior geometry must be that of Schwarzschild. Hence the particle on the surface has the same equation of motion as a spaceship with a rocket motor in an empty Schwarzschild spacetime — we may treat the pressure with which the star pushes outward on the particle as mathematically analogous to the thrust of the rocket motor. The surface of the star must, therefore, necessarily fall towards the singularity $r = 0$, just as the spaceship does (such a surface is said to be a *trapped surface*). Gravitational collapse is inevitable.

In the rest frame of a falling astronaut the amount of proper time needed to enter a black hole and crash into the singularity at $r = 0$ is not only finite, but also quite short. For a typical worldline, the proper time between $r \sim r_S$ and $r = 0$ is of the order of r_S/c. This means that once an astronaut crosses the horizon of a black hole with $M \sim M_\odot$, he only has $\sim 10^{-5}$ sec to live.

Although the Schwarzschild coordinates suffer from the inconvenience of a singularity in the t-time at $r = r_S$ (see Fig. 9.2), this does not affect the calculation of the *proper time* needed to fall into a black hole and into the singularity. Eq. (8.38) gives us $dr/d\tau$ for an astronaut in free fall. For purely radial motion $A = 0$ (see Eq. (8.35)) and hence Eq. (8.38) reduces to

$$\left(\frac{dr}{d\tau}\right)^2 = \frac{r_S}{r} - (1 - B^2)$$

This can be integrated:

$$\tau = \text{constant} + \frac{r[r_S/r - (1 - B^2)]^{1/2}}{1 - B^2}$$

$$+ \frac{r_S}{(1 - B^2)^{3/2}} \tan^{-1}\left[\frac{r_S/r - (1 - B^2)}{1 - B^2}\right]^{1/2} \tag{18}$$

For a finite change in r, the corresponding change in proper time is finite. For example, suppose the astronaut is initially at rest at the radius $r_0 =$

$r_S/(1 - B^2)$; this radius is larger than r_S and hence outside of the black hole. It then only takes him a proper time

$$\Delta\tau = r_S \frac{\pi/2}{(1 - B^2)^{3/2}} = r_S \frac{\pi}{2} \left(\frac{r_0}{r_S}\right)^{3/2} \qquad [19]$$

to fall freely all the way to the radius $r = 0$.

EXERCISE 3. Use Eq. (18) to check this.

If the initial value r_0 is near r_S, then $r_0/r_S \sim 1$ and $\Delta\tau \sim r_S$, as we claimed above.

Although horizons are physical, coordinate-independent properties of a spacetime, it is often easier to recognize the presence, or absence, of a horizon by particular, cleverly chosen coordinates. Thus, the absence of a horizon in flat spacetime is obvious in $x^\mu = (t,x,y,z)$ coordinates, but not so obvious in the rotating x'^μ coordinates used to construct the metric (15). For the case of the Schwarzschild geometry, the coordinates that make things easy are the Kruskal coordinates, which we will describe in the next section.

9.3 THE MAXIMAL SCHWARZSCHILD GEOMETRY

The Schwarzschild coordinates t, r, θ, ϕ are singular at $r = r_S$. There exist several coordinate systems which are well behaved at $r = r_S$. But the price we must pay is that in the new coordinates the metric will not appear static, not even in the exterior region.

A very convenient set of coordinates are the *Kruskal coordinates*[1] u, v, θ, ϕ, defined as follows in the interior of the black hole:

$$u = \sqrt{1 - r/r_S} \; e^{r/2r_S} \sinh t/2r_S \qquad (r < r_S) \qquad [20]$$

$$v = \sqrt{1 - r/r_S} \; e^{r/2r_S} \cosh t/2r_S \qquad [21]$$

For the outside region, the definitions are

$$u = \sqrt{r/r_S - 1} \; e^{r/2r_S} \cosh t/2r_S \qquad (r > r_S) \qquad [22]$$

$$v = \sqrt{r/r_S - 1} \; e^{r/2r_S} \sinh t/2r_S \qquad [23]$$

The inverse transformation is given (implicitly) by

$$(r/r_S - 1) \; e^{r/r_S} = u^2 - v^2 \qquad (r < r_S) \qquad [24]$$

$$t = 2r_S \tanh^{-1} u/v \qquad [25]$$

and

$$(r/r_S - 1)\, e^{r/r_S} = u^2 - v^2 \qquad (r > r_S) \qquad\qquad [26]$$

$$t = 2r_S \tanh^{-1} v/u \qquad\qquad [27]$$

EXERCISE 4. Show that Eqs. (24)–(27) agree with (20)–(23).

The interval in Kruskal coordinates is

$$ds^2 = (4r_S^3/r)\, e^{-r/r_S}\, (dv^2 - du^2) - r^2 d\theta^2 - r^2 \sin^2 \theta d\phi^2 \qquad [28]$$

where r is to be regarded as a function of u, v, as given by Eq. (24).

The coordinate u is spacelike; v is timelike. Note that the new metric has no singularity at $r = r_S$. There is, of course, a singularity at $r = 0$; this is unavoidable since the curvature becomes infinite at that point.

EXERCISE 5. Derive Eq. (28).

Let us now disregard displacements in the θ- and ϕ-directions and concentrate on the radial and time dependence of the geometry. Fig. 9.3 shows spacetime with u-, v-coordinate axes and the corresponding values of r, t. In this diagram r and t are measured in units of r_S; thus, $r = 1.5$ means $r = 1.5r_S$, etc. The curves $r = $ constant are hyperbolas, and the curves $t = $ constant are straight lines through the origin. It is obvious from the diagram that the coordinate transformation from u, v to r, t is singular at $r = r_S$. Thus, the *line u = v* is mapped into the *point r = r_S, t = ∞*. The elimination of the Schwarzschild "singularity" is of course accomplished precisely by the singular character of the coordinate transformation.

One of the nice features of the Kruskal coordinates is that the worldlines of light signals are straight lines at 45° with respect to the u, v-axes.

EXERCISE 6. Show that the worldline of a light signal must have $du = \pm dv$. (Assume $d\theta = d\phi = 0$).

However, the worldlines of other particles are not straight lines. For example, the dashed line in Fig. 9.3 indicates a typical worldline for an astronaut falling into the black hole.

The surface $r = r_S$ divides spacetime into two regions: the black hole (region II) and the asymptotically flat space surrounding it (region I). The existence of an *event horizon* at $r = r_S$ is immediately obvious from Fig. 9.3: a light signal sent out by the astronaut after he crosses $r = r_S$ will proceed at 45° and ultimately hit $r = 0$. The light signal will never reach $r = r_S$; the signal always moves towards *decreasing* values of r. Signals

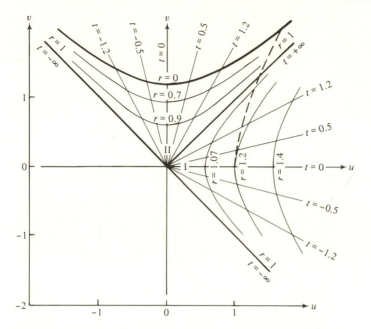

Fig. 9.3 *The Schwarzschild geometry in Kruskal coordinates.*
The u-, v-*coordinates are dimensionless. The* r-, t-
coordinates are given in units of r_S. *The dashed line*
is the worldline of a particle falling towards r = 0.

other than light must necessarily lie within the forward light cone and
will be even less able to escape. Signals from outside the black hole can
of course cross $r = r_S$ with no trouble, i.e., the surface of the black hole
acts as a "one-way membrane."

Note that the singularity $r = 0$ in Fig. 9.3 is *not* a pointlike singularity. Rather, $r = 0$ is a spacelike surface. As seen in u, v-coordinates,
the singularity happens at different points of space at different times. The
singularity happens at the point u when the v-time is

$$v = \sqrt{1 + u^2} \qquad\qquad [29]$$

i.e., at large values of u, the singularity is delayed. We recall that, as seen
in Schwarzschild coordinates, the singularity happens at the same time
($r = 0$) at all points of space ($-\infty < t < \infty$). This difference in behavior
simply reflects a difference in the definition of simultaneity.

The diagram of Fig. 9.3 has been left blank below the line $u = -v$.
It is obvious that the only region that can be explored by an astronaut
who starts his expedition in the exterior (region I) of the black hole is
the region $u > -v$. The region $u < -v$ is necessarily outside of his light
cone. However, the latter region could still have a physical significance

if signals sent by somebody or something in this region cross the line $u = -v$ and reach an astronaut in $u > -v$. Examination of the Kruskal metric shows that in fact the line $u = -v$ is in no way singular and geodesics can cross it without any trouble. This indicates that Fig. 9.3 is incomplete.

A more complete diagram is given in Fig. 9.4. This diagram is as complete as can be. Technically, a manifold is said to be *maximal* if every geodesic is either of infinite length in both directions (has no beginning and no end), or else ends, or begins, on a singularity. If all geodesics are of infinite length, then the manifold is said to be (geodesically) *complete*. In physical terms: in a complete spacetime manifold, every (stable) particle has been in the spacetime forever, and remains in it forever. In a maximal spacetime, particles can be created, or annihilated, only at singularities. We will always demand that spacetime be maximal because it makes no sense at all that the worldline of a particle should suddenly come to an end (i.e., leave or enter our universe) in the middle of nowhere. The spacetime shown in Fig. 9.3 is maximal, but not complete; that shown in Fig. 9.2 is neither maximal nor complete.

Note that several sets of r, t-coordinates are needed to cover all parts of spacetime in Fig. 9.4, but only one set of u-, v-coordinates is enough. The spacetime has *two* singularities $r = 0$; one is at $v = \sqrt{1 + u^2}$,

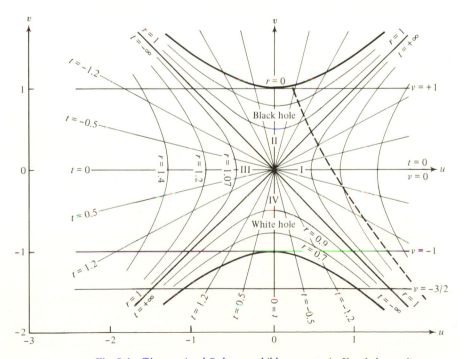

Fig. 9.4 *The maximal Schwarzschild geometry in Kruskal coordinates.*

the other at $v = -\sqrt{1 + u^2}$. The spacetime also has *two asymptotically flat regions,* one at $u \to +\infty$, the other at $u \to -\infty$; these two regions are labeled I and III in Fig. 9.4.

EXERCISE 7. Write down the coordinate transformations from r-, t- to u, v-coordinates that hold in regions III and IV.

To understand what this means, let us look at the time evolution of the geometry using v as our time variable. Several lines $v = $ constant have been drawn through Fig. 9.4. For example, at time $v = -3/2$, we have a space that is asymptotically flat (as $u \to \infty$) with a singularity at $u = \sqrt{5/2}$. At the same time we have a *second* asymptotically flat space $(u \to -\infty)$ with a singularity at $u = -\sqrt{5/2}$. This second space is in a different universe; alternatively, it may be regarded as a region in our own universe if this second region of our universe is very far away from the first. At the time $v = -1$, both of these singularities coalesce and we obtain two asymptotically flat spaces joined together by a region in which the geometry deviates greatly from flatness. Such a region is called a *wormhole.** Whether we regard this wormhole as connecting two universes, or as connecting two remote parts of the same universe, the topology in either case is non-Euclidean. Between the times $v = -1$ and $v = +1$ there is *no singularity,* only a wormhole in which the curvature, although perhaps large, is everywhere finite. At time $v = +1$ the connection between the two asymptotically flat spaces disappears, and we are left with two separate spaces, each with a singularity. These spaces continue to evolve parallel in time. Note that the time evolution described here is perfectly symmetric under time reversal (reversal of v to $-v$).

The second asymptotically flat space (region III) is outside of the forward light cone of the first asymptotically flat space (region I), and conversely. This implies that communication between the two universes is impossible. However, if astronauts from both universes jump into the black hole (region II), then they can meet, embrace, and die together.

We can say that what prevents signals from passing through the wormhole is the time evolution of the interior geometry. The wormhole "pinches off" so fast that not even a light signal can get through before the singularity happens. If we want to interpret the two asymptotically flat regions as remote parts of the same universe, then the impossibility of signal transmission through the wormhole is crucial for the preservation of causality.[2] It is obvious that if remote parts of the universe could communicate via a (short) wormhole, the effective signal velocity would exceed the speed of light.

Region IV is called a *white hole.* According to the above discussion such a white hole may be regarded as the time reverse of a black

---

* Also called an Einstein-Rosen bridge or a Schwarzschild throat.

hole. Signals from region IV can go out to the exterior regions I and III, but no signal from the latter regions can ever be sent into region IV. This must not be interpreted as implying that a white hole is intrinsically luminous. We are dealing with *vacuum solutions* of Einstein's equations; light signals are to be regarded as *test* signals. Light will come out of a white hole only if somebody places a light source in the white hole. There is no reason to suspect that anybody has done this or could do this. The white hole portion of the complete Kruskal solution is probably as irrelevant to our universe as are the advanced solutions of Maxwell's equations. In electromagnetism, the mathematical existence of these solutions is required by the time reversal symmetry of the Maxwell equations, but their physical existence would require some very peculiar boundary conditions at the beginning of the universe (incoming waves that converge on particles and are *exactly absorbed* by them).

So far we have concentrated on the spacetime geometry in the *u*-, *v*- (or *r*-, *t*-) plane. Let us now look at the purely spatial part of the geometry, i.e., the geometry at a given fixed time. We are then dealing with a curved three-dimensional space with coordinates *r* (or *u*), θ, and ϕ. Since it is impossible to make a drawing of three-dimensional curved space, we prefer to look at the geometry in the two-dimensional curved "plane" $\theta = \pi/2$. (Because of spherical symmetry, the geometry is the same in all "planes" passing through the center of symmetry). Fig. 9.5a shows this curved surface at the "time" $v = 0$.* The curved two-dimensional space *r*, ϕ is shown embedded in a three-dimensional space *r*, ϕ, *z*; the *z*-coordinate is an artificial coordinate which has nothing to do with the *z*-coordinate of real space. The shape of the surface has been chosen in such a way that the distances measured along the curved surface with, say, a string, agree with the distances measured in the *r*-, ϕ-plane of the Schwarzschild geometry.

EXERCISE 8. For $v = t = 0$ and $\theta = \pi/2$, the Schwarzschild geometry in the *r*-, ϕ-plane is given by

$$dl^2 = \frac{dr^2}{1 - r_S/r} + r^2 d\phi^2$$

In order to embed this curved two-dimensional surface in flat three-dimensional space, we introduce an extra coordinate *z* and write (*r*, ϕ, *z* are regarded as cylindrical coordinates):

$$dl^2 = dz^2 + dr^2 + r^2 d\phi^2$$

$$= dr^2 \left[1 + \left(\frac{dz}{dr} \right)^2 \right] + r^2 d\phi^2$$

---

* Note that in general the times $v = $ constant and $t = $ constant do not coincide (see Fig. 9.3). The case $v = 0$, $t = 0$ is exceptional; these times do coincide.

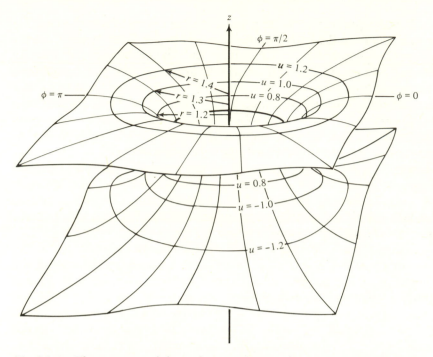

Fig. 9.5(a) *The geometry of the surface* v = 0, θ = π/2 *(or, equivalently,* t = 0, *θ = π/2). Only* u *and* φ *(or, equivalently,* r *and* φ*) vary along this sur-face. The surface has been constructed in such a fashion that the dis-tances measured along the surface with a string directly give the true distances that would be measured in the Schwarzschild geometry. (From Fuller and Wheeler,* Phys. Rev. **128,** *919 (1962).)*

By comparing these two expressions, we obtain the following differential equation for the surface:

$$\left(\frac{dz}{dr}\right)^2 = \frac{1}{1 - r_S/r} - 1$$

Solve this equation to find the shape $z = z(r)$ of the surface of revolution shown in Fig. 9.5a.

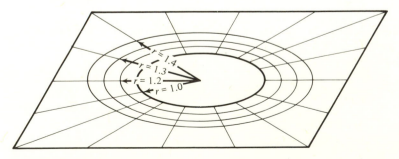

Fig. 9.5(b) *The geometry of the same surface when the mass is zero (flat surface).*

The *upper* half of the figure ($u > 0$) may be thought of as showing the deformation of the flat r-, ϕ-plane which we start out with in the absence of gravitation. To be precise, imagine that the r-, ϕ-plane of flat space (Fig. 9.5b) has the hole $r < 1$ cut out of it (the region $r < 1$ is nonexistent at "time" $v = 0$; in Fig. 9.4, the surface $v = 0$ does not contain values of r smaller than $r = 1$). Then gravitation deforms this plane into the funnel-like structure that is the upper part of the curved surface shown in Fig. 9.5a. The lower part arises from a similar deformation of the r-, ϕ-plane of the second universe.[2]

The two universes of Fig. 9.5a are joined along the circle $0 \leq \phi \leq 2\pi$. If we want to take into account the θ-coordinate, then we must imagine similar circles for different values of θ. This means that the two universes are actually joined along a spherical surface $0 \leq \phi \leq 2\pi$, $0 \leq \theta \leq \pi$ (a "two-sphere").

At later times ($v > 0$) or earlier times ($v < 0$) the "wormhole" is narrower (in Fig. 9.4, the surface $v > 0$ contains values of r smaller than $r = 1$) and at time $v = \pm 1$ the wormhole pinches off at the center and we are left with two separate surfaces, each with a cusp. Fig. 9.6 gives schematic diagrams that take us through a sequence of values from $v < -1$ to $v > 1$. The middle diagram is a cross section through Fig. 9.5a; the other diagrams indicate how the structure shown in Fig. 9.5a evolves.

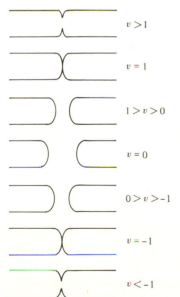

$v > 1$

$v = 1$

$1 > v > 0$

$v = 0$

$0 > v > -1$

$v = -1$

$v < -1$

Fig. 9.6 *Evolution of the Schwarzschild geometry in v-time (schematic).*

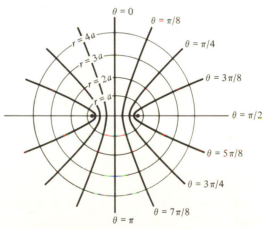

Fig. 9.7 *Spheroidal coordinates in the x-, z-plane. The surfaces $r = $ constant are confocal ellipses with foci at $x = \pm a$; the surfaces $\theta = $ constant are confocal hyperbolas. The spheroidal coordinates in three dimensions are obtained by rotating this pattern about the z-axis.*

Finally, it must be kept in mind that although the Kruskal diagram describes a complete (or at least as complete as possible) solution of Einstein's equations, only a part of this solution will usually be relevant in the real world. For example, in the formation of a black hole by gravitational collapse of a star, the Schwarzschild vacuum solution applies only to those regions of spacetime outside of the star. If the dashed line in Fig. 9.3 is the worldline of the surface of the collapsing star, then the Schwarzschild solution is valid everywhere to the *right* of the dashed line. The region to the *left* of the dashed line is occupied by the matter of the star and is described by some new solution of Einstein's equations with $T^{\mu\nu} \neq 0$; the Kruskal diagram does not apply in this region. Obviously regions III and IV are never of any relevance to the problem of gravitational collapse. Since the complete Kruskal diagram is irrelevant to black holes formed by gravitational collapse, it can only apply to black holes created jointly with the universe, i.e., black holes that have existed ever since the beginning of time.

9.4   THE KERR SOLUTION

In the linear approximation, the stationary solution for the gravitational field surrounding a rotating mass was given in Section 3.10. The corresponding exact solution of the nonlinear Einstein equations is given by the following expression for the interval:[4,5,6]

$$ds^2 = dt^2 - \frac{\rho^2}{\Delta}dr^2 - \rho^2 d\theta^2 - (r^2 + a^2)\sin^2\theta d\phi^2$$

$$- \frac{2GMr}{\rho^2}(dt - a\sin^2\theta d\phi)^2 \quad [30]$$

where ρ^2, Δ are functions of r and θ,

$$\rho^2 \equiv r^2 + a^2\cos^2\theta \qquad\qquad [31]$$

$$\Delta \equiv r^2 - 2GMr + a^2 \qquad\qquad [32]$$

and a, M are constants. This is known as the *Kerr solution*. It is straightforward, and very tedious, to verify that Einstein's equations *in vacuo* are satisfied by this metric.

The Kerr solution is stationary, but not static, that is, the metric is not a function of t, but fails to be invariant under time reversal because of the presence of a term $\sim d\phi dt$. The solution is rotationally symmetric about the z-axis, i.e., invariant under the transformation $\phi \rightarrow \phi + \Delta\phi$.

In the limiting case of large r, Eq. (30) reduces to

$$ds^2 \simeq \left(1 - \frac{2GM}{r}\right) dt^2 - \left(1 + \frac{2GM}{r}\right) dr^2 - r^2 d\theta^2$$

$$- r^2 \sin^2 \theta d\phi^2 - \frac{2GMa^2}{r} \sin^4 \theta d\phi^2 + \frac{4GMa}{r} \sin^2 \theta d\phi dt \quad [33]$$

In this expression all terms $\sim 1/r$ have been kept and all terms $\sim 1/r^2$ omitted. The next to last term on the right side of Eq. (33) can be neglected because it is smaller by a factor $2GMa^2 \sin^2 \theta/r^3$ than the term that precedes it. To recognize the last term, we express it in rectangular coordinates. With $x = r \sin\theta \cos\phi$, $y = r \sin\theta \sin\phi$, we obtain

$$r^2 \sin^2 \theta d\phi = xdy - ydx \quad [34]$$

and therefore Eq. (33) is the same as

$$ds^2 \simeq \left(1 - \frac{2GM}{r}\right) dt^2 - \left(1 + \frac{2GM}{r}\right) dr^2 - r^2 d\theta^2$$

$$- r^2 \sin^2 \theta d\phi^2 + \frac{4GMa}{r^3} (xdy - ydx)dt \quad [35]$$

The first four terms in Eq. (35) are exactly the same as in the Schwarzschild case and permit us to identify M as the total mass of the system. The last term corresponds to an off-diagonal component in the metric,

$$g_{01} = g_{10} = -2GMa \frac{y}{r^3}$$

$$g_{02} = g_{20} = +2GMa \frac{x}{r^3} \quad [36]$$

in rectangular coordinates. Comparison with the weak field solution for a rotating mass (Eq. (3.168)) shows that the quantity Ma is to be identified as the spin angular momentum of the system,

$$S_z = Ma \quad [37]$$

Hence the parameter a is the angular momentum per unit mass. If $a > 0$ ($a < 0$), then the direction of rotation is counterclockwise (clockwise) around the $+z$ axis. Note that in Eq. (30) the time reversal $t \rightarrow -t$ has exactly the same effect as the change $a \rightarrow -a$; this simply corresponds to reversal of the direction of rotation.

The Kerr solution is not the only possible solution for the field surrounding a rotating mass. In general, such solutions will depend on the exact shape of the mass; in the limiting case of large r, different mass distributions will generate different gravitational multipole moments and these will show up in the metric. However, the Kerr solution is of crucial importance because it is the only possible stationary solution for a rotating black hole, i.e., for a rotating system that has collapsed inside its horizon. The gravitational collapse of a rotating, electrically neutral star will ultimately lead to the Kerr geometry.

The coordinates t, r, θ, ϕ used in Eq. (30) are Schwarzschild-like. In fact, in the limit $a \to 0$, the expression (30) for the Kerr solution agrees exactly with the expression (1) for the Schwarzschild solution.

EXERCISE 9. Check this.

However, if $a \neq 0$, then the coordinates t, r, θ, ϕ are not spherical co-ordinates, but *quasi-spheroidal* coordinates. These coordinates are defined in terms of rectangular coordinates by[4]

$$x = (r^2 + a^2)^{1/2} \sin\theta \cos [\phi - f(r)] \tag{38}$$

$$y = (r^2 + a^2)^{1/2} \sin\theta \sin [\phi - f(r)] \tag{39}$$

$$z = r \cos \theta \tag{40}$$

where*

$$f(r) \equiv a \int_\infty^r \frac{dr}{r^2 - 2GMr + a^2} + \tan^{-1} \frac{a}{r} \tag{41}$$

If $M = 0$, then $f(r) \equiv 0$ and the quasi-spheroidal coordinates defined by Eqs. (38)–(40) become ordinary spheroidal coordinates. In the limit $r \to \infty$, the coordinates r, θ, ϕ approach spherical coordinates.

Plotted in the rectangular axes x, y, z, the surfaces $r = $ constant are confocal ellipsoids, and the surfaces $\theta = $ constant are hyperboloids of one sheet. The surfaces $\phi = $ constant are rather complicated "bent planes" which gradually become flat at large values of r (in ordinary spheroidal coordinates, these surfaces are ordinary planes).[5] Fig. 9.7 (p. 323) shows the lines $r = $ constant and $\theta = $ constant in the x-, z-plane. At large values of r, the ellipses tend to circles and the hyperbolas straighten out to give us ordinary polar coordinates. Note that Fig. 9.7 adopts the convention that if the upper part of a given hyperbola is characterized by the parameter θ, then the lower part is characterized by $\pi - \theta$. This makes the correspondence between spheroidal and spherical coordinates more obvious; the discontinuity in θ at $r = 0$ is of course a purely mathematical device with no physical implications whatsoever. Finally, it

---

* The integral appearing in Eq. (41) may be done explicitly, but the result depends on whether $GM > a$ or $GM < a$.

must be emphasized that Fig. 9.7 is only intended to display the relationship between x-, y-, z- and r-, θ-, ϕ-coordinates; this figure does not in any way display the geometry of the curved space.

9.5 HORIZONS AND SINGULARITIES IN THE ROTATING BLACK HOLE

The Kerr geometry is much more complicated than the Schwarzschild geometry, and depends drastically on whether $GM > |a|$, $GM = |a|$, or $GM < |a|$. We will first discuss the case $GM > |a|$. One of the least serious complications of the Kerr geometry is the presence of an infinite redshift surface when $g_{00} = 0$. According to Eq. (30) this corresponds to

$$g_{00} = 1 - \frac{2GMr}{\rho^2} = 0 \tag{42}$$

i.e.,

$$r^2 + a^2 \cos^2 \theta - 2GMr = 0$$

$$r = GM \pm \sqrt{G^2M^2 - a^2 \cos^2 \theta} \tag{43}$$

There are two distinct infinite redshift surfaces; they have been drawn (in quasi-spheroidal coordinates) in Fig. 9.8. As in the Schwarzschild case, these surfaces do not correspond to any physical singularity. The vanishing of g_{00} only tells us that a particle cannot be at rest ($dr = d\theta = d\phi = 0$) at these surfaces; only a light signal emitted in the radial direction can be at rest.

In the region between these two surfaces, the value of g_{00} is negative. Hence t is not a timelike coordinate and the t independence of the metric does not necessarily imply that the geometry is truly time independent. At a radius $r < GM - \sqrt{G^2M^2 - a^2 \cos^2 \theta}$, the coordinate t again reverts to the character of a time coordinate.

In the Kerr geometry, the infinite redshift surfaces do not coincide with event horizons. We can calculate the (coordinate) velocity of a light signal by setting $ds^2 = 0$ in Eq. (30). For a signal in the equatorial plane ($\theta = \pi/2$, $d\theta = 0$) we obtain

$$\left(\frac{dr}{dt}\right)^2 = \frac{\Delta}{\rho^2} \left\{ 1 - (r^2 + a^2) \left[1 + \frac{2GMra^2}{\rho^2(r^2 + a^2)} \right] \left(\frac{d\phi}{dt}\right)^2 \right.$$
$$\left. - \frac{2GMr}{\rho^2} + \frac{4GMr}{\rho^2} a \frac{d\phi}{dt} \right\} \tag{44}$$

If we evaluate this on the infinite redshift surfaces,

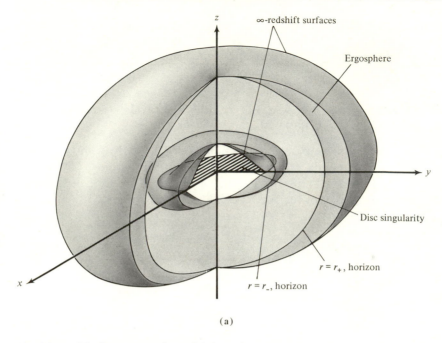

(a)

Fig. 9.8(a)   *The horizons and ∞-redshift surfaces of a rotating black hole*
with **GM** = *1.2a represented in* x-, y-, z-*coordinates. The horizons*
r = r₊ *and* r = r₋ *are ellipsoids.*

$$\left(\frac{dr}{dt}\right)^2 = \frac{\Delta}{2GMr}\left\{2a\frac{d\phi}{dt} - (r^2 + a^2)\left[1 + \frac{a^2}{(r^2 + a^2)}\right]\left(\frac{d\phi}{dt}\right)^2\right\} \qquad [45]$$

where it is understood that r has one of the values given by Eq. (43).
For a signal in the radial direction $d\phi/dt = 0$, and hence $dr/dt = 0$; such
a signal cannot escape.

For a light signal in the *x-, y-*plane, with $d\phi/dt \neq 0$, Eq. (45) does
not determine dr/dt as a function of r. The velocity of the signal must then

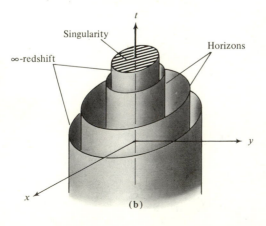

(b)

Fig. 9.8(b)   *The horizons and ∞-*
redshift surfaces of a
rotating   black   hole
represented in x-, y-, t-
coordinates.

be calculated by integrating the geodesic equation so as to obtain dr/dt and $d\phi/dt$. Eq. (45) is a first integral of the equation of motion and may be regarded as a constraint on the initial velocities. It is the *only* constraint. Hence, we can arbitrarily assign dr/dt a positive initial value and adjust $d\phi/dt$ so as to satisfy Eq. (45) (note that excessively large values of dr/dt are not acceptable because we are dealing with a quadratic equation for $d\phi/dt$, and must obtain a *real* solution). Obviously a positive value of $(ad\phi/dt)$ is required. This implies that signals can escape from the infinite redshift surfaces if sent out with a component of velocity in the direction of rotation of the Kerr field.

Where are the event horizons? Their location is given by the condition that $dr/dt = 0$ for arbitrary values of $d\phi$ and $d\theta$. If we set $ds^2 = 0$ in Eq. (30) and solve for $(dr/dt)^2$, we obtain an expression similar to Eq. (44), but with some extra terms involving θ and $d\theta/dt$. In any case, Δ appears as an overall factor on the right side, just as it does in Eq. (44). Hence the condition $dr/dt = 0$ becomes

$$\Delta = 0 \tag{46}$$

i.e.,

$$r^2 - 2GMr + a^2 = 0$$
$$r = GM \pm \sqrt{G^2M^2 - a^2} \tag{47}$$

It is customary to define

$$r_+ \equiv GM + \sqrt{G^2M^2 - a^2} \tag{48}$$

$$r_- \equiv GM - \sqrt{G^2M^2 - a^2} \tag{49}$$

At $r = r_+$ and $r = r_-$ a light signal necessarily has zero velocity in the radial direction, the light cones lie along the surfaces $r = $ constant, and light cannot escape from these surfaces. These horizons are shown in Fig. 9.8; at the poles ($\theta = 0$, $\theta = \pi$), the infinite redshift surfaces meet the horizons. The one-way membrane $r = r_+$ defines the outer boundary of the rotating black hole. The one-way membrane $r = r_-$ is a second, inner boundary.

In the region between the two event horizons, $g_{11} > 0$ and hence r is a timelike coordinate. As in the Schwarzschild case, the r dependence of the metric therefore implies that the metric in this region is necessarily dynamic.

Fig. 9.9 shows the light cones at different values of the radius;* this figure should be compared with Fig. 9.2 for the Schwarzschild case.

---

* The values of GM/a used in Fig. 9.9 have no special significance; they have been picked for convenience in plotting.

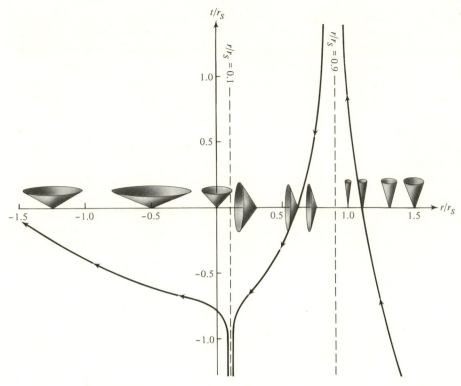

Fig. 9.9(a)   *The forward light cones on the axis of rotation near and inside a rotating black hole with* GM = 1.666 a. *The curve is the worldline of an ingoing light signal.*

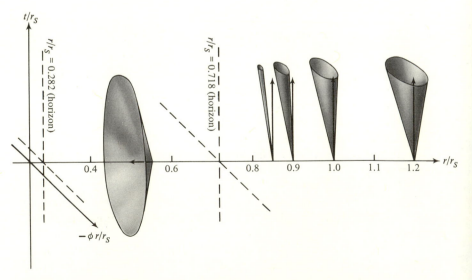

Fig. 9.9(b)   *The forward light cones in the equatorial plane near and inside a rotating black hole with* GM = 2.222 a. *The rotation is counter-clockwise.*

Neither at the infinite redshift surfaces nor at the horizons does the Kerr geometry develop a local singularity; in freely falling geodesic coordinates, the curvature remains finite. The only true singularity occurs at $r = 0$. At this location the curvature tensor diverges.

To understand how this singularity arises, let us examine g_{00} in the vicinity of $r = 0$. According to Eq. (30) we have

$$g_{00} = 1 - \frac{2GMr}{r^2 + a^2 \cos^2 \theta} \qquad [50]$$

In Fig. 9.10, g_{00} is plotted as a function of $z = r \cos \theta$ for the special case $\theta = 0$ (points on the positive z-axis) and $\theta = \pi$ (points on the negative z-axis). Note that at $r = 0$, $\partial g_{00} / \partial r$ is not zero. This implies that the value of $\partial g_{00} / \partial z$ depends on how we approach the point $z = 0$. If we approach from above ($z > 0$), then $\partial g_{00} / \partial z < 0$; if we approach from below ($z < 0$), then $\partial g_{00} / \partial z > 0$. Thus, $\partial g_{00} / \partial z$ has a discontinuity at $z = 0$. It follows that the second derivative $\partial^2 g_{00} / \partial z^2$ does not exist at $z = 0$; more precisely, the second derivative diverges at $z = 0$.[*] The behavior of the second derivatives of g_{11}, g_{33}, and g_{03} is similar. Since the curvature tensor involves these second derivatives, we can conclude that the curvature tensor is singular at $r = 0$. This singularity is a physical singularity which cannot be blamed on our choice of coordinates; the spheroidal coordinates r, θ, ϕ and the metric (30) are perfectly regular when $r = 0$ (except when $\theta = \pi/2$).

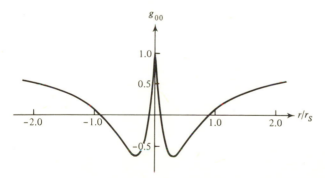

Fig. 9.10 *The function g_{00} evaluated along the z-axis.*

Note that in our spheroidal coordinates, $r = 0$ is a disc centered on the origin. This can be seen from Eqs. (38)–(40); in rectangular coordinates $r = 0$ corresponds to $x^2 + y^2 = a^2 \sin^2 \theta$, $z = 0$. Since θ varies between 0 and $\pi/2$, this is equivalent to $x^2 + y^2 \leq a^2$, $z = 0$. The edge of the disc is at $x^2 + y^2 = a^2$, i.e., it is at the foci of the ellipses in Fig. 9.7. If we take the limit $a \to 0$, then the disc singularity shrinks and ultimately reduces to the Schwarzschild singularity $r = 0$.

---

[*] The second derivative at $z = 0$ behaves as a delta function.

What is the physical meaning of the disc singularity? Einstein's equations tell us that an infinite value of the curvature and of $R_{\mu\nu} - \frac{1}{2}g_{\mu\nu}R$ corresponds to an infinite value of the energy-momentum density. Hence, the disc $r = 0$ contains an infinite density of matter. The Kerr solution can then be interpreted as giving the gravitational fields in the space surrounding a thin (rotating) disc of matter. The Kerr solution also has an alternative interpretation without such a disc of matter, but this interpretation requires that our asymptotically flat space be joined, at the disc $r = 0$, to another asymptotically flat space. This extension of the Kerr geometry will be presented in Section 9.6.

The region between the outer surface of infinite redshift and the outer horizon is called the *ergosphere*[6] (see Fig. 9.8). A procedure discovered by Penrose[7] makes it possible to extract energy from the rotating black hole by means of particles that are moved in and out of the ergosphere. The key to this procedure is that some orbits in the ergosphere have a *negative* total energy, i.e., the gravitational binding energy exceeds the sum of rest mass and kinetic energies. We recall that according to Eq. (3.86) the energy is given by

$$P_o = \frac{\partial L}{\partial u^0}$$

With $L = \sqrt{g_{\mu\nu}u^\mu u^\nu}\, m$ we obtain

$$P_o = \frac{m}{\sqrt{g_{\mu\nu}u^\mu u^\nu}}\, g_{o\alpha}u^\alpha = mg_{o\alpha}u^\alpha \qquad [51]$$

This energy is conserved since the gravitational field is independent of x^0 (see Eq. (3.79)). With the values of $g_{o\mu}$ of Eq. (30),

$$P_o = m(g_{00}u^0 + g_{03}u^3) \qquad [52]$$

In the ergosphere $g_{00} < 0$ and $u^0 > 0$ (for the latter, see Fig. 9.9b and note that u^μ must be in the forward light cone). Furthermore, $g_{03} > 0$ and $u^3 > 0$ (see Fig. 9.9b). Hence the two terms in Eq. (52) are of opposite sign and the energy will be negative if the first term dominates.

EXERCISE 10. Find a set of values for the components of u^μ such that $u^\mu u_\mu = 1$ and $P_0 < 0$. Assume that $\theta = \pi/2$ and $2GM > r > r_+$.

To extract energy from a black hole, proceed as follows: Send a particle (with positive energy) from infinity into the ergosphere and there make the particle split into two pieces in such a way that one piece moves with negative energy.* Since the total energy is conserved,

---

* The splitting of the particle and the injection of one piece into an orbit of negative energy will require some expenditure of internal energy; we ignore this.

the other piece will then have a larger positive energy. This piece carries its energy out to infinity and completes the extraction process. Where does the energy come from? One part of the energy is simply the rest mass and kinetic energy of the piece that was ejected in the ergosphere; the remainder comes from the rotational kinetic energy of the black hole, i.e., the total mass-energy of the black hole decreases.

It was shown by Christodolou and Ruffini[8] that there is an upper limit on the amount of energy that can be extracted from a black hole. The limit is

$$\Delta M = M - M_{\text{ir}} \qquad [53]$$

where M_{ir} is the *irreducible mass:*

$$M_{\text{ir}} = \frac{1}{2G} [(GM + \sqrt{G^2M^2 - a^2})^2 + a^2]^{1/2} \qquad [54]$$

The upper limit given by Eq. (53) corresponds to a complete extraction of the rotational energy; one is then left with a Schwarzschild black hole of mass M_{ir}.

These results follow directly from Hawking's theorem: *In any process involving one or more black holes, the total area of the horizon surfaces can never decrease.*[9] The proof of this theorem uses certain global topological techniques that are beyond the scope of our discussion. Let us forget about the proof, but use the theorem. In the case of the Kerr black hole, the area of the horizon $r = r_+$ is

$$A = 8\pi GM(GM + \sqrt{G^2M^2 - a^2}) \qquad [55]$$

and hence any changes in M and a are subject to the constraint that the expression (55) must not decrease.

EXERCISE 11. Show that Eq. (55) gives the area of the surface $r = r_+$. (Hint: The area is given by the integral $\int\int \sqrt{g_{22}} \sqrt{g_{33}} \, d\theta d\phi$, where g_{22} and g_{33} are evaluated at $r = r_+$.)

It is easy to show that for arbitrary (infinitesimal) changes δM and δa, the corresponding changes in A and M_{ir} are related as follows:

$$\delta M_{\text{ir}}^2 = \frac{1}{16\pi G^2} \delta A \qquad [56]$$

EXERCISE 12. Derive Eq. (56).

Since Hawking's theorem tells us that $\delta A \geq 0$, it follows that $\delta M_{\text{ir}} \geq 0$. The black hole mass M can be written in the form

$$M^2 = M_{\text{ir}}^2 + S^2/(4GM_{\text{ir}})^2 \qquad [57]$$

where $S = aM$ is the spin of the black hole. Since M_{ir} must either remain constant or increase, any decrease in M must come from a decrease of S and the minimum attainable value of M is M_{ir}.

Energy extraction via the Penrose process is not likely to occur spontaneously in nature. However, there exists a quantum process which does occur spontaneously and which may cause a black hole to lose energy. This process involves the spontaneous creation of particle-antiparticle pairs by quantum fluctuations in the vacuum. Normally, the creation of pairs is forbidden by energy conservation. But in the ergosphere one member of the pair can enter an orbit of negative energy and the other an orbit of positive energy in such a way that the total energy is conserved. The former then falls into the black hole while the latter escapes to infinity. The net result is that particles steadily stream out of the ergosphere while the energy of the black hole steadily decreases. Note that the energy loss of the black hole is not due to particles emerging from the horizon; rather, it is due to the capture of negative energy particles that originate in the ergosphere. The rate of energy emission turns out to be small if the mass of the black hole is large. For any black hole of a mass larger than $\sim 10^{16}$ g, the energy loss is insignificant even over a time of 10^{10} years (the age of the universe). Hence this quantum process is only of interest in the case of black mini-holes of mass below 10^{16} g; the corresponding Schwarzschild radius is less than 10^{-12} cm. Perhaps such black mini-holes were created when the universe began.[27,29]

The above mentioned quantum process takes place in the ergosphere. Since a nonrotating black hole has no ergosphere, particles cannot be created by the above process in such a black hole. According to Hawking,[30] there is an additional quantum process which can occur equally well in rotating and nonrotating black holes. This process creates particles from the time-dependent gravitational fields associated with the gravitational collapse that, at some time in the past, formed the black hole. Remarkably, the particles continue to emerge at a constant rate even a long time after the collapse. The energy spectrum of the emerging particles (photons, neutrinos, electrons, etc.) is that of a blackbody radiator with a temperature $T \simeq (M_\odot/M) \times 10^{-6}$ °K. The temperature decreases as the mass increases and is very low if the mass is large. Hence, this process is again only of importance for black mini-holes.

9.6 THE MAXIMAL KERR GEOMETRY

The Kerr geometry has some absolutely horrible features which emerge when one tries to trace worldlines through it. Take an astronaut encapsulated in a spaceship with powerful rocket motors and send him into a rotating black hole. If his rocket motors are strong enough, he can travel on any worldline he chooses, subject only to the requirement that the worldline be timelike ($ds^2 > 0$) and in the forward light cone. Assuming that the motion is along the axis of rotation of the black hole ($\theta = 0$), a possible worldline for the astronaut is shown in Fig. 9.11. The worldline begins outside of the black hole, crosses the horizons $r = r_+$ and $r = r_-$, and approaches near $r = 0$. After a short rest at this position, the astronaut moves in the outward direction and approaches the surface $r = r_-$ from the inside. He reaches this surface in a finite proper time. Then disaster strikes: the astronaut cannot continue across $r = r_-$ because this is a one-way membrane. Physically, we cannot accept that the worldline of the astronaut simply comes to an end — there is no singularity at the horizon. To avoid this geodesic incompleteness, we have only one alternative: we must suppose that upon crossing $r = r_-$ in the outward direction, the astronaut finds himself in another universe in which the surface $r = r_-$ belongs to a *white hole* rather than a black hole. Thus, in this second universe the surface is still a one-way membrane, but it now is one-way-out rather than one-way-in. Of course, the white hole must have the same mass and spin as the black hole. The astronaut then continues towards $r = r_+$, which is to be regarded as another one-way-out membrane, and finally emerges from the white hole and reaches the asymptotically flat regions of the second universe.

If two universes are bad, worse is to come. In a universe with a

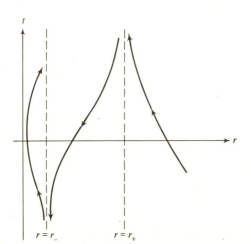

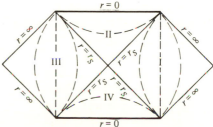

Fig. 9.11 *A possible worldline for an astronaut moving along the axis of a black hole.*

Fig. 9.12 *Conformal representation of the maximal Schwarzschild geometry. The infinite (open) regions on the right and left of Fig. 9.4 have been compressed into the square boxes I and III. The value of* r *is constant along each of the dashed lines. (After Carter, Phys. Rev. 141, 1242 (1966).)*

rotating black hole one can introduce Kruskal-like coordinates. The explicit coordinate transformations are very messy and we will not try to write them down. Once these coordinates have been introduced, one can argue, just as in the Schwarzschild case, that geodesic completeness requires that a universe that contains a black hole must also contain a corresponding white hole and vice versa (compare Fig. 9.4). Returning to our description of the travels of the astronaut, we conclude that the universe *cum* white hole in which at last report he was wandering must also contain a black hole. Assume the astronaut heads for this black hole and drops in. He will then emerge in a third universe and . . . etc., etc.

These arguments suggest that the complete Kerr geometry has a very complicated structure. A careful analysis of the geometry shows that the geodesically most complete extension of the Kerr solution consists of an infinite sequence of universes, each containing a rotating black hole and white hole. The universes are joined by one-way "tunnels" whose entrances and exits are these black and white holes, respectively.

Fig. 9.13 shows part of this infinite sequence.[11] To fit all the universes on the page, the asymptotically flat regions have been compressed so as to appear finite. For comparison, Fig. 9.12 shows the Kruskal diagram subjected to the same type of compression. (Technically, a conformal transformation has been used to map the infinite spacetime into a finite area; diagrams showing the compressed spacetime as in Figs. 9.12 and 9.13 are called Penrose diagrams.) The squares labeled $I(m,n)$ in Fig. 9.13 are asymptotically flat universes in which the coordinate r lies in the range $r_+ < r < \infty$ (this coordinate is spacelike in this range). The squares labeled $II(k,k)$ correspond to the region $r_- < r < r_+$ that lies between the inner and outer horizons (r is timelike in this range). Note that $II(1,1)$ is a black hole as seen from the universes $I(1,0)$ and $I(0,1)$, while $II(2,2)$ is a white hole as seen from $I(3,2)$ and $I(2,3)$. The triangles $III(m,n)$ correspond to the region $0 < r < r_-$.

Some geodesics of freely falling particles are shown in Fig. 9.14. The shape of the geodesic depends on the energy. Note that only geodesics of sufficiently high energy can reach $r = 0$; the others have a turning point at $r > 0$.

Any geodesic that hits the disc singularity $r = 0$ will of course terminate. Is there any way we can avoid having this singularity in our geometry? It turns out to be possible to eliminate the singularity by continuation of the metric to *negative values of r*. If we insist that r is always positive, then the derivative of g_{00} has a discontinuity at $r = 0$ (see Fig. 9.10). But if we extrapolate Eq. (50) to negative values of r, then there is no discontinuity (see Fig. 9.15). Thus, we assume that when a worldline crosses $r = 0$, it enters a space with negative values of r. Conversely, a worldline that approaches $r = 0$ from the negative side

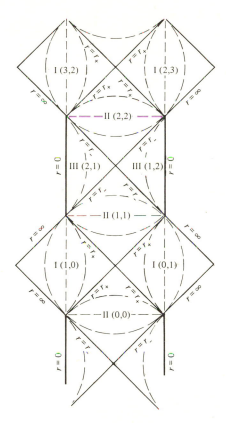

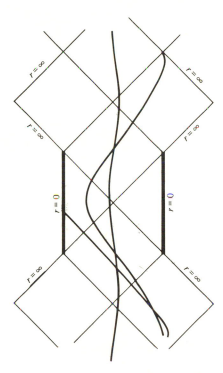

Fig. 9.13 *Conformal representation of the maximal Kerr geometry; GM > a. (After Carter, Phys. Rev. 141, 1242 (1966).)*

Fig. 9.14 *Typical worldlines in the Kerr geometry. (After Carter, Phys. Rev. 141, 1242 (1966).)*

enters into the region of positive r. Of course, the space with negative values of r must form a separate universe, distinct from ours; the universes are joined at $r = 0$.

Fig. 9.16 shows the structure of the complete Kerr geometry, including the spaces of negative values of r. In these spaces r has the range $0 > r > -\infty$. It is obvious from Eq. (30) that as $r \to -\infty$, the space-time becomes flat. Since the values of r given by Eqs. (43) and (47) are *positive*, there are no infinite redshift surfaces or horizons in these new universes.

Note that these diagrams only show the two-dimensional r,t space-time; they can only be used to analyze the motion of a particle that stays on the axis of symmetry. Particles that are not on the axis can also go from one universe to the other provided they cross the disc $r = 0$; that is,

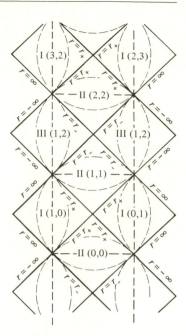

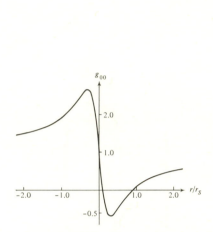

Fig. 9.15 *The function g_{00} extrapolated to negative values of* r *along the z-axis.*

Fig. 9.16 *Conformal represen-*
tation of the Kerr
geometry extended to
negative values of r.

their orbit must thread through the ring singularity $r = 0$, $\theta = \pi/2$ (see Fig. 9.8). An orbit that avoids this disc remains within the same universe.

To arrive at a physical interpretation of the negative values of r, we note that in Eq. (30), a change in the sign of r is equivalent to a change in the sign of M. An observer in the new universe will of course insist on interpreting his observations in terms of a positive value of r, and he will therefore measure M to be negative. A particle of positive mass in this universe will feel a repulsive gravitational force that pushes it away from $r = 0$.

It must be emphasized that the continuation to negative values of r is nothing but an exotic — and very unlikely — speculation. This speculation should probably be discarded on physical grounds. Negative values of the mass (or total energy) are very objectionable. If particles of negative mass could be created, the vacuum would be extremely unstable because quantum mechanical vacuum fluctuations could continuously create an endless number of pairs of positive and negative mass particles. In the absence of negative mass such an instability is prevented by energy conservation.

There is also another objection to a continuation to negative values of r: in some regions of negative r, the worldline of a particle can be made to close upon itself, i.e., the particle can return into its own past.[10] This can be seen from Eq. (30) by examining the coefficient of $d\phi^2$:

$$g_{33} = -(r^2 + a^2)\sin^2\theta\left(1 + \frac{2GMra^2}{\rho^2(r^2 + a^2)}\sin^2\theta\right)$$

The term in brackets becomes negative if

$$r < -\frac{(r^2 + a^2\cos^2\theta)(r^2 + a^2)}{2GMa^2\sin^2\theta} \qquad [58]$$

When this condition is satisfied ϕ becomes a timelike coordinate. Hence a circle around the symmetry axis, with $r =$ constant, $\theta =$ constant, $t =$ constant, is a possible particle worldline. If a particle moves along this circle, it returns to its starting point when ϕ increases by 2π. Note that the motion proceeds without any increase in t-time, that is, not only is the "orbit" (in three dimensions) closed, but also the worldline (in four dimensions).

Although it is necessary that any closed timelike worldline enter the region (58), it is not necessary that such a worldline lie entirely within this region. For example, consider a worldline that emerges from the region (58), proceeds to the asymptotically flat region ($r \rightarrow -\infty$), plunges back into the region (58), and there is made to close upon itself. Since, in the asymptotically flat region, t necessarily increases along a timelike worldline, we must arrange the worldline in the region (58) in such a way that t *decreases* along it. This is possible because there exist timelike worldlines with $a\, d\phi/dt > 0$. Note that in order to decrease t, the worldline must go around the axis of rotation in a direction opposite to that of the rotation.

The worldline described in the preceeding paragraph has the paradoxical property that the particle emerges from the Kerr hole before it enters it. The time interval between emergence and entrance can be made arbitrarily large by completing many orbits around the symmetry axis before leaving the region (58). Thus, the Kerr hole can act as a "time machine"; by passing through the machine, a particle travels into the past.

Such a "time machine" violates causality and results in a logical contradiction. To put the paradox in a very glaring form, suppose we regard the particle as a signal (e.g., a signal rocket) that is emitted, at time $t = 0$, by an apparatus located in the asymptotically flat universe and is received by this apparatus at some earlier time, say $t = -2$. Suppose the apparatus is programmed with the following instructions: emit the signal if the signal is not received before $t = 0$, do not emit the signal if the signal is received before $t = 0$. These instructions are in logical contradiction with emission at $t = 0$ and reception at $t = -2$.

This causality paradox is a strong argument against the existence of regions of negative *r*. If we suppose that only the regions with positive *r* in Fig. 9.13 exist, then the paradox disappears.

Incidentally, there is another causality violation which is sometimes discussed in connection with Fig. 9.13. Since the regions I(1,0) and I(3,2), for instance, have exactly identical geometries, it is tempting to identify them. An astronaut who drops into the rotating black hole would then simply emerge from the rotating white hole in the *same* universe. Such identification would be helpful in reducing the number of universes. But we run into the causality problem: some timelike worldlines that enter the black hole would emerge from the white hole at an earlier time. This type of causality violation has been called *trivial*, because it is the consequence of unfortunate and unnecessary topological identifications. In constrast, the causality violation at negative *r* is nontrivial.[10]

So far our discussion of the Kerr geometry only touched on the case $|a| < GM$. In the case $|a| = GM$, the two horizons coincide (see Fig. 9.17) and in the case $|a| > GM$, there is no horizon. Of course, there

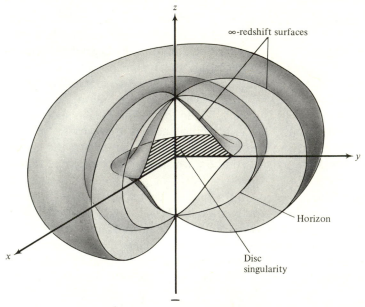

Fig. 9.17 *The horizons and ∞-redshift surfaces of a rotating black hole with*
  GM = a.

still is a ring singularity at $r = 0$, $\theta = \pi/2$. A singularity *not* surrounded by a horizon is called a *naked singularity*. A rotating "hole"* of large spin, $|a| > GM$, has such a naked singularity. Figs. 9.18 and 9.19 show the geometry for these two cases.

Although naked singularities do exist mathematically as solutions of the Einstein equations, Penrose has conjectured that the actual ap-

---

* If there are no horizons, we will not call it a *black* hole.

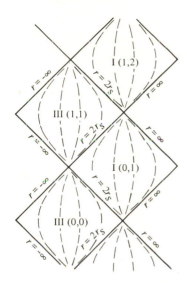

Fig. 9.18 *Conformal representation of the maximal Kerr geometry;* GM = a. *(After Carter, Phys. Rev.* **141,** *1242 (1966).)*

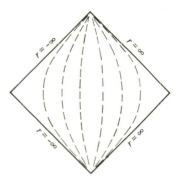

Fig. 9.19 *Conformal representation of the maximal Kerr geometry;* GM < a. *(After Carter, Phys. Rev.* **141,** *1242 (1966).)*

pearance of naked singularities is always forbidden in the real world. This conjecture is known as the *cosmic censorship hypothesis.* The naked singularities in the Kerr solution could perhaps be forbidden by an instability: the solution might be unstable for $|a| \geq GM$, and therefore states with naked singularities never form in gravitational collapse. Whether or not the Kerr solution is in fact stable or unstable is not yet known.

From a practical point of view, the complete Kerr geometry is as irrelevant to physics as the complete Schwarzschild geometry. If a rotating black hole forms by gravitational collapse, then the interior region will be occupied by matter and the pure vacuum Kerr solution does not apply. Hence the plurality of universes in the Kerr geometry has nothing to do with reality unless a rotating black hole was — somehow — created when the universe began. And even if such an object was created, it is unlikely that it has survived: there is evidence that the "tunnels" and white holes of the maximal Kerr geometry are violently unstable.

Finally, note that in the case of collapse of a spherical, nonrotating star, the Schwarzschild solution fails to hold in the interior, but the Birkhoff theorem guarantees that it will at least hold in the exterior. In the case of collapse of a rotating star, there is no such theorem and the exterior solution need not be that of Kerr. Only a long time after the collapse, when everything becomes stationary, will the exterior fields tend towards those of the Kerr solution.

9.7 GRAVITATIONAL COLLAPSE AND THE FORMATION OF BLACK HOLES

In normal stars, such as the sun, the compressional gravitational forces are held in equilibrium by the thermal pressure of the gas. This thermal pressure will be sufficient to resist the pull of gravity only if the star is hot enough. The star can therefore remain in equilibrium as long as the thermonuclear reactions in its core supply enough heat, i.e., as long as the energy released in these reactions compensates for the energy lost by radiation at the surface. In a star that has exhausted its supply of nuclear fuel, the thermal pressure will ultimately disappear and the star will collapse under its own weight. The collapse may be sudden (implosion) or gradual (contraction), but in any case it can only be halted when an alternative mechanism for generating sufficient pressure becomes available at high density.

In white dwarf stars and in neutron stars such an alternative mechanism is operative: these stars are so dense that the quantum mechanical zero-point pressure becomes dominant. Essentially, a degenerate Fermi gas of electrons supplies the equilibrium pressure in a white dwarf and a Fermi gas of neutrons that in a neutron star.

Fig. 9.20 shows equilibrium configurations of matter at high densities as calculated by numerical integration of the relativistic hydrostatic equations.[6] These curves give the mass of the star as a function of the density at the center. The separate curves of Fig. 9.20, labeled "Oppenheimer-Volkoff," "Hagedorn," "Pandharipande," correspond to somewhat different assumptions about the equation of state of matter at high densities. In the calculation of the curves it has been assumed that the matter has reached the endpoint of thermonuclear evolution—the star is completely burnt out and cold.

At the white dwarf densities ($\rho \geq 10^5$ g/cm³) the electrons are

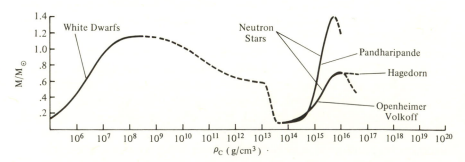

Fig. 9.20 *Total mass as a function of central density for cold stars. Different curves result from different assumptions about the equation of state. (Courtesy of Prof. R. Ruffini, Institute for Advanced Study.)*

detached from their nuclei and move quite freely throughout the volume of the star. The star consists of interpenetrating gases of electrons and nuclei. The zero-point pressure of the electron gas gives the main contribution to the pressure, and the nuclei give the main contribution to the (mass) density. The equation of state (pressure as function of density) based on this model permits equilibrium configurations provided that the total mass is below a certain upper limit. If the mass exceeds this limit, then the electron pressure cannot support the star. The trouble is this: the Fermi energy increases with density and ultimately the electron gas becomes relativistic. The relativistic Fermi gas, in contrast to the non-relativistic gas, can not supply enough pressure for equilibrium.

The following rough calculation will help to make this clear. Let us pretend that the gas has uniform density throughout the star. The Fermi momentum for a gas of N electrons in a volume V is

$$p_F = (3\pi^2 \hbar^3 N/V)^{1/3} \qquad [59]$$

In the nonrelativistic case, the Fermi energy is $p_F^2/2m_e$ (where m_e is the electron mass); in the extreme relativistic case the Fermi energy is equal to p_F. The average energy per electron is of the same order of magnitude as the Fermi energy, and hence we may regard the quantity

$$N(3\pi^2 \hbar^3 N/V)^{2/3}/2m_e \qquad \text{(nonrelativistic)}$$

$$[60]$$

$$N(3\pi^2 \hbar^3 N/V)^{1/3} \qquad \text{(extreme relativistic)}$$

as an estimate of the total zero-point energy. The Newtonian gravitational binding energy is of the order of magnitude of

$$-\frac{GM^2}{R} \qquad [61]$$

where M is the mass of the star and R the radius. Hence, the total energy, ignoring rest masses, is roughly

$$\frac{N^{5/3}(9\pi\hbar^3/4)^{2/3}}{2m_e R^2} - \frac{GM^2}{R} \qquad \text{(nonrelativistic)}$$

$$[62]$$

$$\frac{N^{4/3}(9\pi\hbar^3/4)^{1/3}}{R} - \frac{GM^2}{R} \qquad \text{(extreme relativistic)}$$

where we have taken $V = 4\pi R^3/3$. We must now try to adjust the density, or the radius R, so that a star of given N and M is placed in an equilibrium configuration. For stable equilibrium, the energy (62) should have a minimum as a function of R. In the nonrelativistic case, a minimum exists.

But in the extreme relativistic case, there is no minimum. Essentially, the trouble is that although in the extreme relativistic case the Fermi energy increases with decreasing R, it does not increase strongly enough. The relativistic gas is too soft to resist the gravitational compression.

EXERCISE 13. Check that the energy (62) has a minimum at

$$R = \frac{N^{5/3}(9\pi\hbar^3/4)^{2/3}}{GM^2 m_e} \qquad\qquad [63]$$

in the nonrelativistic case and that it has no minimum in the extreme relativistic case.

The critical mass at which the equilibrium disappears may be estimated by asking when the electron gas turns relativistic. This happens when $p_F \simeq m_e c$. Show that according to Eq. (59) this corresponds to

$$R = \frac{(9\pi\hbar^3 N/4)^{1/3}}{m_e c} \qquad\qquad [64]$$

By comparing Eqs. (63) and (64), show that the critical mass is given by

$$M_{\text{crit}}^2 = \frac{cN^{4/3}(9\pi\hbar^3/4)^{1/3}}{G} \qquad\qquad [65]$$

If the nuclei in the white dwarf are helium, then $N \simeq M/2M_p$. Show that this leads to a critical mass

$$M_{\text{crit}} = \sqrt{\frac{9\pi}{4^3}} \left(\frac{\hbar c}{GM_p^2}\right)^{3/2} M_p \qquad\qquad [66]$$

Evaluate this numerically; the result, $M_{\text{crit}} \simeq M_\odot$, of this rough calculation agrees fairly well with the result of a more precise calculation.

The limiting value of the mass, beyond which the zero-point pressure of electrons becomes insufficient for equilibrium, is known as the *Chandrasekhar limit*. For a realistic model of white dwarfs, in which the nuclear matter is in the form of helium nuclei, a precise calculation gives the value $1.44M_\odot$ for this limit.

For the ideal white dwarfs of Fig. 9.20, in which the nuclear matter is assumed to have reached the state of least nuclear energy,* the Chandrasekhar limit is somewhat lower.

The equation of state based on the zero-point pressure of electrons is subject to a correction which further lowers the value of the limiting mass. For the sake of simplicity let us ignore the strong forces that bind protons and neutrons into nuclei. We are then dealing with a gas of elec-

---

* The nucleus with the least nuclear energy is Fe^{56}.

trons, protons, and neutrons. At high densities the electrons will react with the protons and form neutrons. Normally these neutrons are unstable (β-decay) and significant numbers of neutrons do not accumulate. However, when the Fermi energy exceeds the mass difference $M_n - M_p - m_e$, neutrons become stable. (Under these conditions all the electron states that are energetically accessible to the β-decay electron are already occupied, and hence the decay becomes impossible.) Thus at extreme densities matter consists of a mixture of electrons, protons, and neutrons in β-equilibrium; the equilibrium shifts in favor of neutrons as the density increases.

In Fig. 9.20, the configurations between $\rho_0 \sim 10^8$ to $\rho_0 \sim 10^{13}$ g/cm^3 correspond to an increasing fraction of neutrons. (Incidentally, in these configurations the equilibrium is unstable and therefore of little practical interest.) At a density above $\sim 10^{13}$ g/cm^3 we reach the realm of the neutron stars in which neutrons are by far the most abundant kind of particle. According to Fig. 9.20, the masses of stable neutron stars lie in the range from $\sim 0.3 M_\odot$ to $\sim 1.4 M_\odot$.

Fig. 9.21 (p. 346) shows the structure of a neutron star.[12] The density is comparable to nuclear densities ($\sim 2 \times 10^{14}$ g/cm^3) and hence the star may be described as a single giant nucleus. The bulk of the star consists of a Fermi gas of neutrons, with some very few electrons and protons. At the center is a small core with heavy particles (hyperons). Near the surface is a layer of heavy nuclei and a crust of iron crystals. A plasma atmosphere surrounds the star.

For our purposes, the most important feature of Fig. 9.20 is the absence of any equilibrium configuration with a mass larger than $\sim 1.4 M_\odot$. A star of mass larger than this cannot be supported by the zero-point pressure of electrons or neutrons. The details of Fig. 9.20 depend on the equation of state. One might wonder if, at high densities, the nuclear forces between neutrons could somehow "stiffen" the material. To some extent this is possible — equations of state that try to take into account nuclear interactions tend to permit stable neutron stars of slightly higher mass. But no matter how much the equation of state at high densities is modified by interactions, it can be shown that there always is an upper limit to the mass that can be supported. The reason for this is that the stiffness of a material is directly related to the speed of sound waves in the material — the stiffer the material, the higher the speed of sound. The requirement that the speed of sound be no more than the speed of light then sets a limit on the stiffness of the material. According to Rhoades and Ruffini[13] this leads to the conclusion that the mass of a neutron star can never exceed $3.2 M_\odot$, independent of any assumption about the details of the equation of state at high densities.

The absence of equilibrium configurations for massive burnt-out stars has drastic consequences for stellar evolution. A star with a mass below the Chandrasekhar limit evolves along a sequence of equilibrium

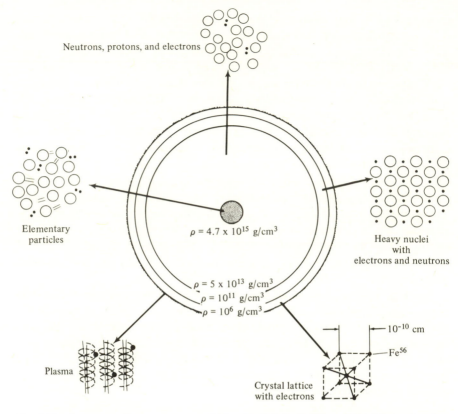

Neutrons, protons, and electrons

Elementary particles

$\rho = 4.7 \times 10^{15}$ g/cm^3

Heavy nuclei with electrons and neutrons

$\rho = 5 \times 10^{13}$ g/cm^3
$\rho = 10^{11}$ g/cm^3
$\rho = 10^6$ g/cm^3

Plasma

10^{-10} cm

Fe^{56}

Crystal lattice with electrons

Fig. 9.21 *Structure of a neutron star. (After Ruffini, in DeWitt and DeWitt, eds.* Black Holes)

configurations. When the nuclear fuel is exhausted it settles down in a white dwarf state. The luminosity of such a white dwarf is entirely due to the residual thermal energy; as the energy is lost by radiation the star grows dimmer and dimmer, ultimately becoming a "black star" (the cooling time is typically of the order of 10^9 years).

On the other hand, when the nuclear fuel is exhausted in a star with a mass above the Chandrasekhar limit, the pressure will become insufficient to prevent gravitational collapse. At late stages of its evolution, a massive star will have a dense core, and the collapse begins in this core. The core implodes. If the mass of the core is sufficiently small, then the implosion can be halted when neutron star densities are reached. The sudden halt of the implosion creates a shock wave which blows away the outer layers of the star. The outward explosion of these layers shows up on the sky as a supernova. The famous Crab pulsar is a neutron star that was formed in such a supernova explosion; the Crab nebula, which surrounds it, consists of hot gases which once constituted the outer layers of the original massive star (of a mass of $\sim 4M_\odot$). If the mass of the collapsing core exceeds the critical mass for a neutron star, then the implosion cannot be halted. The complete gravitational collapse of the core produces a black hole.

Supernova explosions are one possible mechanism for the formation of black holes. Mass exchange in a close binary star system is another mechanism. Consider a neutron star in an orbit near the surface of a more or less normal star. If the neutron star is near enough, it will pull material off the surface and gradually increase in mass. Once the mass increases beyond the critical limit, the neutron star will collapse to a black hole. The binary system Centaurus X-3 (see next section) is believed to contain such a neutron star which is accreting matter from its companion; ultimately, this neutron star will come to a bad end.

In a sense the collapse never ends. An outside observer would say that the surface of the collapsing star asymptotically approaches the Schwarzschild radius, but never reaches it in a finite time. In this sense, the object may be appropriately called a "frozen star" rather than a black hole. However, the "frozen star" very soon becomes practically indistinguishable from a black hole. The intensity of any light emitted by the collapsing surface sharply decreases while the redshift increases (the luminosity decreases exponentially with time; the characteristic "half-life" is of the order of GM).[23] Furthermore, the gravitational and electromagnetic fields surrounding the collapsing mass asymptotically approach those corresponding to a black hole. The evolution of the initial gravitational and electromagnetic fields into the final black hole fields will in general be accompanied by radiation. For example, calculations by Price[23] show that if the initial fields deviate slightly from the spherically symmetric Schwarzschild field, then these deviations (multipoles) get radiated away as the collapse proceeds. Within a finite time the electromagnetic and gravitational fields surrounding the collapsing mass approach so close to those surrounding a black hole that the quantum uncertainties in the measurement process prevent an outside observer from detecting any difference at all.[14]

It is believed that when the external gravitational and electromagnetic fields of a collapsing mass ultimately reach a stationary state, the fields will be entirely determined by three parameters: mass (M), spin angular momentum (S), and electric charge (Q).* This conjecture has been rigorously proved in some special cases: there exists one theorem (Israel's theorem) which shows that if $S = 0$, then the stationary solutions are uniquely characterized by M and Q; there exists another theorem (Robinson's) which shows that if $Q = 0$, then the stationary solutions are uniquely characterized by M and S.[28] In the general case, with $S \neq 0$ and $Q \neq 0$, the conjecture appears very plausible. Any two black holes with the same mass, angular momentum, and charge are then absolutely identical (as far as an outside observer is concerned). If anything is

---

* The analytic solution corresponding to $S \neq 0$, $Q \neq 0$ is known as the Kerr-Newman solution; the special case $S = 0$, $Q \neq 0$ is the Reissner-Nordstrøm solution (see ref. 26).

dropped into a black hole, there is no way of telling from the field surrounding the black hole at a later time what was dropped into the black hole. When a black hole swallows up some matter, only its mass, spin angular momentum, and charge change. All other properties of matter, such as intrinsic multipole moments, baryon number, lepton number, etc., are "forgotten" by the black hole. The absence of distinctive fields — such as, e.g., independent gravitational or electromagnetic multipole fields, or nuclear force fields—in the region surrounding a black hole led Wheeler to remark that "a black hole has no hair."[6]

Although the collapse of a mass never quite ends for an outside observer, it does end very soon for an observer who rides along with the collapsing mass. Within a finite proper time, this observer hits a singularity. In the special case of spherical symmetry or axial symmetry, the existence of the Schwarzschild singularity ($r = 0$) or the Kerr singularity ($r = 0$) follows directly from the solution of Einstein's equations. In the general case, the existence of singularities is indicated by the Hawking-Penrose theorem which roughly says that a singularity will develop whenever the spacetime contains a *closed trapped surface*.[3]

By a closed trapped surface is meant the following: Consider a closed surface S (see Fig. 9.22a). At some time, let this surface suddenly become luminous for an instant. At any later time, the ingoing and outgoing wavefronts will then form surfaces S_1 and S_2 (see Fig. 9.22a). In flat spacetime, the area S_1 will of course be smaller than S, while the area S_2 will be larger than S. In a strongly curved spacetime, *both* S_1 and S_2 can be smaller than S; if so, then S is called a trapped surface (see Fig. 9.22b). Note that any kind of signal emitted by a trapped surface falls towards the interior. For example, the surface $r = $ constant $ < 2GM$ is a closed trapped surface in the Schwarzschild geometry.

It is believed that trapped surfaces will be formed in almost all

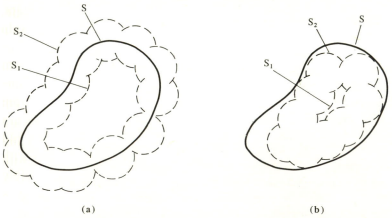

(a)   (b)

Fig. 9.22   *The solid line shows a closed, luminous surface. The broken lines show the wavefronts at a later time. In (a) the surface is not trapped. In (b) the surface is trapped.*

cases of gravitational collapse. The Hawking-Penrose theorem then tells us that singularities will also be formed.

What is the nature of the singularity? In the context of the above theorem, a singularity is simply a region in which geodesics terminate. What is not known is how they terminate. The most obvious possibility is that the curvature becomes infinite; this is, of course, exactly what happens at $r = 0$ in the Schwarzschild geometry. In any case, the presence of a singularity signals the breakdown of classical geometrodynamics. It is tempting to conjecture that the Hawking-Penrose theorem of classical geometrodynamics is as irrelevant to the real world as the Earnshaw theorem of classical electrodynamics[24] according to which atoms are unstable and must collapse to a singularity. It may well be that when quantum effects are taken into account, the collapsed matter attains some nonsingular final state.

9.8   THE SEARCH FOR BLACK HOLES

Since black holes are an end product of stellar evolution, they could be quite abundant. It is even possible that in some galaxies most of the mass exists in the form of black holes. For example, in elliptical galaxies the observed ratio of total mass to luminosity is ~ 70 times that of the sun. Since the sun is a quite typical star, most of the matter in these galaxies must be in a state of very low luminosity. This implies the presence of diffuse matter (gas, dust), stars of very low mass, evolved (dark) stars, or black holes. Wolfe and Burbidge[15] conclude that the most likely alternatives are that these galaxies contain either a very large population of stars of very low mass ($\sim 5 \times 10^{12}$ stars of mass $\lesssim 0.1 M_{\odot}$) or else black holes. The observed light distribution and the observed velocity dispersion in bright elliptical galaxies may be used to set limits on the amount of mass that can be in the form of black holes. If there is a large central black hole, then its mass can not exceed $10^{10} M_{\odot}$. However, a much larger amount of mass can exist in the form of a large number of somewhat smaller black holes ($\sim 10^{3} M_{\odot}$ each) distributed over the galaxy.

The birth of a large black hole by gravitational collapse in the center of a galaxy would be accompanied by violent fireworks in the form of optical and radio emissions. The spectacular explosions seen in the nuclei of some galaxies (see Fig. 9.23) and the enormous amounts of energy radiated by the quasi-stellar objects or quasars* (see Fig. 9.24) may be associated with the formation of large black holes. The energy released by an exploding galaxy or a quasar is estimated as $10^{7} M_{\odot}$ to $10^{9} M_{\odot}$ and is likely to be of gravitational origin. Possible mechanisms for the release of this energy are collapse of a superstar ($10^{8} M_{\odot}$–$10^{9} M_{\odot}$), star collisions, or multiple supernova explosions in a small volume.[16]

---

* We suppose that the quasi-stellar objects are at the cosmological (large) distances indicated by their redshift.

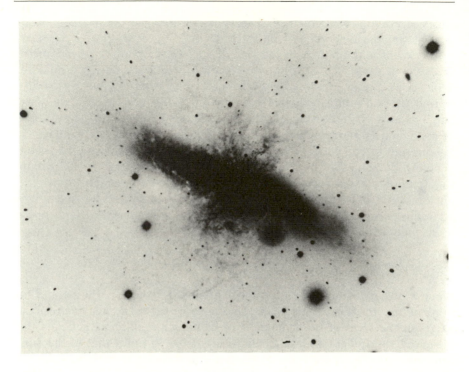

Fig. 9.23   *The exploding galaxy M82. Filaments of hydrogen gas, ejected by the explosion, extend 25,000 light years outwards from the nucleus of the galaxy. This is a negative print of a photograph taken with the 200-inch reflector on Mt. Palomar. (Hale Observatories photograph.)*

Independent of the details of the process, what is involved is a large amount of mass (up to $10^9 M_\odot$) confined in a small volume (no more than a few light years across). Such a configuration is unstable and will evolve into a black hole in a time that is short by galactic standards.[15]

Lynden-Bell[17] has suggested that quasars are galaxies with a large central black hole and that the intense electromagnetic emission from these quasars results from the capture of interstellar gas by the black hole. As the gas becomes exhausted, the brightness of the quasar diminishes and it gradually turns into a normal galaxy. In this view, many of the ordinary galaxies are old, or dead, quasars and they should contain large black holes ($10^7 M_\odot$ to $10^8 M_\odot$) in their nuclei. This could explain the origin of the radio emissions that are often associated with galactic nuclei. Our own galaxy produces such radio emissions which could be accounted for by a hole of $\sim 3 \times 10^7 M_\odot$ at its nucleus.

Besides black holes in galaxies, there also may exist some in intergalactic space. There are some clusters of galaxies in which the visible (luminous) mass is insufficient to give gravitational binding. The total mass needed to hold the cluster together may be 5 to 10 times as large

Fig. 9.24 *The quasi-stellar object 3C 147 is indicated by a wedge. Note that the image has a starlike appearance. However, this object is at a distance of about 10¹⁰ light years and it is also a source of radio waves. (Hale Observatories photograph.)*

as the visible mass. This suggests that these clusters are held together by a large amount of invisible mass. One explanation is that large black holes are distributed over the space between the galaxies. For example, the Coma cluster, which contains about 1,000 galaxies, may also contain 10^6 black holes, each of a mass of about $10^9 M_\odot$, distributed over the space between the galaxies.[18]

The preceding arguments offer only rather circumstantial evidence for the existence of black holes. Can we obtain some direct evidence? An isolated black hole does not emit light and is nearly unobservable. In principle, if a black hole is on, or very near, the line of sight from the earth to a star, then the deflection of light brings about a large change in the apparent position of the star. However, the probability for such an alignment is very small and even if it happens we would not be likely to notice.

The most favorable conditions for the observation of black holes are found in close binary systems. If a black hole or neutron star is in an orbit very near to a normal star, it can capture some matter from this star. As the matter falls towards the black hole it suffers heating by friction and compression. Temperatures of 10^8 to 10^9 °K are reached and intense thermal X-rays are emitted.

The observation of X-ray sources from the earth is complicated by atmospheric absorption. The X-ray telescopes must be placed on satellites or high-flying rockets. Much of our knowledge of X-ray sources is due to the work of Giacconi et al.[19] with the UHURU satellite. Among these sources are several close binary systems in which the X-ray source appears to orbit a more or less normal star and is periodically eclipsed by it. The orbital period is typically a few days. These sources are very intense; the X-ray luminosity is of the order of 10^{37}–10^{38} erg/sec.*

What makes these binary systems most interesting is that the mass of the primary and secondary components can be determined from the observational data. For example, if the spectral class and luminosity of the normal star are known, astrophysics gives its mass; orbital data

---

* For X-rays with an energy between 2 and 10 KeV.

(period, velocity, and angle of inclination between the orbit and our line of sight*) then determines the mass of the secondary star.

The binary X-ray sources fall into two groups.[20] The first group consists of *pulsating* sources in which the X-rays are emitted as a sequence of regular pulses with intervals of a few seconds. These regular X-ray pulsations are very similar to the radio pulsations found in the case of pulsars. In fact, it is believed that the mechanism behind the pulsations is the same in both cases: the sources are rotating neutron stars. The extremely large (10^{37} erg/sec) output of X-rays is evidence in favor of a neutron star interpretation of these sources. Ordinary radio pulsars obtain their transmission power from their rotational kinetic energy. However, the power emitted by the above X-ray sources is too large to arise from conversion of rotational kinetic energy. The X-ray emission is powered by *accretion*. The neutron star captures material from its companion, and as this material falls on the neutron star it heats up and emits X-rays. This mechanism for X-ray production is only possible for an object as dense as, or denser than, a neutron star, since only very intense gravitational fields can supply the required energy. The pulsation of the X-ray sources is probably due to the presence of a strong magnetic field. If the poles of the magnetic field are inclined off the axis of rotation, the periodic variation of the magnetic field will deflect particles first to one part of the surface, then to another; it is not unlikely that a pulsating X-ray flux would result.

Table 9.1 gives the data for two pulsating X-ray sources.[20,25] The masses are given in units of the solar mass. Note that the masses for the X-ray emitter (M_X) are consistent with the neutron star interpretation.

TABLE 9.1   PULSATING X-RAY SOURCES

X-ray source	Pulsation period (sec)	Orbital period (days)	Primary star	Mass of primary star	M_X
Hercules X-1	1.24	1.7	Hz Herculis	~ 2	0.2–1.2
Centaurus X-3	4.84	2.1	(blue giant)	~ 20	0.2–0.3

The second group of binary X-ray sources consists of *flickering* sources in which the X-rays are emitted in the form of irregular bursts. These bursts happen on a very short time scale; large changes in intensity have been observed in times as short as 10^{-3} sec. Such sudden variations in intensity indicate that the source is probably quite compact.

---

* The velocity can be determined from Doppler shifts; the angle of inclination can be determined from eclipse data (in an eclipsing binary) or from some model that makes assumptions about the structure of the binary system.

EXERCISE 14. Suppose that the emitting region has a roughly spherical shape (with radius R), and a roughly uniform surface brightness. If such a source is suddenly turned off (or on), then an observer at some distance will first receive the change in intensity contributed by that part of the source that is nearest to him; the change in intensity contributed by the farthest part of the source will arrive with a time delay $\sim R/c$. Hence, even if the source switches off (or on) instantaneously, the change in intensity is spread out over a time R/c as seen by the observer. Show that the change in the observed intensity in a time interval $\Delta t \ll R/c$ is small compared with the initial (or final) intensity. Hence, a large change in intensity in a time of 10^{-3} sec implies a size no larger than $R \sim c\Delta t \sim 100$ km for the source. (Hint: the observed intensity is proportional to the amount of luminous surface visible to the observer.)

The mechanism for X-ray production in these systems is presumably the same as for the pulsating systems: the compact object in orbit around the normal star draws gas off the latter and X-rays are emitted as the gas falls towards the compact object. This model may also explain the time scale of the bursts. Before disappearing in the black hole, the captured gas spirals around the black hole and forms an equatorial disc (accretion disc) centered on the black hole (see Fig. 9.25). The orbital period at the inner edge of the accretion disc is expected to be about a millisecond.

EXERCISE 15. The smallest stable circular orbit around a black hole has a radius of $6GM$ (see Problem 8.9). Use Kepler's law $(T/2\pi)^2 = r^3/GM$ to show that the period T for a circular orbit of radius $6GM$ is of the order of 10^{-3} sec for $M \sim 6M_\odot$. This period does not have to be corrected for redshift because it is already expressed in t-time (see Problem 8.10).

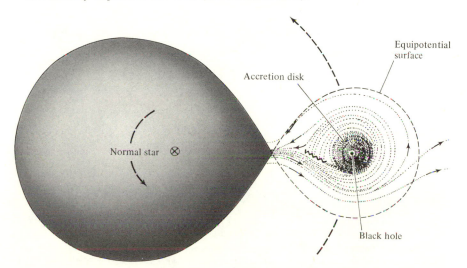

Fig. 9.25 *Primary star, accretion disc, and black hole. (After Eardley and Press,* Annual Review of Astronomy and Astrophysics, *1975.)*

Table 9.2 gives the data for some of the best known flickering X-ray sources.[20,21,22,25,27] Fig. 9.26 is a photograph of the primary star associated with the source Cygnus X-1.

TABLE 9.2   FLICKERING X-RAY SOURCES

X-ray source	Time scale of bursts	Orbital period (days)	Primary star	Mass of primary star	M_X
Cygnus X-1[a]	~ 0.001 sec	5.6	HDE 226868	$30M_\odot$	$\geq 6\ M_\odot$
SMC X-1	~ 1 min	3.89	Sk 160	20	~ 6
Vela X-1	~ 0.1 sec	8.9	HD 77581	~ 15	~ 1.5

[a] Cygnus X-1 is not an eclipsing binary. The orbital period is determined from the velocity curve of the primary star (obtained from Doppler shifts).

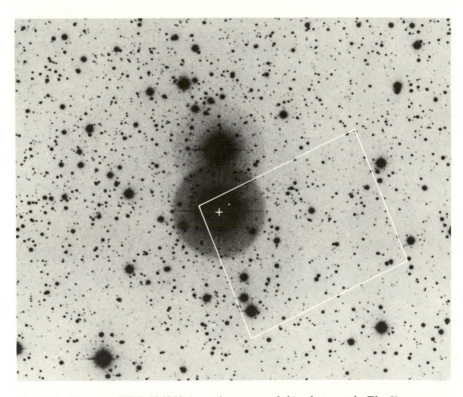

Fig. 9.26 *The star HDE 226868 is at the center of this photograph. The X-ray source Cygnus X-1 is believed to orbit this star; the box indicates the uncertainty in the position of the X-ray source. The system also emits radio waves; the cross indicates the uncertainty in the position of the radio source. The distance from the earth is about 10^4 light years and hence the size of the image is not related to the size of the star; rather, it is due to the effects of overexposure. (Hale Observatories photograph made by J. Kristian. Kindly supplied by Prof. K. S. Thorne.)*

According to Table 9.2, the mass of the compact object (M_X) in Cygnus X-1 and SMC X-1 is much in excess of the upper limit for the mass of a neutron star; in Vela X-1 the excess is marginal. Thus, the first two sources are certainly not neutron stars and the last is probably not a neutron star. Since the large X-ray flux and the short time scale of the bursts tell us that the sources are very compact, it seems that we are dealing with black holes.

FURTHER READING

An excellent and quite comprehensive presentation of the physics of Schwarzschild and Kerr black holes will be found in Misner, Thorne, and Wheeler, *Gravitation;* the general black hole with both spin and electric charge is also treated there. Rees, Ruffini, and Wheeler, *Black Holes, Gravitational Waves, and Cosmology,* also contains much information on these topics.

A more advanced treatment is given in C. DeWitt and B. S. DeWitt, eds., *Black Holes.* This contains the lectures delivered at Les Houches in the summer of 1972, as well as numerous reprints of relevant articles. It includes:

Hawking, "The Event Horizon" (general definition of horizons and of black holes).

Carter, "Black Hole Equilibrium States" (analytic and geometric properties of the Kerr solution with both angular momentum and electric charge; integration of the geodesic equation; general theorems regarding black holes).

Bardeen, "Rapidly Rotating Stars, Disks, and Black Holes" (the Kerr solution treated as the limiting case of the field surrounding a rotating disk of matter).

Novikov and Thorne, "Astrophysics of Black Holes" (physics of the accretion disc formed by matter as it falls into the hole).

Ruffini, "On the Energetics of Black Holes" (critical mass for the gravitational collapse of a neutron star, emission of gravitational and electromagnetic radiation by a particle orbiting or falling into a black hole).

Hawking and Ellis, *The Large Scale Structure of Space-Time,* presents a rigorous mathematical discussion of the topology of the Schwarzschild, Kerr, and other solutions and general theorems (Hawking-Penrose theorems) on the formation of singularities in spacetime; this book presupposes some knowledge of differential geometry and topology.

The equilibrium (or lack of equilibrium) of large masses under the influence of their gravitational self-attraction is described at length in Harrison, Thorne, Wakano, and Wheeler, *Gravitation Theory and Gravitational Collapse.* A short, very simple, and very clear presentation of the essential physics behind the equilibrium problem is given in the article "The Superdense Star and the Critical Nucleon Number" by Wheeler (*in* Chiu and Hoffmann, eds., *Gravitation and Relativity*).

Our knowledge of what happens in gravitational collapse is summarized in the article "Non-spherical Gravitational Collapse: A Short Review" by Thorne (*in* Klauder, ed., *Magic Without Magic: John Archibald Wheeler*).

The article "Astrophysical Processes near Black Holes" by Eardley and

Press (see ref. 27) gives a comprehensive review of the formation of black holes and of the observable effects of the black hole on its environment.

X-ray binaries and the evidence for the existence of black holes in these systems are reviewed in:

Blumenthal and Tucker, "Compact X-ray Sources" (see ref. 22).

Eardley and Press, "Astrophysical Processes near Black Holes" (see ref. 27).

Gursky and van den Heuvel, "X-Ray-Emitting Double Stars" (*Scientific American,* March, 1975).

Ruffini and Gursky, eds., *Neutron Stars, Black Holes, and Binary X-ray Sources.*

There are several excellent articles addressed to a wider audience which discuss gravitational collapse and black holes in a more or less qualitative way:

Chandrasekhar, "The Black Hole in Astrophysics: The Origin of the Concept and Its Role" (*Contemp. Physics* **14,** 1, 1974).

Penrose, "Gravitational Collapse: The Role of General Relativity" (*Nuovo Cimento,* vol. I, special number, 1969).

Penrose, "Black Holes" (*Scientific American,* May, 1972).

Ruffini, "Neutron Stars and Black Holes in Our Galaxy" (*N.Y. Academy of Sciences* **35,** 196, 1973).

Ruffini and Wheeler, "Introducing the Black Hole" (*Physics Today,* January, 1971).

Thorne, "Gravitational Collapse" (*Scientific American,* November, 1967).

Thorne, "The Search for Black Holes" (*Scientific American,* December, 1974).

REFERENCES

1. M. D. Kruskal, Phys. Rev. **119,** 1743 (1960).
2. R. W. Fuller and J. A. Wheeler, Phys. Rev. **128,** 919 (1962).
3. For a precise statement and proof of the theorem see S. W. Hawking and G. F. R. Ellis, *The Large Scale Structure of Space-Time* (Cambridge University Press, 1973), p. 266.
4. R. P. Kerr, Phys. Rev. Lett. **11,** 237 (1963).
5. R. H. Boyer and R. W. Lindquist, J. Math. Phys. **8,** 265 (1967).
6. M. Rees, R. Ruffini, and J. A. Wheeler, *Black Holes, Gravitational Waves, and Cosmology* (Gordon and Breach, New York, 1974).
7. R. Penrose, Nuov. Cim. Ser. I, **1,** 252 (1969).
8. D. Christodolou, Phys. Rev. Lett. **25,** 1596 (1970); D. Christodolou and R. Ruffini, Phys. Rev. D**4,** 3552 (1971).
9. See ref. 3, p. 318; also, see S. W. Hawking, Proc. N.Y. Acad. Sci. **224,** 268 (1973).
10. B. Carter, Phys. Rev. **174,** 1559 (1968).
11. B. Carter, Phys. Rev. **141,** 1242 (1966).
12. R. Ruffini, *in* C. DeWitt and B. S. DeWitt, eds., *Black Holes* (Gordon and Breach, New York, 1973).
13. C. E. Rhoades and R. Ruffini, Phys. Rev. Lett. **32,** 324 (1974).
14. H. C. Ohanian and R. Ruffini, Phys. Rev. D**10,** 3903 (1974).
15. A. M. Wolfe and G. R. Burbidge, Ap. J. **161,** 419 (1970).

16. For different theories that try to account for the energy requirements of quasars, see G. Burbidge and M. Burbidge, *Quasi-Stellar Objects* (Freeman, San Francisco, 1967).
17. D. Lynden-Bell, Nature **223**, 690 (1969).
18. P. J. E. Peebles, *Physical Cosmology* (Princeton University Press, 1971).
19. R. Giacconi, Ann. N.Y. Acad. Sci. **224**, 149 (1973).
20. R. Ruffini and H. Gursky, *Neutron Stars, Black Holes, and Binary X-ray Sources* (Reidel, Amsterdam, 1975).
21. S. Chandrasekhar, Contemp. Phys. **14**, 1 (1974).
22. G. R. Blumenthal and W. H. Tucker, *in* G. R. Burbidge, D. Layzer, and J. G. Philipps, eds., *Annual Review of Astronomy and Astrophysics* (Annual Reviews Inc., Palo Alto, 1974), vol. 12.
23. K. S. Thorne, *in* J. R. Klauder, ed., *Magic Without Magic: John Archibald Wheeler* (Freeman, San Francisco, 1972).
24. J. Jeans, *The Mathematical Theory of Electricity and Magnetism* (5th ed., Cambridge University Press, 1960), pp. 168, 631, 632.
25. H. Gursky and E. P. J. van den Heuvel, Scientific American **232**, 24 (1975).
26. C. W. Misner, K. S. Thorne, and J. A. Wheeler, *Gravitation* (Freeman, San Francisco, 1973).
27. D. M. Eardley and W. H. Press, *in* G. R. Burbidge, D. Layzer, and J. G. Philipps, eds., *Annual Review of Astronomy and Astrophysics* (Annual Reviews Inc., Palo Alto, 1975).
28. D. C. Robinson, Phys. Rev. Lett. **34**, 905 (1975).
29. W. G. Unruh, Phys. Rev. **D10**, 3194 (1974).
30. S. W. Hawking, Nature **248**, 30 (1974).

PROBLEMS

1. *Estimate* the tidal force (stretching force) that acts on an astronaut of normal size and mass as he falls, feet first, through the horizon of a black hole of mass $M_\odot = 2 \times 10^{33}$ g.

2. A flash of light is emitted at $r = 2r_S$, $\theta = \pi/2$ in a Schwarzschild field. Make a polar plot of the coordinate velocity $[(dr/dt)^2 + r^2(d\phi/dt)^2]^{1/2}$ at $r = 2r_S$ as a function of the "emission angle." The "emission angle" θ is defined by

$$\tan \theta = r d\phi/dr$$

3. A spaceship enters a Schwarzschild black hole on a radial line with a radial velocity

$$\frac{dr}{dt} = -(1 - r_S/r)[(r/r_S)^2 - (r/r_S) + 1]^{-1/2}$$

(a) Show that the worldline of the spaceship is timelike for the given dr/dt (therefore this dr/dt is permissible). (b) How much *proper time* elapses as the spaceship moves from $r = 4r_S$ to $r = 0$? (c) In order to move with the given dr/dt, must the spaceship use its rocket motor? (Hint: Solve this problem in r, t-coordinates).

4. An astronaut in command of a spaceship equipped with a powerful rocket motor enters the horizon $r = r_S$ of a Schwarzschild black hole. Prove that in a proper time no larger than $(\pi/2)r_S$, the spaceship reaches the singularity $r = 0$. Prove that in order to avoid the singularity as long as possible, it is best to move in a purely radial direction. (Hint: For purely radial motion, with $dr < 0$ and $dt = d\theta = d\phi = 0$, the increment in proper time is

$$d\tau = -\frac{dr}{\sqrt{r_S/r - 1}}; \quad r \le r_S$$

Integrate this between $r = r_S$ and $r = 0$ to obtain

$$\Delta\tau = \frac{\pi}{2} r_S$$

Finally, check that if dt, $d\theta$, $d\phi$ are different from zero, then the increment $d\tau$, for a given value of $-dr$, is necessarily smaller than the value given above.)

Show that in order to achieve this longest proper time the astronaut must use his rocket motor in the following way: outside the horizon he must brake his fall so as to arrive at $r = r_S$ with (nearly) zero radial velocity; inside the horizon he must shut off his motor and fall freely. (Hint: Show that $\Delta t = (\pi/2)r_S$ corresponds to free fall from $r = r_S$, see Eq. (19).)

5. Suppose that a black hole is surrounded by gas which, initially at rest at a large distance, falls radially into the black hole. For the purpose of this problem, pretend that gravitation can be described with sufficient precision by the Newtonian $1/r^2$ field and that the supply of gas at large distance is inexhaustible. Show that the density of the gas increases as $r^{-3/2}$ as it falls. (Hint: At any given radius the density is constant in time because gas leaving is replaced by gas arriving, i.e., $\partial\rho/\partial t = 0$. Show that this implies $(1/r^2)(\partial/\partial r)r^2\rho v = 0$ where v is the velocity at radius r.)

6. Show that in *classical mechanics* the angular velocity of rotation of a rigid body can be expressed as

$$\omega_x = \frac{\partial E}{\partial S_x}, \quad \omega_y = \frac{\partial E}{\partial S_y}, \quad \omega_z = \frac{\partial E}{\partial S_z}$$

where E is the energy (kinetic energy of rotation) of the body.

This suggests that we define the angular velocity of a Kerr black hole as follows:[*]

$$\omega_z = \frac{\partial M}{\partial S_z}$$

Differentiate Eq. (57) with respect to S (holding M_{ir} constant) and show that the angular velocity of rotation of the black hole is

$$\omega = \frac{a}{r_+^2 + a^2}$$

---

[*] See ref. 26 for a rigorous justification of this definition.

7. Given a surface $f(x^0, x^1, x^2, x^3) = 0$ in a four-dimensional space, we can define the *normal* to the surface by

$$n_\mu = \frac{\partial f}{\partial x^\mu}$$

The surface is said to be a *null surface* if $n_\mu n^\mu = 0$. Show that the surface $r - r_s = 0$ is a null surface in the Schwarzschild geometry. Show that the surfaces $r - r_\pm = 0$ are null surfaces in the Kerr geometry.

8. (a) Use Eq. (63) to calculate the radius of a white dwarf of mass $M = M_\odot$. Assume that the nuclear matter is in the form of helium nuclei.

   (b) By carrying out a calculation similar to that given in Eqs. (59)–(63), show that the radius of a neutron star is of the order of

   $$R = \frac{N^{5/3}(9\pi\hbar^3/4)^{2/3}}{GM^2 M_n}$$

   Evaluate this numerically for $M = M_\odot$.

10. COSMOLOGY

We are all in the gutter, but some of us are looking at the stars.

Oscar Wilde, *Lady Windermere's Fan*

As we begin the study of the universe, a fundamental question immediately comes to mind: is it legitimate to apply the laws of physics that are known to hold on and near the earth to faraway regions of the universe? In a broad sense the answer must necessarily be in the affirmative. For suppose that the laws of physics in two regions of the universe are different; say, in one region gravitation is a tensor field and in another a scalar field. Then at the boundary between these regions physics makes no sense—there is an unacceptable discontinuity.

Such arguments based on continuity cannot rule out the possibility that our laws contain some parameters with a smooth and gradual variation in space or time. In general terms, functions of space and time are *fields* and the question therefore is: do the laws of physics contain a dependence on some cosmological, or "background," fields? If there are such extra fields then we will not say that the laws of physics are different in different regions of the universe; rather we will include the description of these fields in our laws and simply say that physics is more complicated than we were wont to believe.

One possible manifestation of a cosmological field may be a change

360

of "constants." Thus, the gravitational constant or the fine structure constant may vary with position or with time. However, observation indicates that the general features of the universe are pretty much independent of the direction in which we look. The universe is quite *isotropic*. This constitutes strong evidence against a dependence of "constants" on position. For example, regions of the universe with different values of G would expand at different rates and we would observe an anisotropy in the expansion rate and a consequent anisotropy in the density of galaxies. Likewise, a dependence of atomic constants on position would affect the characteristics of the emitted light and galaxies at different positions of the sky would display different, and mysterious, spectral lines.

The evidence against a dependence of "constants" on time has already been presented in Section 5.1.

A cosmological field may also lead to so-called preferred-frame effects.[1] By this is meant that a measurement of the cosmological field can serve to distinguish between different Lorentz frames. In particular, one could use such (local) measurements to determine the velocity of the earth relative to the universe. If the cosmological field is an absolute object (independent of the state of matter), then it violates Lorentz invariance. But if it is a dynamical object, with a field equation that relates it to, say, the mass distribution of the universe, then Lorentz invariance and even general invariance could remain valid. Thus, "preferred-frame" effects do not per se tell us whether there is a fundamental violation of Lorentz invariance; we need to know something about the field equation. In fact, even in geometrodynamics there exist "preferred-frame" effects since, in principle, the curvature tensor can be measured and used to discriminate between different Lorentz frames.

The observational evidence against certain kinds of "preferred-frame" effects has led to the rejection of some gravitational theories with extra cosmological fields.[1,2] For our study of cosmology we will suppose that there are no extra cosmological fields and that Einstein's equations are valid everywhere in the universe.

Another fundamental question that comes to mind is related to a peculiar property of the universe: there is one and only one universe. As long as we are dealing with physical systems smaller than the universe, we can regard differential equations as a suitable expression of the laws of nature. Differential equations, such as, e.g., Maxwell's equations, are summaries of our knowledge of the dynamics of physical systems. We can test these differential equations against experiment by observing systems which have been started with different initial conditions. But if we are dealing with the universe, the initial conditions are beyond our control and we cannot carry out such a test. The question is then, would it not be better to give up any attempt at formulating a differential equation of motion of the universe? Could it not be that the law of motion of the universe is something completely unrelated to Einstein's differential equations? The answer is that there really is no choice. Physics on a small scale determines physics on a large scale. Given that Einstein's equations hold in all regions of, say, the size of the solar system, it then follows that they also hold across the boundaries of these regions and throughout the universe. Einstein's equations therefore not only determine the local gravitational disturbances produced by the sun and planets, but also the gravitational effects of the universe as a whole.

There is, however, one extra detail that we have to take into account when we write Einstein's equations for the universe. A small cosmological term (see Eq. (7.45)) would have no appreciable effects inside the solar system, but could have very important effects on the evolution of the universe. We will therefore include such a term in Einstein's equations for the universe.

10.1 ISOTROPY AND HOMOGENEITY OF THE UNIVERSE

The stars around us that are visible to the naked eye are not uniformly distributed over the sky. These stars are part of a disc-shaped aggregate called the galaxy of the Milky Way. This galaxy contains about 10^{11} stars and measures $\sim 3 \times 10^4$ pc* across. There are many other galaxies; about 10^{10} can be seen with the 200-inch telescope. Even this large number represents only a small fraction of the total of all galaxies—there must be many that are too small, or too far away, to be seen.

The galaxies are not uniformly distributed over the universe. Rather, galaxies tend to cluster. Our galaxy is part of a cluster called the Local Group. This cluster contains about twenty galaxies and measures $\sim 10^6$ pc across. As clusters go, the Local Group is not very large. For example, the Virgo cluster contains over 2,000 galaxies, the Coma cluster about 1,000, and there are many clusters of several hundred galaxies (see Fig. 10.1). Distances between clusters are typically 10^7 pc.

The distribution of clusters is predominantly random. There may be some evidence that some of the clusters tend to be associated in superclusters, but this effect involves at most a small fraction of all clusters. On a scale of a few times 10^7 pc, the most remarkable feature of the universe is its lack of distinguishing features. On a large scale, the universe is quite homogeneous. The assertion that all positions in the universe are essentially equivalent, except for local irregularities, is known as the *Cosmological Principle*.

In discussing the distribution of galaxies, or clusters, we should really differentiate between the angular distribution over the celestial sphere and the distribution in depth, as a function of distance. Fig. 10.2 shows the distribution of clusters of galaxies over a small part of the sky. Obviously, the distribution of nearby (large angular size) clusters

---

* The pc (parsec) is a unit of distance; 1 pc $\cong$ 3.26 light years.

Fig. 10.1   *The Hercules cluster. The fuzzy objects are galaxies. This cluster contains a total of about 75 galaxies. (Hale Observatories photograph.)*

is far from uniform on the scale of this figure. But the distribution of faraway (small angular size, faint) clusters is remarkably uniform, even on a scale of a few degrees. Thus, the distribution is essentially isotropic, except for local irregularities.

There is a connection between isotropy and homogeneity: unless the earth occupies a special position in the universe, the isotropy of the distribution of galaxies implies homogeneity. The reason is that isotropy about the position of the earth requires that the density of galaxies be some function $n(r)$, independent of angular variables. Unless $n(r)$ is a constant, this would imply that the universe has a unique center of symmetry ($r = 0$) and the earth is at this center. If we make the likely assumption that the earth does *not* occupy such a privileged position, then the function $n(r)$ must be a constant and we have homogeneity.

We can also test the large-scale homogeneity of the universe by simply counting the number of galaxies of a given brightness. If the number of galaxies per unit volume is constant, then the total number within a distance l increases as l^3. Consider those galaxies that have a given intrinsic luminosity $\mathscr{L}$ (erg/sec); the brightness, or incident flux measured at the earth, is

$$S = \frac{\mathscr{L}}{4\pi l^2} \tag{1}$$

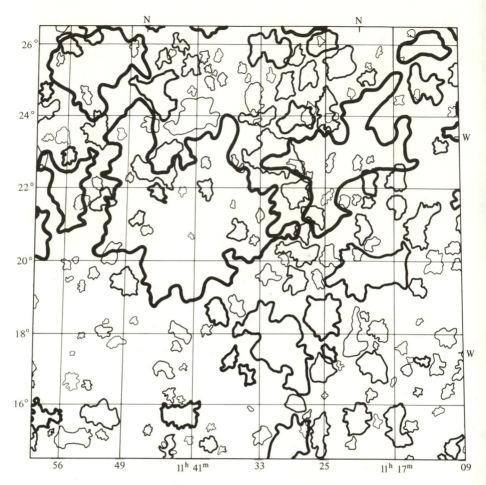

Fig. 10.2 *Clusters of galaxies in a small patch (12° × 12°) of the sky in the constellation Leo. The outlines of the nearest clusters are indicated by the heaviest lines, those of the farthest clusters by the lightest lines. (After Zwicky and Herzog,* Catalogue of Galaxies and of Clusters of Galaxies.)

This formula is based on the assumption that space is flat; the effects of the curvature of the universe only become noticeable at distances larger than 10^8 or 10^9 light years. According to Eq. (1), $l^3 \sim S^{-3/2}$ and therefore the number of such galaxies with a brightness in excess of S is proportional to $S^{-3/2}$,

$$N(>S) \propto S^{-3/2} \qquad [2]$$

The proportionality does not depend on the value of $\mathscr{L}$ and hence applies to the totality of all galaxies. Thus, a count of numbers of galaxies as a function of brightness can be used as a direct observational test of homogeneity.

For the brighter (hence nearer) galaxies, the counts are in good agreement with Eq. (2).[3] For very faint (far away) galaxies, Hubble found deviations from Eq. (2), the counts for the faintest galaxies being too low by a factor of about two. It is now believed that these deviations are due to systematic measurement errors. Peebles[4] has remarked that in any case Hubble's data do prove that even for distances as large as 10^9 pc the density of galaxies shows no drastic decrease; that is, there is no indication of any edge of the universe or end in the galactic distribution.

For radio sources, by which are meant both galaxies that are strong radio emitters (radio galaxies) and quasi-stellar objects (quasars), the counts are not in agreement with Eq. (2). There is an excess of sources of low brightness. This is believed to be an evolutionary effect. A source at a distance l is seen by us as it was at a time l/c ago. If the luminosity of the sources or the density of the sources is time dependent, then the relation (1) will be modified. An increase of $N(>S)$ results if we make the assumption that, on the average, the sources had a higher luminosity or a higher density (or both) in the past.

The evidence for evolutionary effects in radio sources can be used as an argument against the steady-state theory of the universe of Bondi, Gold, and Hoyle. In this theory the universe has no beginning and no end and the average properties of the universe are not only constant in space, but also constant in time. In such a universe new radio sources must be continually created to replace those that fade away or move away and the average luminosity and density must remain constant.

The conclusion that radio source counts contradict the steady-state theory was first reached by Ryle and Scheuer in 1955. Later, more precise counts by Ryle and Pooley and others have confirmed this conclusion.[5,6,31]

For quasars, the analysis of counts has also been carried out using their *optical,* rather than radio, brightness. According to Schmidt,[7] the counts indicate that the density of quasars must have been much higher in the past than it is now. This again contradicts the steady-state theory.

10.2 COSMIC DISTANCES

The methods used by astronomers to measure distances to objects in the sky vary with the distance involved. For the nearest stars (up to ~30 pc), triangulation with the diameter of the earth's orbit as base line is adequate. Some other geometrical methods give the slightly larger distances to nearby clusters of stars (e.g., the Hyades cluster at ~41 pc). Beyond a few hundred pc, distances are determined by the apparent brightness of a standard light source. If a light source, such as a star, has a luminosity $\mathscr{L}$ (erg/sec), then the apparent brightness S (erg/cm² sec) measured at the earth is

$$S = \frac{\mathscr{L}}{4\pi l^2}$$ [3]

where l is the distance. A measurement of the flux S therefore gives the distance l provided the luminosity $\mathscr{L}$ is known.

For distances within our galaxy ($\sim 10^5$ pc) main-sequence stars can be used as standard light sources of known luminosity. In main-sequence stars the luminosity and surface temperature, or, what amounts to the same thing, spectral type, are directly related. The relationship is given by the Hertzsprung-Russell diagram of Fig. 10.3.

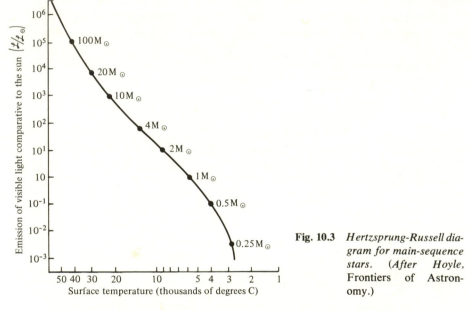

Fig. 10.3   *Hertzsprung-Russell diagram for main-sequence stars.* (*After Hoyle, Frontiers of Astronomy.*)

Inspection of the spectrum of light from a star will tell us the spectral type; the diagram then gives us $\mathscr{L}$ and Eq. (3) gives us the distance.

Distances within our local group ($\sim 10^6$ pc) can be measured by using variable stars of the Cepheid type as standard sources. These stars are very bright, 10^3 to $10^4 \mathscr{L}_\odot$, and they are not on the main sequence. The luminosity of these stars oscillates with a period of 2 to 40 days. What makes them useful as distance indicators is the relation between period and luminosity: the long-period Cepheids are intrinsically the brightest, the short-period Cepheids are the least bright. The period-luminosity relation is plotted in Fig. 10.4. This plot can be used for distance determinations in much the same way as Fig. 10.3.

To measure distances to nearby clusters of galaxies ($\sim 10^7$ pc), standard sources brighter than the Cepheids are needed. It has been found that in the galaxies of the local group, the brightest stars in a

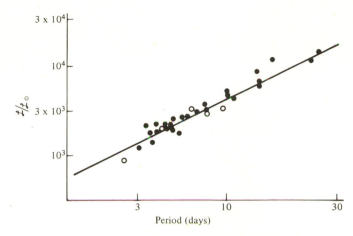

Fig. 10.4   *The period-luminosity relation for Cepheid stars. (After Kraft, Ap. J. **134**, 616 (1961).)*

galaxy usually have a luminosity of $\sim 4 \times 10^5 \mathscr{L}_\odot$; hence, the brightest stars in a galaxy may be regarded as a standard light source and their apparent brightness directly gives the distance. The brightest globular star cluster (see Fig. 10.5 for an example) in a galaxy may be used similarly.[8] The luminosity of the brightest globular star cluster in the Andromeda galaxy M31 is $\sim 10^6 \mathscr{L}_\odot$* and it is assumed that the brightest clusters in other, more distant, galaxies have the same luminosity.

Another method that can be used for this range of distances is based on measurement of the size of large luminous clouds of hydrogen that are present in galaxies (HII regions). Sandage and Tammann[10] found that the size of the clouds in a given galaxy is related to the luminosity of the galaxy. Hence, measurement of both apparent (angular) size of these clouds and apparent brightness of the galaxy will determine the distance.

Finally, if distances beyond those of the nearest clusters are sought, nothing will serve as standard light source except entire galaxies. For example, all galaxies of the spiral type ScI (see Fig. 10.6 for an example) have nearly equal luminosities (differences in luminosity do not exceed 50%). But in order to penetrate the most remote depths of space even ScI galaxies are not bright enough. It becomes necessary to use the brightest galaxy in a cluster of galaxies; these galaxies are usually of the elliptical type E (see Fig. 10.7 for an example of an elliptical galaxy). The luminosity of the brightest galaxy in the Virgo cluster is $\sim 8 \times 10^{10} \mathscr{L}_\odot$ ($M_B = -21.68$) and this is used as a standard luminosity for the brightest galaxy in remote clusters.

Fig. 10.8 summarizes different methods used for distance deter-

---

* This luminosity is only a rough number. Actual distance determinations are based directly on the photographic blue magnitude (M_B). This magnitude depends on the luminosity in a narrow range of wavelengths rather than on the total luminosity.

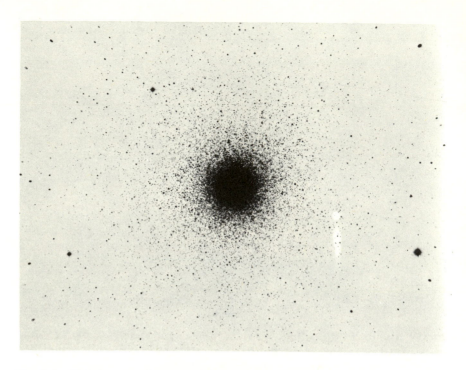

Fig. 10.5 *Globular star cluster (NGC 5272) in Canes Venatici. (Hale Observatories photograph.)*

Fig. 10.6 *Spiral galaxy (NGC 5457) in Ursa Major. (Hale Observatories photograph.)*

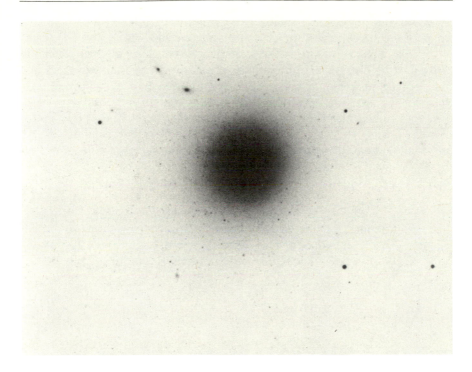

Fig. 10.7 *Elliptical galaxy (NGC 4486) in Virgo. (Hale Observatories photograph.)*

mination.[9] It must be emphasized that the measurement of large distances depends on the validity of the methods used for the lesser distances; the calibration of each method relies on all the methods that are to the left of it in Fig. 10.8. Thus, the method of "brightest galaxy" is calibrated

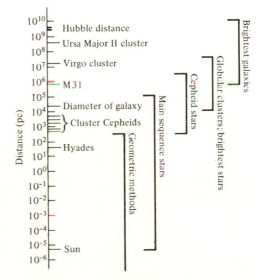

Fig. 10.8 *The cosmic distance ladder. The brackets indicate the range for which a particular method may be used. Each method depends on all the other methods that are to the left of it in the diagram. (After Weinberg, Gravitation and Cosmology.)*

by comparing it with the method of "brightest globular cluster" in the range of distances where the two methods overlap, etc.

Any mistake in one of the calibrations will affect all the determinations of larger distances. For example, it was shown by Baade in 1952 that the period-luminosity relation for Cepheids that had been in use until then was in error. A recalibration indicated that the determinations of distances by this method made previously had to be increased by a factor of two. This implied that all extragalactic distance determinations also had to be increased by this factor.

10.3 THE COSMOLOGICAL REDSHIFT AND THE EXPANSION OF THE UNIVERSE

The spectral lines of almost all galaxies display a shift towards lower frequencies, i.e., a redshift. The amount of redshift is directly proportional to the distance, except for some random irregularities. Fig. 10.9 is

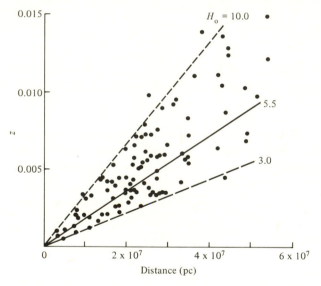

Fig. 10.9 *Observed redshift of galaxies as a function of distance. The straight line corresponds to $H_0 = 5.5 \ cm \ sec^{-1} \ pc^{-1}$. (After Sandage and Tammann, Ap. J. **196**, 313 (1975).)*

a plot of the redshift of some galaxies versus their distance. The redshift z is defined as

$$z = \frac{\lambda - \lambda_0}{\lambda_0} \qquad\qquad [4]$$

where λ is the wavelength received at the earth and λ_0 the wavelength emitted by the atoms in the remote galaxy.

The physical interpretation of the redshift is that it is a Doppler shift produced by the motion of the galaxies away from the earth. For velocities low compared to c, the fractional change in wavelength is given by the Doppler formula

$$\frac{\lambda - \lambda_0}{\lambda_0} = \frac{v}{c} \qquad [5]$$

A redshift directly proportional to the distance therefore indicates a recession velocity directly proportional to the distance. Mathematically, we can express this as

$$v = H_0 l \qquad [6]$$

where H_0 is known as *Hubble's constant*; Eq. (6) is *Hubble's law*. According to a recent determination by Sandage and Tammann,[10] the value of H_0 is

$$H_0 = (5.5 \pm 0.5) \text{ cm sec}^{-1} \text{ pc}^{-1} \qquad [7]$$

The uncertainty ± 0.5 cm sec^{-1} pc^{-1} only takes into account random errors in the data. Systematic errors could be much larger and could possibly change the value of H_0 by a factor of two. All we can say with confidence is that H_0 is somewhere between 3 and 12 cm sec^{-1} pc^{-1}.

Although according to (6) galaxies are moving radially away from us, this does not mean that the earth occupies a preferred position. Let us write (6) in vector notation:

$$\mathbf{v} = H_0 \mathbf{r} \qquad [8]$$

Consider now an observer on some galaxy at a position $\mathbf{r}'$. Relative to the earth this observer has a velocity

$$\mathbf{v}' = H_0 \mathbf{r}' \qquad [9]$$

Hence, relative to this observer, some other galaxy has a velocity

$$\mathbf{v} - \mathbf{v}' = H_0 (\mathbf{r} - \mathbf{r}') \qquad [10]$$

which shows that there is a general motion of galaxies radially away from this observer, just as there is such a motion away from an observer on earth. Hence Eq. (8) indicates a general expansion of the universe, with distances increasing in proportion.

Of course, Eq. (8) and the proportional increase in distances only apply to the average motion of galaxies. Individual galaxies may have a recession velocity somewhat larger or smaller than that given by Hubble's law (see Fig. 10.9). It should also be noted that the distances within a galaxy or within a (bound) cluster of galaxies do not increase.

The motion of expansion has all the appearances of having been generated by a gigantic explosion or *big bang*. It seems that initially all the matter was collected in a very dense fireball.* The violent internal pressures blew this fireball apart. As the fireball expanded and cooled, slight fluctuations in density grew into local condensations which evolved into gravitationally bound systems (galaxies and bound clusters of galaxies). These bound systems continued the average motion of expansion. Thus, the general motion of recession of the galaxies is directly related to the initial velocities that these galaxies had when they were first formed.

An important question is how the recession velocities change in time. Since the attractive gravitational forces oppose the expansion, we expect a decrease in the expansion rate. If the recession velocities decrease strongly enough, then the universe will reach a state of maximum size at some time in the future and then begin to contract.

Let us calculate the expected deceleration in the expansion rate. Since we are only interested in the average motion of the galaxies, we can approximate the mass distribution of the universe by an average uniform mass density. Suppose that in such a universe with a uniform mass density ρ we remove all the mass from a small spherical volume. Since the remaining mass distribution has spherical symmetry (isotropy) around the empty volume, the gravitational field in this volume will then be exactly zero. It now follows that if the mass is put back into the spherical volume, the motion of the mass will be completely unaffected by the rest of the universe. Consider then a particle on the surface of the sphere of mass; the acceleration of this particle is

$$\frac{dv}{dt} = -\frac{Gm}{l^2} \qquad [11]$$

where l is the distance to the center, and m the mass in the sphere. With $m = 4\pi l^3 \rho/3$, we obtain

$$\frac{dv}{dt} = -\frac{4\pi G\rho}{3}l \qquad [12]$$

This equation shows that the change in the relative velocity of two galaxies is proportional to the distance between them. Since Hubble's law tells us that the present velocities are proportional to the distance,

---

* This fireball should not be thought of as having been located at some preferred position in the universe; rather, it filled the entire universe.

it follows that the velocities remain proportional to the distance at all times. We can, therefore, write

$$v(t) = H(t)l(t) \qquad [13]$$

where $H(t)$ is the Hubble "constant" at time t. (H_0 is the Hubble constant at the present time.)

The rate of change of $H(t)$ is directly related to the mass density. To obtain this relation, we need only differentiate Eq. (13):

$$\frac{dv}{dt} = Hv + \frac{dH}{dt} l$$

$$= \left(H^2 + \frac{dH}{dt}\right) l \qquad [14]$$

It is customary to define a *deceleration parameter*,

$$q = -\left(1 + \frac{1}{H^2}\frac{dH}{dt}\right) \qquad [15]$$

so that

$$\frac{dv}{dt} = -qH^2 l \qquad [16]$$

Note that if the expansion is slowing down, then $q > 0$. If we compare Eqs. (12) and (16) we find

$$-qH^2 = -\frac{4\pi G}{3}\rho \qquad [17]$$

i.e.,

$$q_0 = \frac{4\pi G}{3H_0^2}\rho_0 \qquad [18]$$

where the subscript $_0$ indicates that the quantities are to be evaluated at the present time.

Note that Eq. (18) is equally valid in Newtonian theory and in geometrodynamics. The derivation of this equation depends on the absence of gravitational fields in a spherical cavity surrounded by a spherical mass distribution. According to Birkhoff's theorem (see Section 8.4), this is just as true in geometrodynamics as it is in Newtonian theory. Furthermore, if we assume that the cavity is of small dimension, then the velocity and acceleration will be small and therefore the non-relativistic equation of motion will be a valid approximation.

There is, however, one modification of Eq. (18) which we may want to consider. If the Einstein equations contain a cosmological constant, then we must take this into account. We know from Section 7.4 that the cosmological term has the effect of a uniform mass density $-\Lambda/4\pi G$ (see Eq. (7.46)). Thus, the mass density ρ in Eq. (12) must be replaced by $\rho - \Lambda/4\pi G$:

$$\frac{dv}{dt} = -\frac{4\pi G}{3}\left(\rho - \frac{\Lambda}{4\pi G}\right)l \qquad [19]$$

and our final result becomes

$$q_0 = \frac{4\pi G}{3H_0^2}\rho_0 - \frac{\Lambda}{3H_0^2} \qquad [20]$$

This equation will play an important role in the comparison of the theoretical models of the universe with observational data (see Section 10.12).

The direct measurement of the deceleration parameter is very difficult. It is possible to determine q_0 by looking for systematic deviations from Hubble's law at large distance. Since we see galaxies at large distances as they were at the time when they emitted the light that reaches us now, galaxies at large distances indicate the state of the universe a long time ago. If the expansion rate has been decreasing, we expect remote galaxies to show velocities in excess of those given by Hubble's law.

In the analysis of the data it must be taken into account that for large values of the redshift ($z \sim 1$), the connection between z and v is not linear. Even more important: the inverse square law (3) relating brightness to distance must be modified to take into account the curvature of spacetime and the decrease of the energy of light brought about by the reduction of the frequency of the light wave as it moves through the expanding universe (see Section 10.5). The precise form taken by these corrections depends on the model of the universe that one assumes. For a certain class of models, known as Friedmann models (see Section 10.9) Gunn and Oke[11] find

$$q_0 = 0.3 \pm 0.7 \qquad [21]$$

This value of q_0 is based on the assumption that there are no significant evolutionary effects in the luminosities of the galaxies used as distance indicators. If the luminosity of a galaxy decreases as it ages, then q_0 could be considerably smaller, perhaps as low as $q_0 \simeq -2$.[11]

A negative value of q_0 would indicate that the rate of expansion of the universe is *accelerating*. According to Eq. (20), a negative value of q_0 requires a positive value of Λ. For example, if $q_0 = -2$, then

$$\Lambda = 6H_0^2$$
$$\simeq 2 \times 10^{-35}/\text{sec}^2$$

where ρ_0 has been neglected. Since we are reasonably confident that q_0 is not much less than -2, we can regard $10^{-35}/\text{sec}^2$ as a rough upper limit on Λ (see Section 7.4).

An alternative method for the determination of q_0 makes use of the fact that in a Friedmann model with a given value of H_0, the value of q_0 is uniquely related to the age of the universe (i.e., the Friedmann model is uniquely characterized by *two* parameters). Using an age of 15×10^9 years (see Section 10.4), Sandage and Tammann[10] obtain the result

$$q_0 = 0.10 \begin{cases} +0.16 \\ -0.08 \end{cases} \tag{22}$$

Unfortunately, this result is quite sensitive to systematic errors in the age; for example, if the age is $\sim 10 \times 10^9$ years, then one would obtain $q_0 \simeq 1$.

Finally, some remarks about quasars and their redshifts. The largest redshift observed for a galaxy is $z = 0.46$. However, the quasi-stellar objects have redshifts much larger than this—redshifts two or four times as large are quite common. The largest quasar redshifts yet observed are $z = 3.40$ and $z = 2.88$ for the quasars QSO(OH 471) and 4C 05.34, respectively. If $z = 3.40$, the received wavelength differs from the emitted wavelength by the factor of

$$\frac{\lambda}{\lambda_0} = 1 + z = 4.40$$

Obviously, for such a large redshift we cannot use the approximate Doppler formula (5). If we use the formula

$$\frac{\lambda}{\lambda_0} = \left(\frac{1+v}{1-v} \right)^{1/2}$$

given by special relativity,* then $v = 0.90$ and Hubble's law gives $l = 5 \times 10^9$ pc. For such large distances Hubble's law is actually not valid, but in any case it is clear that the quasars are at very large distances.

The flux radiated by a quasar is enormous—at least 10^{48} erg/sec. Furthermore, large changes in brightness have been observed to occur in a relatively short time. For example, 3C 446 varies in brightness by a factor of two on a time scale as short as one day. This implies that the

---

* This means we ignore the curvature of spacetime. The exact redshift formula for the Friedmann model is given in Section 10.11.

object is quite small, no more than one light day across (see Exercise 9.14).

The reliability of quasar redshifts as distance indicators has been the subject of much controversy. Obviously, if the quasars are nearby objects their energy requirements would be much reduced. But if the redshift is not related to the general motion of recession of the universe, it must involve some other mechanism. For example, the quasars may be fairly small nearby objects that have been ejected at relativistic speeds from the nucleus of our galaxy, or perhaps the redshift has nothing to do with speed and is of gravitational origin.

There is now some very good evidence against such a local hypothesis for quasars. Gunn and Oke[12] have discovered several cases of quasars associated with clusters of galaxies. These quasars have the *same* redshift as the galaxies in the cluster. Hence in these cases there can be no doubt that the redshifts are cosmological, that is, they arise from the motion of recession of the universe and indicate a large distance.

If we want to use quasars to test for deviations from Hubble's law, we must first find a method of distance determination that does not depend on the redshift. Hills and Bahcall[13] have suggested that the brightest quasar (at a given redshift) can be used as a standard light source. The results obtained by this method are consistent with $q_0 \simeq 1$ but the uncertainties are very large.

10.4 THE AGE OF THE UNIVERSE

The present expansion of the universe indicates that the average density of matter is decreasing. If we extrapolate the expansion backwards in time, then in the past the universe must have been much more dense. Under the assumption that the recession velocity v of Eq. (6) remains constant for any given galaxy, the time needed for complete collapse is $v/l = H_0^{-1}$. This time

$$H_0^{-1} = \frac{1}{5.5} \sec \frac{\text{pc}}{\text{cm}} = 1.8 \times 10^{10} \text{ years} \qquad [23]$$

is called the *Hubble age*. The true age of the universe is probably somewhat less than (23) because the recession velocities in the past were somewhat larger than they are now.

As stated in the preceding section, the observed expansion suggests an explosive origin of the universe. Some finite time ago the universe began as a very dense and very hot fireball which then flew apart in a primordial explosion; the universe began with the *big bang*. Galaxies and stars were gradually formed as the gas expanded and cooled.

Steady-state theories, such as that of Bondi, Gold, and Hoyle, avoid the singular origin of the universe demanded by the big bang.

In these theories it is postulated that matter is created, out of nothing, in the empty space between galaxies or perhaps in the very active nuclei of some galaxies. It is assumed that the rate of creation is such that the mean density of the universe remains constant in spite of the expansion. Extrapolation of the expansion backwards in time does then not lead to a denser state of the universe because the spontaneously created matter disappears.

Such a steady-state universe lasts forever and forever, with no essential changes. On the macroscopic scale the universe is then time independent and hence has symmetry under time translations (homogeneity in time). Although in a superficial way this is appealing, in a more fundamental way time symmetry has been lost. The steady-state theory violates energy conservation and such a violation implies that the fundamental laws of physics must have an explicit time dependence — that is, the laws of physics lack symmetry under time translations.* Thus, the steady-state theory achieves symmetry on the macroscopic scale by giving it up on the microscopic scale. Since conservation laws, and in particular the conservation of energy, are very fundamental to physics, we prefer not to give them up unless forced to do so by the evidence. As was mentioned in Section 10.1, the evidence from radio source counts is against, rather than in favor of, steady-state theories.

Although the age of the universe need not agree exactly with H_0^{-1}, we may guess that the expansion rate has not been changing very drastically and that therefore the age agrees roughly with H_0^{-1}. How old, then, is the universe?

Radioactive dating methods give for the age of solid bodies in the solar system (rocks on earth and moon, meteorites) a value of $(4.6 \pm 0.1) \times 10^9$ years.[14] As an illustration of these dating methods, we briefly discuss the decay of U^{238}. This uranium isotope decays, by a series of α and β disintegrations, into Pb^{206}, a stable isotope of lead. The decay rate† for this process is $\lambda = 1.55 \times 10^{-10}$/year. If we designate the abundance, or number of atoms in a sample, by a square bracket, then the abundances of U^{238} and Pb^{206} are related by

$$[Pb^{206}]_0 = [Pb^{206}]_t + (e^{\lambda t} - 1)[U^{238}]_0 \tag{24}$$

In this equation time is reckoned backwards from the present. Thus $t = 0$ is now and $t = t$ is the age of the sample.

EXERCISE 1. Derive Eq. (24). Assume that uranium decays directly into lead with a decay rate λ; this is a valid approximation because all the intermediate

---

* The arguments presented in the appendix show that energy is conserved unless the Hamiltonian has an explicit time dependence.

† The relation between decay rate λ and half-life $T_{1/2}$ is $\lambda = 0.693/T_{1/2}$.

steps in the decay chain proceed very fast (nearly instantaneously) compared with the first step.

The measurable quantities are the isotopic abundance ratios. It is therefore convenient to divide Eq. (24) by the abundance of Pb^{204}; this is a stable isotope which is not produced by radioactive decay whose abundance is therefore a constant, $[Pb^{204}]_0 = [Pb^{204}]_t$. Then

$$\left[\frac{Pb^{206}}{Pb^{204}}\right]_0 = \left[\frac{Pb^{206}}{Pb^{204}}\right]_t + (e^{\lambda t} - 1)\left[\frac{U^{238}}{Pb^{204}}\right]_0 \qquad [25]$$

Because the isotopes Pb^{206} and Pb^{204} have the same chemistry, all minerals formed at time t will have the same value of $[Pb^{206}/Pb^{204}]_t$; thus, this relative abundance depends on t but not on the type of mineral. Measurement of $[Pb^{206}/Pb^{204}]_0$ and $[U^{238}/Pb^{204}]_0$ in several types of minerals of the same age will determine the "unknowns" $[Pb^{206}/Pb^{204}]_t$ and $e^{\lambda t} - 1$ in Eq. (25); the age then follows from the known value of λ.

The age of solid bodies in the solar system only sets a lower, although very rigorous, limit on the age of the universe. Another limit is set by the age of the chemical elements. Almost all elements, with the exception of hydrogen, some of the helium, and a few other light elements, were formed by nuclear reactions in the interior of stars. In particular, the uranium and thorium found on the earth were produced by massive stars that lived, and died, before the solar system was born. Astrophysical theory allows one to calculate the relative abundance with which U^{238} and Th^{232} must have been produced in the nuclear reactions:

$$\left[\frac{Th^{232}}{U^{238}}\right] = 1.9 \pm 0.3 \qquad [26]$$

At present, the measured abundance ratio of these isotopes is

$$\left[\frac{Th^{232}}{U^{238}}\right] = 3.9 \pm 0.5 \qquad [27]$$

The increase of the relative abundance is due to the lower decay rate of Th^{232} as compared to U^{238}; the decay rate of the former is 4.95×10^{-11}/year, of the latter 1.55×10^{-10}/year. If it is assumed that all of the uranium and thorium were produced very suddenly and only changed by radioactive decay ever since, it is a simple matter to calculate the time needed to change (26) into (27),

$$t_0 = 6.8 \times 10^9 \text{ years} \qquad [28]$$

However, it is possible that production took place gradually, so that, for a while, the abundance was determined by the balance between generation and decay. This complicates the analysis somewhat and can increase the age given by (28) by a factor of about two.

The ages of some other elements may be calculated similarly; numbers comparable to the preceding are obtained. These calculations place the age of elements in our galaxy somewhere between 6×10^9 and $\sim 15 \times 10^9$ years.[14]

The age of globular star clusters in our galaxy gives us another clue to the age of the universe. The theory of stellar evolution permits us to calculate the age of the stars in these clusters from their observed temperature and luminosity; both of these quantities change in a predictable way with the age of the star. Ages obtained by this method lie somewhere between 8×10^9 and 18×10^9 years.[14]

Although the universe must be older than the stars and the chemical elements, it is believed that star formation and nucleosynthesis began soon ($\sim 10^8$ years) after the big bang. Hence the age of globular clusters and the age of elements may be regarded as equal to the age of the universe, to within the stated uncertainties. It is very satisfying that the age of the universe arrived at by these arguments agrees with the Hubble age, just as the big bang theory leads us to expect. On the other hand, in the steady-state theory, the Hubble "age" is not the age of anything, and the agreement is nothing but a (very unlikely) numerical coincidence.

10.5 THE COSMIC BLACKBODY RADIATION

At a very early epoch, the dense and hot universe must have contained thermal blackbody radiation at a very high temperature. As the universe expanded, the density of this radiation would have decreased but the radiation should still be present, filling the interstellar and intergalactic space around us even now.

The effect of expansion on the thermal radiation amounts to a reduction of the temperature of the radiation. To see what happens to the thermal radiation in some region of space, imagine that this region, of size much smaller than the size of the universe, is enclosed in a container with perfectly reflecting walls. Suppose that this container expands at the same rate as the universe. Under these assumptions the radiation in the region inside the container will behave in the same way as the radiation in some region outside the container. The reason is that while in the case of a region surrounded by walls, photons are reflected and cannot escape, in the case of a region without walls, any photons that do escape are, on the average, simply replaced by similar photons which enter the region from the surrounding space.

It is now easy to calculate what happens to the radiation in the expanding container. At temperature T, the number of photons in some normal mode of oscillation of the container is given by statistical mechanics as

$$n(\nu) = \frac{1}{e^{h\nu/kT} - 1} \qquad [29]$$

where ν is the frequency of the mode. If the linear dimension of the container increases by a factor ξ, then the wavelength of the mode must increase by the same factor and hence the frequency decreases to the value $\nu' = \nu/\xi$. If the expansion proceeds sufficiently slowly (adiabatically), the number of photons remains constant as their frequency changes. Hence after expansion by the factor ξ, Eq. (29) will give the number of photons at the frequency ν'

$$n(\nu') = \frac{1}{e^{h\nu/kT} - 1}$$

$$= \frac{1}{e^{\xi h\nu'/kT} - 1} \qquad [30]$$

Obviously (30) still corresponds to a spectrum of the same shape but with the new temperature

$$T' = T/\xi \qquad [31]$$

This shows that expansion by, say, a factor of two changes thermal radiation of a given temperature into thermal radiation of one-half the initial temperature.

In the first detailed calculation of nucleosynthesis in the early universe, Alpher, Bethe, and Gamow[15] estimated the temperature of the primordial fireball as $\sim 10^9$ °K at a time of ~ 200 sec after the big bang. From this Alpher and Herman[16] estimated the present temperature of the remaining blackbody radiation at ~ 5 °K. Such a low temperature implies that the maximum in the spectral distribution of the radiation (Planck's law) is in the millimeter wavelength range.

Radiation of this kind was discovered in 1965 by Penzias and Wilson[17] with a very sensitive horn antenna originally constructed for a satellite communication system. The radiation was identified as the cosmic fireball radiation by Dicke, Peebles, Roll, and Wilkinson.[18] These first measurements were carried out at a wavelength of several cm and they gave the radiation temperature of ~ 3° K.

Since then many measurements of the intensity of the radiation have been performed at many different wavelengths. For the range of wavelengths between ~ 100 and 0.3 cm, ground-based radiometers may be used, but for wavelengths below 0.3 cm the radiometer must be carried above the atmosphere by a rocket or balloon because the atmosphere absorbs the radiation at these wavelengths.

It is also possible to obtain the blackbody radiation temperature of interstellar space by a measurement of the absorption of starlight by interstellar clouds of molecules, such as cyanogen (CN). These clouds are in thermal equilibrium with the blackbody radiation. At zero temperature all the molecules would be in the ground state, but at nonzero temperature some of the excited states are populated and this affects the absorption of starlight by these molecules. This method is mainly useful for setting an upper limit on the temperature.

Fig. 10.10 summarizes the measured intensities as a function of wavelength.[19] The curve drawn through the data points is the Planck spectrum for a blackbody temperature of $T = 2.7\ °K$; the good agreement is evidence that we are in fact observing thermal radiation.

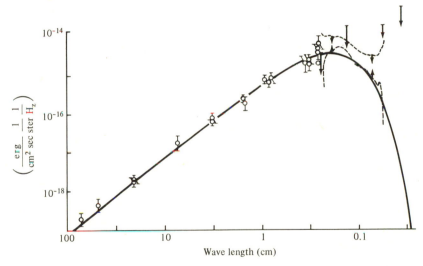

Fig. 10.10 *The energy spectrum of the cosmic microwave radiation. The quantity i_ν is the flux per unit frequency interval and unit solid angle. The arrows indicate upper limits that have been obtained from data on the absorption of starlight by interstellar gas. The dashed lines indicate limits obtained with infrared detectors. The solid line corresponds to a blackbody spectrum with $T = 2.7\ °K$.*

The angular distribution of the radiation has also been measured. Here one is interested in both small scale and large scale anisotropies. On a small scale, anisotropy may be due to the presence of hot spots in the early universe; the extra thermal radiation emitted by such spots may still be around and would be reaching us from some particular direction. Measurements have shown that angular irregularities in the blackbody temperature amount to less than 1% at an angular scale of a few minutes of arc.[20] This means that the radiation is even more isotropic than the distribution of galaxies.

Large-scale anisotropies are expected if the universe on the whole is anisotropic. Furthermore, the motion of the earth through the blackbody radiation background would result in an angular dependence of the temperature: radiation incident from forward would have a higher effective temperature. A deviation of the velocity of the earth from that corresponding to the uniform expansion of the universe is very likely since the solar system moves at ~250 km/sec around the center of the galaxy; furthermore, the galaxy is falling towards the Andromeda nebula (M31) which is part of our Local Cluster; and finally, the entire Local Cluster is presumably attracted by, and falling towards, the Virgo cluster. The uniform expansion law of Hubble is only valid if these peculiar velocities of galaxies, and clusters of galaxies, are averaged out. The observed large scale anisotropies (on a scale of 180° to 360°) in the temperature amount to no more than ~0.1%. From this one can deduce that the velocity of the earth is no larger than ~300 km/sec.[21] Incidentally, this measurement of the velocity of the earth does not conflict with the principle of relativity. What is measured is not an absolute velocity, but rather a velocity *relative* to the radiation background.

The existence of blackbody radiation with the expected temperature lends support to the big-bang theory. In a steady-state universe we would not expect any thermal radiation. Of course, there might be some microwave radiation, emitted by suitable radio sources, with a spectrum sufficiently close to the blackbody spectrum of Fig. 10.10 so as to mislead us. It is hard to believe that nature is so malicious. Furthermore, it is unlikely that discrete sources, distributed perhaps in the same way as galaxies, could give us microwave radiation with a nearly perfect isotropy over the sky.

10.6 THE MASS DENSITY

In order to apply Einstein's equations to the dynamics of the universe, we need to know the energy-momentum tensor of the universe and in particular the mass density. In view of the large-scale homogeneity of the universe, it will be sufficient to deal with the *mean* mass density, i.e., the mass density averaged over volumes large compared to $\sim (10^7)^3$ pc^3.

The most visible contribution to the mass density is that due to galaxies. The mass of spiral galaxies, in which stars and gas move in nearly circular orbits about the galactic center, can be determined by measuring the orbital velocity of stars and gas as a function of radius ("rotation curve"). The mass density and total mass of the galaxy are then obtained by fitting the density so that there is agreement between the theoretical and observed rotation curves.

In the case of binary systems of galaxies (pairs of galaxies in a tightly bound orbit about each other), the mass can also be determined by using the virial theorem. For two galaxies of equal mass, this theorem takes the form

$$\tfrac{1}{4} M \overline{v^2} = \tfrac{1}{2} G M^2 \left(\overline{\frac{1}{r}} \right) \tag{32}$$

The relative velocity v can be obtained by Doppler shift measurements and the distance r between the galaxies follows from the angular size and distance of the system. The equation is valid in a statistical sense for an ensemble consisting of a large number of binary systems, hence the averages $\overline{v^2}$ and $(\overline{1/r})$ must be used in this equation.*

It is usual to express the mass in terms of a mass-to-light ratio $M/\mathscr{L}$ (mass divided by luminosity). For spiral galaxies the value of this ratio obtained by the above methods is typically 1 to 20 times $M_\odot/\mathscr{L}_\odot$. For elliptical galaxies it is about $50 M_\odot/\mathscr{L}_\odot$. For all galaxies taken together, the average value of the mass to light ratio is $\sim 20 M_\odot/\mathscr{L}_\odot$.

To obtain the mass density in the universe, we must multiply the mass-to-light ratio by a factor of (number density of galaxies) × (average luminosity per galaxy). According to a survey by Oort, the value of this factor is $1.6 \times 10^{-10} \mathscr{L}_\odot/pc^3$ and therefore the mean galactic mass density is[22]

$$\begin{aligned} \rho_0 &\simeq 3 \times 10^{-9} M_\odot/pc^3 \\ &\simeq 2 \times 10^{-31} \ g/cm^3 \end{aligned} \tag{33}$$

This density could be subject to systematic errors. One source of such errors is the mass-to-light ratio. If one applies the virial theorem, in a form somewhat more general than (32), to *clusters* of galaxies, one finds mass-to-light ratios which are considerably larger than the values given above. Thus, for the Coma cluster one finds $M/\mathscr{L} \sim 200 M_\odot/\mathscr{L}$. Even in our own Local Group, measurements of distances and velocities give a mass-to-light ratio $M/\mathscr{L} \sim 100 M_\odot/\mathscr{L}_\odot$ for the group as a whole.

Ostriker and Peebles[29,30] have suggested that for spiral galaxies the typical value of the mass-to-light ratio is actually $\sim 200 M_\odot/\mathscr{L}_\odot$. This would mean that there is a large amount of invisible mass in or near such galaxies. A likely explanation is that these galaxies are surrounded by extensive "halos," consisting of a large number of stars of very low mass (red dwarfs with $M \simeq 0.1 M_\odot$) which are distributed over a large sphere concentric with the galaxy. Such stars would give off next to no visible light and they would therefore increase the mass of the galaxy, as measured from outside the halo, but not the luminosity. The existence of halos is also suggested by a study of the dynamical stability of galaxies.

* Since the angle between r and our line of sight is not known, it is also necessary to average over this angle.

By numerical integration of the equations of motion, it was found that a rotating disclike distribution of stars is unstable — within a short time the stars collect in a shapeless clump. But if the rotating disc is enclosed in a spherically symmetric mass distribution, of a mass and size at least as large as the disc, then the rotating disc remains stable.[29]

It is also possible that there exists a fair amount of matter in the intergalactic space within clusters. Such intergalactic matter could be in the form of dispersed stars, or dwarf galaxies (neither of which would show up on photographic plates), or perhaps black holes.

One can place a limit on the mass density of intergalactic stars. Although their light would be too faint to leave an image on a photographic plate, the light would make a contribution to the average brightness of the night sky. Observations have set a limit on the extragalactic contribution to this brightness and indicate that the mass density of starlike intergalactic objects is no higher than ~ 5 times the value given by (33).[22]

Black holes in the intergalactic space within clusters would be unobservable. If their mass is larger than $\sim 10^{10} M_\odot$ they could disrupt galaxies by collision, but if their mass is less than that their presence in a cluster would only show up as an increase of the mass-to-light ratio of the cluster. If the large mass-to-light ratios of clusters are entirely blamed on the presence of such invisible forms of matter, then the mass of this matter may come to ~ 10 times the visible galactic mass.

Deep intergalactic space (between clusters) contains a certain amount of electromagnetic energy. For example, the blackbody radiation at $2.7°K$ contributes 4×10^{-34} g/cm^3, and radio waves, starlight (optical), and X-rays contribute smaller amounts. Besides, there are cosmic rays with a density no larger than $\sim 10^{-35}$ g/cm^3.

Other contributions to the mass are highly speculative. Intergalactic atomic hydrogen, perhaps left over from the big bang, would show up by its absorption of 21-cm radio waves in the spectrum of an extragalactic radio source. The absence of such an absorption line permits one to set a limit of 2×10^{-30} g/cm^3 on the hydrogen mass density (this corresponds to $\sim 10^{-6}$ atoms/cm^3). It is also possible to look for the ultraviolet Lyman-α absorption line in the spectrum of the light reaching us from quasars. This test is much more sensitive than that using the 21-cm line and sets a limit of 5×10^{-36} g/cm^3 on the hydrogen mass density.[22] A similar limit can be set on the density of molecular hydrogen. Of course, these results depend on quasars as the source of the light that is supposedly absorbed by hydrogen; if the quasars are not cosmological (not at the distances indicated by their redshifts), then these results are not valid.

It is possible that intergalactic hydrogen does not exist in the form of neutral atoms, but rather in the form of a totally ionized plasma at $\sim 10^6$ °K. Ultraviolet radiation and cosmic rays emitted by the galaxies could have heated the hydrogen to this temperature. Such a plasma would not signal its presence by producing absorption lines. If dense enough, the plasma could be detected because Bremsstrahlung would lead to the emission of soft X-rays. The available experimental limits on the X-ray flux therefore set a limit of 10^{-29} g/cm^3 on the plasma density.[22]

Table 10.1 lists the known and conjectural contributions to the mean mass density of the universe. It is also conceivable that neutrinos and gravitational radiation are present with densities much in excess of the numbers given in the table. Because neutrinos and gravitational radiation interact so weakly with matter, they are very hard to detect. Furthermore, deep intergalactic space could contain a large density of black holes; these would be quite undetectable. (Actually, if the mass of a black hole is very large, of the order of $10^{11}M_\odot$, then it can act as a lens and produce focusing effects on the light emitted by a remote galaxy. No such focusing effects have been observed. This rules out very massive black holes, but leaves open the question of black holes of mass comparable to that of the sun.)

The estimates for the mass density presented in this section, and listed in Table 10.1, are sensitive to the cosmological scale of distances. If our determination of distances is in error, then the mass densities will also be. Since the value of the Hubble constant H_0 may be regarded as a summary of our knowledge of cosmic distances, it is customary to say that the mass densities depend on the value of H_0. For example, the galactic mass density (33) varies as H_0^2; if $H_0 = 5.5$ cm sec^{-1} pc^{-1} then $\rho_0 = 2 \times 10^{-31}$ g/cm^3, but if the true value of H_0 is twice as large,

TABLE 10.1   THE MEAN MASS DENSITY OF THE UNIVERSE

Form of Matter	Mean Mass Density (g/cm^3)[a]
Galaxies	2×10^{-31}
Galactic halos, and intergalactic stars, dwarf galaxies, or black holes (in clusters)	$\lesssim 2 \times 10^{-30}$
Electromagnetic radiation	
blackbody (2.7 °K)	4×10^{-34}
radio waves	10^{-40}
starlight (optical)	10^{-35}
X-rays	10^{-37}
Cosmic rays	10^{-35}
Intergalactic hydrogen (atomic)	$< 4 \times 10^{-36}$ g/cm^3 (2×10^{-30})[b]
Intergalactic hydrogen (molecular)	$< 5 \times 10^{-34}$
Intergalactic plasma ($\sim 10^6$ °K)	$\leq 10^{-29}$
Total protons, neutrons, and other baryons	$\leq 6 \times 10^{-31}$[c]

[a] Densities in this table are based on the value $H_0 = 5.5$ cm sec^{-1} pc^{-1} for Hubble's constant.
[b] The second number does not assume a cosmological redshift for quasars.
[c] From theory of deuterium production.

then the true value of this density is four times as large. The other mass densities listed in Table 10.1 are subject to similar uncertainties.

We can bypass some of the uncertainties in the mass density caused by the uncertainties in H_0 by defining a *density parameter*

$$\sigma_0 = \frac{4\pi}{3}\frac{G\rho_0}{H_0^2}$$

[34]

The density parameter that corresponds to the galactic mass density listed in Table 10.1 is

$$\sigma_0 = 0.02$$

This is independent of the value of H_0.

The galactic mass density of Table 10.1 represents a lower limit on the mass density of the universe. An *upper* limit on the mass density can be obtained from an interesting argument regarding the abundance of deuterium in the universe.[27] Normal main-sequence stars, such as the sun, produce deuterium by the burning of hydrogen. However, the deuterium thus produced is rapidly converted into tritium, He^3, and heavier elements and therefore a normal star will destroy deuterium faster than it creates it—that is, a star that has an initial deuterium abundance will tend to reduce this abundance. For example, our sun started with an abundance of about 20 parts deuterium in 10^6 parts hydrogen (by number) and is now down to 4 parts in 10^6. If stars do not make deuterium, and do not bequeath it to other stars in the way they bequeath heavy elements, then where does the deuterium come from? The answer seems to be that it was around before the stars were born— it was produced in the primordial fireball of the big bang. Between 100 and 1,000 seconds after the big bang, the temperature of the fireball is around 10^9 °K which is the right temperature to cook deuterium out of protons. However, if the density of protons, neutrons, etc., is excessively large at this time, then the deuterium will be quickly converted into tritium, He^3, He^4, etc., and little of it will remain. The observed abundance of deuterium (20 parts per 10^6 hydrogen) therefore sets an upper limit on the mass density of protons and neutrons in the universe. This limit translates into a present density of

$$\rho_0 < 6 \times 10^{-31} \text{ g/cm}^3$$

Of course, all of this hinges on the assumption that the big bang really is the source of the observed deuterium. Supernovae have been suggested as an alternative source, but there are arguments against this possibility.[24,27] Note that the above result is independent of any uncertainties in H_0; but if we want to express our limit on ρ_0 in terms of a limit on σ_0, then the uncertainties in H_0 will come in. Using the smallest value of H_0 (3 cm sec^{-1} pc^{-1}) gives the upper limit

$$\sigma_0 < 0.2$$

on the density parameter.

10.7 COMOVING COORDINATES

In order to write down the Einstein equations for the dynamics of the universe we will ignore the small-scale deviations from uniformity. We want to concentrate on the large-scale features of the metric and we will assume that on the large scale the universe is isotropic and homogeneous. The mass in the universe will be treated as a fluid, with a constant density throughout. The galaxies, or clusters of galaxies, may be regarded as the particles out of which this fluid is made; this assumes that we discount local irregularities in the motion of galaxies and take into account only the motion of uniform expansion.

By taking advantage of the symmetry (homogeneity and isotropy) of the universe, we can simplify the metric considerably. We begin by specifying the coordinates we will use: they will be *comoving* coordinates. By this is meant that the spatial coordinates partake of the uniform motion of expansion of the matter in the universe. If we ignore the local irregularities in the motion of galaxies (irregular deviations from Hubble's law), we may say that each galaxy carries its spatial coordinates with it; the coordinate points move with the galaxies as the latter fall freely in the background gravitational field of the universe. The coordinate interval between any two galaxies remains forever constant and the expansion of the universe results not from a change in the coordinate position of the galaxies, but rather from a change in the metric of the space.

For the time coordinate x^0 we will use the proper time measured by clocks carried by the galaxies. We will assume that these clocks not only run at the same rate, but also that they are *synchronized*. In general synchronization is a very tricky problem, but in a uniform universe a simple operational procedure for synchronization is available. An observer on a galaxy A sends out a flash of light when his clock reads time T_0; an observer on galaxy B also sends out a flash of light when *his* clock reads T_0. The flash from A is received at B when the clock at the latter point reads T_B; the flash from B is received at A when the clock there reads T_A. The clocks are synchronized if $T_A = T_B$. This method depends on the uniformity of the universe; homogeneity and isotropy guarantee that both light signals will have the same travel time. An alternate synchronization method is also available: all observers set their clocks so that they read zero at the instant of the big bang.

We can show that with the above choice of coordinates,

$$g_{00} = 1 \qquad\qquad\qquad\qquad\qquad\qquad\qquad\qquad [35]$$

$$g_{0k} = 0 \qquad k = 1, 2, 3 \qquad\qquad\qquad\qquad\qquad [36]$$

This means that our comoving coordinates are also time orthogonal (see Eq. (7.38)).

Eq. (35) follows from the assumption that the clocks used to measure x^0 are at rest in the coordinates. For such a clock $dx^1 = dx^2 = dx^3 = 0$ and hence

$$d\tau^2 = g_{00}(dx^0)^2$$

This gives (35) because, by definition of x^0, we must have $d\tau = dx^0$.

To derive Eq. (36) we introduce local geodesic coordinates x'^μ, with $g'_{\mu\nu} = \eta_{\mu\nu}$, at the position of some given galaxy. By a Lorentz transformation we can make sure that the origin of geodesic coordinates has the same instantaneous velocity as the galaxy. Hence the galaxy has zero velocity in these coordinates,

$$\frac{dx'^n}{dt} = 0 \qquad\qquad\qquad\qquad\qquad\qquad\qquad\qquad [37]$$

Since $dt = dx^0$ we can also write this as

$$\frac{\partial x'^n}{\partial x^0} = 0 \qquad\qquad\qquad\qquad\qquad\qquad\qquad\qquad [38]$$

Furthermore, the clock of any neighboring galaxy is synchronized with that of the given galaxy. If the coordinate interval between these galaxies is dx^k, then in the local geodesic coordinates synchronization can be expressed as

$$x'^0(x^k + dx^k) = x'^0(x^k)$$

i.e.,

$$\frac{\partial x'^0}{\partial x^k} = 0 \qquad\qquad\qquad\qquad\qquad\qquad\qquad\qquad [39]$$

With the aid of (38) and (39) we can now calculate g_{0k}:

$$g_{0k} = \frac{\partial x'^\mu}{\partial x^0} \frac{\partial x'^\nu}{\partial x^k} g'_{\mu\nu} = \frac{\partial x'^\mu}{\partial x^0} \frac{\partial x'^\nu}{\partial x^k} \eta_{\mu\nu}$$

$$= \frac{\partial x'^0}{\partial x^0} \frac{\partial x'^0}{\partial x^k} - \frac{\partial x'^n}{\partial x^0} \frac{\partial x'^n}{\partial x^k}$$

$$= 0 \qquad\qquad\qquad\qquad\qquad\qquad\qquad\qquad\qquad [40]$$

Since we have chosen our coordinates so that galaxies are for-ever at rest, we must check that this is consistent with the geodesic equation of motion. Given that a galaxy is at rest at some initial time, will it remain at rest? To see that this is so, we only need to write down the geodesic equation of motion for a particle at rest $[dx^\mu/d\tau = (1,0,0,0)]$:

$$\frac{d^2x^k}{d\tau^2} = -\Gamma^k{}_{00} \qquad [41]$$

But

$$\Gamma^k{}_{00} = \tfrac{1}{2}g^{kl}(g_{l0,0} + g_{0l,0} - g_{00,l}) = 0 \qquad [42]$$

according to (35) and (36). Hence the acceleration (41) vanishes.

10.8 THE THREE-GEOMETRY

In view of (35) and (36) the metric of spacetime reduces to

$$ds^2 = (dx^0)^2 - g_{kn}dx^k dx^n \qquad [43]$$

It is customary to write this as

$$ds^2 = dt^2 - dl^2 \qquad [44]$$

where $dt = dx^0$ is simply called the time interval and

$$dl^2 = -g_{kn}dx^k dx^n \qquad [45]$$

is the space interval. In order to emphasize the three-dimensional nature of the space interval, we will write (45) as

$$dl^2 = {}^{(3)}g_{kn}dx^k dx^n \qquad [46]$$

where

$${}^{(3)}g_{kn} = -g_{kn} \qquad [47]$$

The 3×3 tensor ${}^{(3)}g_{kn}$ describes the geometry of space at a given instant of time.

What can we say about the geometry of three-dimensional space at a given instant of time? This space is supposed to be homogeneous and isotropic: the geometry should not distinguish between different points nor between different directions about a point. In order to dis-cover the consequences of these symmetry requirements, let us con-centrate on the curvature tensor ${}^{(3)}R_{mnsk}$ of the three-geometry.

It follows from the isotropy and homogeneity of the three-dimensional space that $^{(3)}R_{mnsk}$ must have the form

$$^{(3)}R_{mnsk} = K \left[^{(3)}g_{ms}\,^{(3)}g_{nk} - \,^{(3)}g_{mk}\,^{(3)}g_{ns} \right] \qquad [48]$$

where K is some constant. Eq. (48) may be justified as follows: At a given point, introduce geodesic coordinates so that the metric becomes $^{(3)}g_{mk}{}' = \delta_m{}^k$. The isotropy of space demands that it must not be possible to distinguish between different directions by their curvature; i.e., the curvature tensor $^{(3)}R'_{mnsk}$ must be left unchanged by rotations of the geodesic coordinates. Since the unit tensor $\delta_m{}^k$ is the only tensor unchanged by rotation, the curvature tensor must be some combination of unit tensors,

$$^{(3)}R'_{mnsk} = K\delta_m{}^s\delta_n{}^k + K_1\delta_m{}^k\delta_n{}^s + K_2\delta_m{}^n\delta_k{}^s \qquad [49]$$

The antisymmetry relation $^{(3)}R_{mnsk} = -^{(3)}R_{nmsk}$ requires that $K = -K_1$ and $K_2 = 0$; hence,

$$^{(3)}R'_{mnsk} = K(\delta_m{}^s\delta_n{}^k - \delta_m{}^k\delta_n{}^s) \qquad [50]$$

Transformation of this equation from geodesic coordinates back to the original coordinates gives the form (48). The quantity K must be a constant (the same everywhere) in order to satisfy the condition of homogeneity. This conclusion also follows from isotropy: if there is some point at which the gradient of K is different from zero, then the direction of this gradient defines a preferred direction in space, in contradiction with isotropy. Thus, isotropy at all points implies that K is a constant, i.e., it implies homogeneity.

EXERCISE 2. Show that

$$^{(3)}R_{ns} = -2K^{(3)}g_{ns} \qquad [51]$$

$$^{(3)}R = -6K \qquad [52]$$

Since the curvature tensor $^{(3)}R_{mnsk}$ can be expressed in terms of the metric $^{(3)}g_{kn}$ and the first and second derivatives of this metric (see Eq. (6.77)), Eq. (48) may be regarded as a differential equation for $^{(3)}g_{kn}$. We want to find a solution of this equation.

(I) THE CASE OF POSITIVE CURVATURE Rather than try to solve Eq. (48) by brute force, we will use geometrical arguments. First consider the analogous *two-dimensional* problem. What two-dimensional space has uniform (homogeneous and isotropic) curvature? Obviously,

the surface of an ordinary sphere has this property; this surface is usually called a two-sphere.*

To obtain a three-dimensional space of uniform curvature we take the surface of a four-dimensional hypersphere; this surface is called a three-sphere. The equation of the surface of a four-dimensional hypersphere is, in rectangular coordinates,

$$(x^1)^2 + (x^2)^2 + (x^3)^2 + (x^4)^2 = a^2 \qquad [53]$$

where a is the "radius" of the sphere. The distance between any two nearby points on the surface is

$$dl^2 = (dx^1)^2 + (dx^2)^2 + (dx^3)^2 + (dx^4)^2 \qquad [54]$$

Note that the coordinate x^4 has nothing to do with time; it is simply an extra, unphysical, coordinate that must be introduced if one wants to pretend that the curved three-dimensional space is a subspace of a flat Euclidean space.**

Eq. (53) can be used to eliminate the unphysical coordinate x^4. This yields a distance

$$dl^2 = (dx^1)^2 + (dx^2)^2 + (dx^3)^2 + \frac{(x^1dx^1 + x^2dx^2 + x^3dx^3)^2}{a^2 - (x^1)^2 - (x^2)^2 - (x^3)^2} \qquad [55]$$

In this expression there appear only the physical coordinates x^1, x^2, x^3. Eq. (55) gives us the desired metric for a space of uniform curvature.

It is easy to check that the curvature tensor for the metric (55) does satisfy Eq. (48) with

$$K = \frac{1}{a^2} \qquad [56]$$

EXERCISE 3. Check this. (Hint: It is sufficient to check Eq. (48) near the origin; since the geometry described by (55) is uniformly curved (by construction), what holds at one point must hold at all. In the vicinity of the origin it is sufficient to approximate

$$^{(3)}g_{kn} \simeq \delta_k{}^n + \frac{x^k x^n}{a^2} \cdot \Big) \qquad [57]$$

---

* Flat two-dimensional space is a special case of a two-sphere of infinite radius.

** The three-dimensional space is *embedded* in the four-dimensional space (compare Exercise 9.8).

A geometry with a positive value of K is said to have *positive curvature*. To study this geometry further, we begin by introducing "polar" coordinates in the usual way:

$$x^1 = r \sin \theta \cos \phi$$
$$x^2 = r \sin \theta \sin \phi$$
$$x^3 = r \cos \theta \qquad\qquad\qquad [58]$$

We can then express the space interval (55) as

$$dl^2 = \frac{dr^2}{1 - r^2/a^2} + r^2 d\theta^2 + r^2 \sin^2 \theta d\phi^2 \qquad\qquad [59]$$

Note that the "rectangular" coordinates x^k and the "radial" coordinate r are actually periodic coordinates. If we start at the "top" of the three-sphere ($x^4 = a$), move off in the x^1-direction, and continue straight ahead, we reach the equator ($x^4 = 0$), then the bottom of the sphere ($x^4 = -a$); if we continue straight ahead we again pass through the equator and finally return to the top. The values of x^1 for the top, the first equatorial crossing, the bottom, and the second equatorial crossing are, respectively, $x^1 = 0$, $x^1 = a$, $x^1 = 0$, $x^1 = -a$, and the values of r are, respectively, $r = 0$, $r = a$, $r = 0$, $r = a$ (see Fig. 10.11). That there are two distinct points with the same value of x^1 and r need not bother us too much; we can resolve the ambiguity by, say, writing coordinates for points in the upper hemisphere in red ink, and points in the lower in blue.

To gain some feeling for our space of positive curvature, let us look at the radius and circumference of a circle placed in this space.

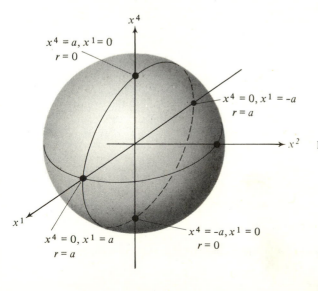

Fig. 10.11 *Section* $x^3 = 0$ *through a 3-sphere. The equation for this section is* $(x^1)^2 + (x^2)^2 + (x^4)^2 = a^2$; *i.e., it is the surface of an ordinary sphere. A path that goes from* $x^4 = a$ *to* $x^4 = -a$ *to* $x^4 = a$ *is shown.*

For convenience take the circle defined by $r = b = $ constant around the origin. This circle has as radius the distance between $r = 0$ and $r = b$,

$$l = (\text{radius}) = \int_0^b \frac{dr}{\sqrt{1 - r^2/a^2}} = a \, \sin^{-1} (b/a) \qquad [60]$$

The circumference of the circle is obtained in the usual way. For example, if the circle is in the plane $\theta = \pi/2$,

$$(\text{circumference}) = \int_0^{2\pi} b \, \sin \theta \, d\phi = 2\pi b \qquad [61]$$

The ratio of radius to circumference is therefore *larger* than $1/2\pi$; this is a familiar property of spaces of positive curvature.

Note that for a radius larger than $\pi a/2$, the circumference *decreases* as the radius increases.

The surface area of the sphere $r = b = $ constant surrounding the origin is

$$(\text{area}) = \int_0^{2\pi} \int_0^{\pi} b^2 \, \sin \theta \, d\theta d\phi = 4\pi b^2 \qquad [62]$$

Hence the ratio of the radius squared to the area is larger than $1/4\pi$. For a radius larger than $\pi a/2$, the area decreases as the radius increases. The volume inside the sphere $r = b$ is

$$(\text{volume}) = \int_0^{2\pi} \int_0^{\pi} \int_0^b \frac{r^2}{\sqrt{1 - r^2/a^2}} \, dr \, \sin \theta \, d\theta d\phi$$

$$= 4\pi \left(\frac{a^3}{2} \sin^{-1} \frac{b}{a} - \frac{ba^2}{2} \sqrt{1 - b^2/a^2} \right) \qquad [63]$$

EXERCISE 4. Show that if $b \ll a$, then Eq. (63) gives a volume $4\pi b^3/3$ and explain why this makes sense.

To obtain the total volume of the three-sphere we must take $b = 0$, $\sin^{-1} b/a = \pi$ ("bottom" of sphere). The total volume is then $2\pi^2 a^3$. Our three-sphere is a *closed space;* it has a finite volume even though it has no boundaries.

Since r is a periodic coordinate, it is convenient to introduce a new angular coordinate χ such that

$$r = a \, \sin \chi, \qquad 0 < \chi < \pi \qquad [64]$$

The coordinate χ has an advantage over the coordinate r: it is single-valued. In terms of χ,

$$dl^2 = a^2[d\chi^2 + \sin^2 \chi(d\theta^2 + \sin^2 \theta d\phi^2)] \tag{65}$$

Note that the radial distance from the origin is, according to (60), simply

$$l = (\text{radial distance}) = a\chi \tag{66}$$

(II) THE CASE OF NEGATIVE CURVATURE   Next we want to find a homogeneous and isotropic geometry of negative curvature ($K < 0$). The metric for such a geometry may be obtained from (55) by simply replacing a^2 by $-a^2$:

$$dl^2 = (dx^1)^2 + (dx^2)^2 + (dx^3)^2 + \frac{(x^1 dx^1 + x^2 dx^2 + x^3 dx^3)^2}{-a^2 - (x^1)^2 - (x^2)^2 - (x^3)^2} \tag{67}$$

The geometrical arguments that precede Eq. (55) are, of course, not applicable when this replacement is made.

Since the metric (55) satisfies the condition (48) on the curvature tensor, the metric (67) will also satisfy this condition, the algebra being the same in both cases. The only difference is that the value of K will now be negative,

$$K = -\frac{1}{a^2} \tag{68}$$

In "polar" coordinates (67) becomes

$$dl^2 = \frac{dr^2}{1 + r^2/a^2} + r^2 d\theta^2 + r^2 \sin^2 \theta d\phi^2 \tag{69}$$

The radius of a circle $r = b$ around the origin is

$$l = (\text{radius}) = a \sinh^{-1} (b/a) \tag{70}$$

This makes the ratio of radius to circumference smaller than $1/2\pi$. Likewise, for a sphere, the ratio of radius squared to area is smaller than $1/4\pi$. The volume of a sphere with $r = b$ is

$$(\text{volume}) = 4\pi \left(-\frac{a^3}{2} \sinh^{-1} \frac{b}{a} + \frac{ba^2}{2} \sqrt{1 + b^2/a^2} \right) \tag{71}$$

As $b \to \infty$, this volume diverges. The space of negative curvature is open and infinite.

We can introduce a dimensionless coordinate χ such that

$$r = a \sinh \chi \tag{72}$$

In terms of this coordinate,

$$dl^2 = a^2[d\chi^2 + \sinh^2 \chi(d\theta^2 + \sin^2 \theta d\phi^2)] \qquad [73]$$

This is the analogue of (65).

(III) *THE CASE OF ZERO CURVATURE*   Obviously, Eq. (48) also has the trivial solution $^{(3)}R_{mnks} = K = 0$. This corresponds to flat, Euclidean space. At first sight, such a flat geometry seems of no interest since we certainly expect that the matter distribution in the universe produces some curvature. However, we must keep in mind that matter generates curvature in the four-geometry; it need not generate curvature in the three-geometry. As we will see, it is possible to construct solutions of Einstein's equations such that the three-geometry is flat at each instant of time, but the four-geometry is curved.*

The flat space is of course open and infinite.

Table 10.2 summarizes the properties of our homogeneous, isotropic three-geometries of constant curvature.

TABLE 10.2   HOMOGENEOUS, ISOTROPIC THREE-GEOMETRIES

	Curvature Constant	Metric^a	Radial Distance from Origin	Three-Volume
(i) Positive curvature	$K = \dfrac{1}{a^2}$	$dl^2 = \dfrac{dr^2}{1 - r^2/a^2} + r^2 d\Omega^2$	$l = a \sin^{-1} \dfrac{r}{a}$	finite, closed
(ii) Negative curvature	$K = \dfrac{-1}{a^2}$	$dl^2 = \dfrac{dr^2}{1 + r^2/a^2} + r^2 d\Omega^2$	$l = a \sinh^{-1} \dfrac{r}{a}$	infinite, open
(iii) Zero curvature	$K = 0$	$dl^2 = dr^2 + r^2 d\Omega^2$	$l = r$	infinite, open

$^a d\Omega^2 \equiv d\theta^2 + \sin^2 \theta d\phi^2$

10.9   THE FRIEDMANN MODELS

Although the three-dimensional geometry, of positive or negative curvature, can be described equally well by the coordinates (x^1, x^2, x^3), (r, θ, ϕ), and (χ, θ, ϕ), these coordinates are not equally good for describing the expanding universe. Only the coordinates (χ, θ, ϕ) may be regarded as comoving with the galaxies. The reason is that isotropy and homogeneity of the expansion require that if $dl(t)$ is the distance between two galaxies

---

* Note that this depends on the choice of coordinates. If we introduce some new coordinates, which mix the old space and time coordinates, then the three-geometry in the new coordinates will in general not be flat.

at time t, then the distance at time $t + \Delta t$ must be proportional to the initial distance.

$$dl(t + \Delta t) = A(t)dl(t) \qquad [74]$$

where the proportionality factor A depends on time, but *not on the position in space*. Eq. (74) simply says that when the distance between a pair of galaxies increases by a factor of A, then the distance between any other pair also increases by the same factor.

In the expressions (55), (59), and (65) for the metric, the only quantity that can change in time is a; an expanding universe will have a time-dependent value of a. Only in the last of these metrics does a enter in such a way that (74) is satisfied. Thus, the coordinates (χ,θ,ϕ) may be regarded as comoving coordinates with respect to which the galaxies are at rest.

The time dependence of a is determined by the Einstein equations. We will solve these equations in each of the cases considered in the last section.

In order to write down the Einstein equations, we must also make some assumptions about the energy-momentum tensor of the matter in the universe. We will assume that the energy-momentum tensor is that of a uniform gas with a density and a pressure. The galaxies may be regarded as the particles out of which this gas is made, and since the velocities of the galaxies do not deviate much from uniform expansion, we can actually neglect the pressure of the gas (in general, kinetic pressure is an indication of an r.m.s. deviation of the velocities from uniform velocity). However, in the very early, very hot universe, the pressure due to radiation and particles must have been very important (see Section 10.13).

We will call the general homogeneous, isotropic models of the universe with mass density, pressure and, possibly, a cosmological term, the *Lemaitre models*. The special models with zero pressure and zero cosmological constant are usually called *Friedmann models*.* We begin with a study of the latter.

(1) POSITIVE CURVATURE   The spacetime interval for the closed isotropic universe is

$$
\begin{aligned}
ds^2 &= dt^2 - dl^2 \\
&= dt^2 - a^2(t)[d\chi^2 + \sin^2 \chi(d\theta^2 + \sin^2 \theta d\phi^2)]
\end{aligned}
\qquad [75]
$$

The solution of the Einstein equations is simplified if we replace the time coordinate t by a time parameter η defined by

---

* There is no general agreement on the naming of cosmological models; the terminology adopted here is as good as any.

$$dt = ad\eta \tag{76}$$

Then

$$ds^2 = a^2(\eta)[d\eta^2 - d\chi^2 - \sin^2\chi\,(d\theta^2 + \sin^2\theta d\phi^2)] \tag{77}$$

and the metric tensor is

$$g_{\mu\nu} = \begin{pmatrix} a^2 & & & 0 \\ & -a^2 & & \\ & & -a^2\sin^2\chi & \\ 0 & & & -a^2\sin^2\chi\sin^2\theta \end{pmatrix} \tag{78}$$

Some of the Christoffel symbols are easy to calculate:

$$\Gamma^0_{00} = \frac{\dot{a}}{a} \qquad \Gamma^0_{kn} = -\frac{\dot{a}}{a^3}\,g_{kn} \qquad \Gamma^k_{on} = \frac{\dot{a}}{a}\,\delta_n{}^k$$

$$\Gamma^0_{k0} = 0 \qquad \Gamma^k_{00} = 0 \tag{79}$$

where

$$\dot{a} \equiv da/d\eta \tag{80}$$

Note that, contrary to the usage of Chapter 8 (see Eqs. (8.31)–(8.34)), the dot does *not* stand for the derivative with respect to proper time. The latter derivative is

$$\frac{da}{d\tau} = \frac{da}{dt} = \frac{1}{a}\frac{da}{d\eta} = \frac{1}{a}\,\dot{a}$$

EXERCISE 5. Check the results given in Eq. (79).

The R_{00} component of the Ricci tensor is entirely determined by (79):

$$R_{00} = \frac{3}{a^2}\,(a\ddot{a} - \dot{a}^2) \tag{81}$$

The R_{kn} components can be easily evaluated by relating them to $^{(3)}R_{kn}$; this saves some labor. We begin with the definition

$$R_{kn} = R^{\alpha}{}_{kn\alpha} = R^0{}_{kn0} + R^l{}_{knl} \tag{82}$$

The term $R^0{}_{kn0}$ only involves Christoffel symbols of the type (79). The term $R^l{}_{knl}$ equals

$$R^l_{knl} = -\Gamma^l_{kn,l} + \Gamma^l_{kl,n} + (\Gamma^0_{kl}\Gamma^l_{on} + \Gamma^s_{kl}\Gamma^l_{sn})$$
$$- (\Gamma^0_{kn}\Gamma^l_{0l} + \Gamma^s_{kn}\Gamma^l_{sl}) \quad [83]$$

In this expression, the Christoffel symbols with one zero index are known from (79); the remaining symbols, with purely spatial indices, add up to the Ricci tensor of 3-space:

$$R^l_{knl} = {}^{(3)}R^l_{knl} + \Gamma^0_{kl}\Gamma^l_{on} - \Gamma^0_{kn}\Gamma^l_{0l} \quad [84]$$

If we insert

$${}^{(3)}R^l_{knl} = \frac{1}{a^2}{}^{(3)}g^{lm}[{}^{(3)}g_{mn}{}^{(3)}g_{kl} - {}^{(3)}g_{ml}{}^{(3)}g_{kn}] = \frac{2}{a^2}g_{kn} \quad [85]$$

according to (48), then everything on the left side of (84) is determined. The result is

$$R_{kn} = \frac{1}{a^4}(2a^2 + \dot{a}^2 + a\ddot{a})g_{kn} \quad [86]$$

The curvature scalar is then given by

$$R = R^0_0 + R^k_k = g^{00}R_{00} + g^{kn}R_{kn} = \frac{6}{a^3}(a + \ddot{a}) \quad [87]$$

The Einstein equation, including the cosmological constant, is

$$R_\mu^\nu - \tfrac{1}{2}\delta_\mu^\nu R = -8\pi GT_\mu^\nu - \Lambda\delta_\mu^\nu \quad [88]$$

We will concentrate on the $_{00}$ component of this equation,

$$R_0^0 - \tfrac{1}{2}R = -8\pi GT_0^0 - \Lambda \quad [89]$$

With (81) and (87) this becomes

$$-\frac{3}{a^4}(a^2 + \dot{a}^2) = -8\pi GT_0^0 - \Lambda \quad [90]$$

Note that in this equation $\ddot{a}$ has canceled; we only have to solve a first-order equation. This equation is sufficient to determine $a(t)$ if T_{00} is given. The other components of the Einstein equation can be ignored because it turns out that they tell us nothing different from (90).

We must now decide what to use for the energy density T_0^0 of matter. According to Table 10.1 the main contribution would be due to "particles" of one kind or another (galaxies, hydrogen gas, etc.). We will neglect the pressure that these particles might exert. The energy-momentum tensor is then

$$T_\mu{}^\nu = \rho u_\mu u^\nu \tag{91}$$

where ρ is the proper mass density. In our comoving coordinates, the matter is at rest and hence

$$T_0^0 = \rho \tag{92}$$

EXERCISE 6. Derive Eq. (92). (Hint: Show that $u^0 = d\eta/d\tau = 1/a$ and $u_0 = a$.)

Since the volume of the universe varies in time as a^3, it follows that the density varies as a^{-3} and we can write

$$\rho(t) = \frac{M}{2\pi^2 a^3} \tag{93}$$

where M is a constant. Since $2\pi^2 a^3$ is the total volume of the universe, we find it convenient to say that M is the total "mass of the universe." Of course, from an operational point of view, it is meaningless to talk of the mass of the universe: in order to measure the mass of a system one must place oneself outside of the system and measure how the system reacts to pushes and pulls; this cannot be done if the system is the whole universe. Furthermore, according to (93), M is really the sum of all the *proper* masses of the particles in the universe; we know from special relativity that such a sum will not give the total mass since it fails to take into account kinetic and binding energies.

Our general dynamical equation (90) now becomes

$$\frac{3}{a^4}(\dot{a}^2 + a^2) = 8\pi G\rho + \Lambda$$

$$= \frac{4GM}{\pi a^3} + \Lambda \tag{94}$$

In the Friedmann models, it is assumed that the cosmological constant is either exactly zero, or else that it is so small that it can be neglected compared to ρ. Under this assumption, Eq. (94) reduces to

$$\frac{3}{a^4}(\dot{a}^2 + a^2) = \frac{4GM}{\pi a^3} \tag{95}$$

This is the differential equation that describes the *closed Friedmann model.*

The equation of motion (95) can also be derived without making explicit use of Einstein's equations. In Section 10.3 we gave a simple argument which led to the following equation of motion for the distance between galaxies (see Eq. (11), but include the correction for the cosmological term):

$$\frac{dl^2}{dt^2} = -\frac{Gm}{l^2} + \frac{\Lambda}{3}\, l \qquad\qquad [96]$$

If this is multiplied by dl/dt,

$$\frac{dl}{dt}\frac{d^2l}{dt^2} = \frac{dl}{dt}\left(-\frac{Gm}{l^2} + \frac{\Lambda}{3}\, l\right) \qquad\qquad [97]$$

i.e.,

$$\frac{d}{dt}\left[\tfrac{1}{2}\left(\frac{dl}{dt}\right)^2\right] = \frac{d}{dt}\left(\frac{Gm}{l} + \frac{\Lambda}{6}\, l^2\right) \qquad\qquad [98]$$

The integral of (98) is

$$\tfrac{1}{2}\left(\frac{dl}{dt}\right)^2 + \text{constant} = \frac{4\pi}{3}\, G\rho l^2 + \frac{\Lambda}{6}\, l^2 \qquad\qquad [99]$$

where the substitution $m = 4\pi\rho l^3/3$ has been made.

To compare this with (94), we note that distances in the universe vary as $a(t)$, i.e., $l(t) \propto a(t)$. Furthermore, by (76) and (80), .

$$\frac{dl}{dt} \propto \frac{da}{dt} = \frac{da}{d\eta}\frac{d\eta}{dt} = \frac{\dot a}{a}, \qquad\qquad [100]$$

which results in

$$\tfrac{1}{2}\left(\frac{\dot a}{a}\right)^2 + \text{constant} = \frac{4\pi G}{3}\, \rho a^2 + \frac{\Lambda}{6}\, a^2 \qquad\qquad [101]$$

This equation has the same form as (94), except that the value of the constant is left undetermined.

The solution of Eq. (95) is

$$a(\eta) = a_*(1 - \cos\,\eta) \qquad\qquad [102]$$

where

$$a_* = 2GM/3\pi \qquad\qquad [103]$$

By integrating (76) we can find the connection between the time t and the time parameter η,

$$t = a_*(\eta - \sin\,\eta) \qquad\qquad [104]$$

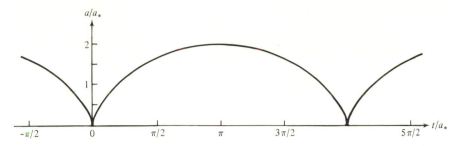

Fig. 10.12   *Radius of curvature of the closed Friedmann universe as a function of time. The curve is a cycloid.*

Eq. (102) and (104) may be regarded as parametric equations for $a(t)$; the curve represented by these equations is a cycloid. Fig. 10.12 shows a plot of a as a function of t. At $t = 0, \pm 2\pi a_*, \pm 4\pi a_*, \ldots,$ etc., $a(t)$ vanishes — that is, the universe contracts to a point. Since the density will become very large when this is about to happen, our approximate expression for the energy-momentum tensor will fail. For example, it will obviously be necessary to take into account the pressure exerted by matter and radiation in the dense, hot universe. We should also keep in mind the possibility that the classical Einstein equations become inapplicable at very high densities; quantum effects will be one obvious source of difficulties. It is therefore not clear exactly what happens at the singular points of Fig. 10.12 and we do not know whether the universe actually has the periodic behavior suggested by this figure.

We will concentrate on one expansion-contraction cycle, with the big bang at $t = 0$, the moment of maximum expansion at $t = \pi a_*$ and total, or nearly so, collapse at $t = 2\pi a_*$.

(II) NEGATIVE CURVATURE   For the case of the open isotropic universe, a calculation similar to that carried out in Eqs. (75)–(94) leads to an equation of motion

$$\frac{3}{a^4}(\dot{a}^2 - a^2) = 8\pi G\rho + \Lambda \qquad [105]$$

$$= \frac{4GM}{\pi a^3} + \Lambda \qquad [106]$$

Here M is defined in exactly the same way as in Eq. (93),

$$\rho = \frac{M}{2\pi^2 a^3}$$

Since the volume of the space is infinite, M cannot be interpreted as a total mass; but we may say that M is the sum of rest masses contained in a volume $2\pi^2 a^3$.

EXERCISE 7. Derive Eq. (106).

Again, we introduce the assumption that Λ can be neglected so that

$$\frac{3}{a^4}\,(a^2 - a^2) = \frac{4GM}{\pi a^3} \tag{107}$$

The model of the universe described by (107) is called the *open Fried-mann model*.

The solution of (107) is

$$a = a_*(\cosh \eta - 1) \tag{108}$$

It follows that

$$t = a_*(\sinh \eta - \eta) \tag{109}$$

Fig. 10.13 shows a as a function of t. The universe begins with a big bang and continues to expand forever. As $t \to \infty$, the universe gradually becomes flat. Again, the state near the singularity (at $t = 0$) is not correctly described by our equations; some of the required modifications are understood, but the closer we approach $t = 0$, the less we understand.

EXERCISE 8. Use Eqs. (108) and (109) to show that the universe becomes flat as $t \to \infty$.

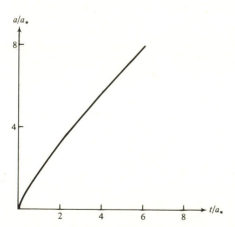

Fig. 10.13 *Radius of curvature of the open Friedmann universe as a function of time. Note that for small values of* t, *the behavior is similar to that shown in Fig. 10.12.*

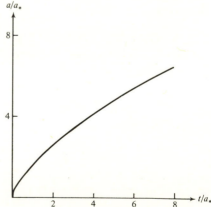

Fig. 10.14 *Radius of curvature of the flat Friedmann universe as a function of time. The behavior at small values of* t *is similar to that shown in Figs. 10.12 and 10.13.*

(III) ZERO CURVATURE   In the case of a universe with a flat three-geometry, we can write the spacetime interval as

$$ds^2 = a^2(\eta)[d\eta^2 - dr^2 - r^2 d\theta^2 - r^2 \sin^2 \theta d\phi^2] \qquad [110]$$

so that

$$g_{\mu\nu} = \begin{pmatrix} a^2 & & & 0 \\ & -a^2 & & \\ & & -a^2 r^2 & \\ 0 & & & -a^2 r^2 \sin^2 \theta \end{pmatrix}$$

The equation of motion can then be derived as in Eqs. (78)–(95) with the result

$$\frac{3}{a^4}(\dot{a}^2) = 8\pi G\rho + \Lambda$$

$$= \frac{4GM}{\pi a^3} + \Lambda \qquad [111]$$

where M is defined as before (see Eq. (93)).

In the Friedmann case $\Lambda = 0$ and with $\dot{a} = a\,da/dt$, Eq. (111) reduces to

$$\left(\frac{da}{dt}\right)^2 = \frac{4GM}{3\pi}\frac{1}{a} \qquad [112]$$

which has the solution

$$a(t) = \left(\frac{3GM}{\pi}\right)^{1/3} t^{2/3} \qquad [113]$$

This function is plotted in Fig. 10.14. As $t \to \infty$, the four-geometry tends to become flat.

10.10   THE EMPTY LEMAÎTRE MODELS

The Lemaître models differ from the Friedmann models in that the cosmological constant is not zero. For the sake of simplicity, we will assume that the cosmological constant is much larger than $8\pi G T^0{}_0$. We can then neglect $T^0{}_0$ altogether and regard the universe as empty. The behavior of the universe will then be controlled by the cosmological term; this term roughly has the same effect as a uniform mass density, constant in space and time (see Eq. (7.46)). A positive value of Λ corresponds to a negative effective mass density (repulsion) and a negative value of Λ corresponds to a possitive mass density (attraction). Hence, we expect that in universes with a positive value of Λ the expansion will tend to accelerate (monotonic behavior), while in universes with a negative value of Λ the expansion will slow down, stop, and reverse (oscillatory behavior). The calculations confirm this.

(I) POSITIVE CURVATURE   If we neglect T^0_0 in Eq. (90), we obtain

$$-\frac{3}{a^4}(a^2 + \dot{a}^2) = -\Lambda$$

With $\dot{a} = ada/dt$, this becomes

$$\left(\frac{da}{dt}\right)^2 = -1 + \Lambda a^2/3 \qquad [114]$$

It is immediately obvious from Eq. (114) that $-1 + \Lambda a^2/3$ cannot be negative. This implies that $\Lambda > 0$, and that the value of a can never be less than $\sqrt{3/\Lambda}$. Thus, the empty Lemaître model with positive curvature requires a positive cosmological constant, and its radius of curvature is never zero (no big bang).

The integration of Eq. (114) is elementary and gives

$$a(t) = \sqrt{\frac{3}{\Lambda}} \cosh\left[\sqrt{\frac{\Lambda}{3}}(t - t_{min})\right] \qquad [115]$$

where t_{min} is the time at which a has its minimum value. Fig. 10.15a gives a plot of $a(t)$. For $t > t_{min}$, the universe expands monotonically, and as $t \rightarrow \infty$ the universe becomes flat.

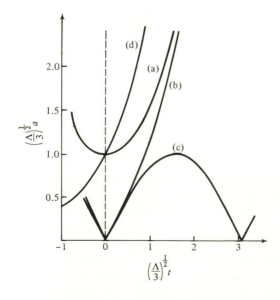

Fig. **10.15**   *Radius of curvature of the empty Lemaître universe as a function of time.*
(a) Closed model, $\Lambda > 0$.
(b) Open model, $\Lambda > 0$.
(c) Open model, $\Lambda < 0$.
(d) Flat model, $\Lambda > 0$.

EXERCISE 9. Integrate (114) and obtain (115).

(II) NEGATIVE CURVATURE   In this case, the equation analogous to Eq. (114) is

$$\left(\frac{da}{dt}\right)^2 = 1 + \Lambda a^2/3$$

which can be integrated to give

$$a(t) = \sqrt{\frac{3}{\Lambda}}\ \sinh\sqrt{\frac{\Lambda}{3}}\ t \qquad \text{for } \Lambda > 0$$

$$a(t) = \sqrt{\frac{3}{|\Lambda|}}\ \sin\sqrt{\frac{|\Lambda|}{3}}\ t \qquad \text{for } \Lambda < 0 \qquad [116]$$

These functions are plotted in Figs. 10.15b and c, respectively. Note that both universes begin with a big bang at $t = 0$. The first one expands monotonically, while the second one oscillates with a period $2\pi\sqrt{3/|\Lambda|}$. Of course, in our actual universe, the mass density near the big bang must have been very large and hence the empty Lemaître models cannot be used to describe the behavior near $t = 0$.

(III) ZERO CURVATURE   Our equation of motion is now

$$\left(\frac{da}{dt}\right)^2 = \Lambda a^2/3 \qquad [117]$$

This equation only makes sense if $\Lambda > 0$, and it then has the solution

$$a(t) = a(0)\ \exp\left(\sqrt{\Lambda/3}\ t\right) \qquad [118]$$

This universe expands exponentially with a characteristic doubling time of $0.693\ \sqrt{3/\Lambda}$ (see Fig. 10.15d). The model described by Eq. (118) is usually called the *de Sitter model*. Note that Eq. (117) also has an exponentially decreasing solution, but this is of no relevance to our (expanding) universe.

Table 10.3 summarizes the characteristics of the Friedmann and empty Lemaître models. The differential equation for the general Lemaître model with both a matter density and a cosmological term cannot be integrated in terms of elementary functions. However, these general models can be roughly described as some combination of the cases listed in Table 10.3.

Consider a universe that begins with a big bang. In the early universe the mass density is very large, and we can neglect the cosmological term—i.e., we have a Friedmann universe. As the universe expands and

TABLE 10.3 ISOTROPIC, HOMOGENEOUS MODELS

	Three-Volume	Time Dependence
Friedmann, $\Lambda = 0$		
(i) Positive curvature	finite	oscillatory
(ii) Negative curvature	infinite	monotonic (for $t > 0$)
(iii) Zero curvature	infinite	monotonic (for $t > 0$)
Empty Lemaître, $\Lambda \neq 0$		
(i) Positive curvature, $\Lambda > 0$	finite	monotonic (for $t > t_{min}$)
(ii) Negative curvature, $\Lambda > 0$	infinite	monotonic (for $t > 0$)
Negative curvature, $\Lambda < 0$	infinite	oscillatory
(iii) Zero curvature, $\Lambda > 0$	infinite	monotonic

the mass density decreases, the cosmological term will become more important. In the Friedmann cases of negative or zero curvature, the expansion and the decrease in mass density are monotonic (see Figs. 10.13 and 10.14) and hence the cosmological term, if any, will ultimately dominate the behavior of the universe. Thus, the universe gradually turns into an empty Lemaître universe of negative or zero curvature. Note that in the case of negative curvature with $\Lambda < 0$, the expansion of the Lemaître universe will stop at some later time (see Fig. 10.15c); the universe reverses and we finally end up in a contracting Friedmann universe of negative curvature.

In the case of a Friedmann universe of positive curvature, the mass density reaches a minimum when one-half the period has elapsed (see Fig. 10.12). Hence, the cosmological term will dominate the behavior of the universe only if it is sufficiently large compared to this minimum mass density. The critical value of Λ is (see Exercise 11)

$$\Lambda_E = (\pi/2GM)^2 \qquad [119]$$

If Λ is larger than Λ_E, then what began as a closed Friedmann universe (see Fig. 10.12) gradually turns into a closed, expanding Lemaître universe (see Fig. 10.15a). A typical example of the function $a(t)$ for $\Lambda > \Lambda_E$ is shown in Fig. 10.16. The transition from the nearly Friedmann to the nearly empty Lemaître universe can be fast or slow depending on the value of Λ. In the exceptional case $\Lambda = \Lambda_E$ the transition is never completed; the universe gets stuck at a constant value of a,

$$a = \text{constant} = 1/\sqrt{\Lambda_E} \qquad [120]$$

The static universe with this constant value of a is known as the *Einstein universe*. However, the equilibrium at the value $a = 1/\sqrt{\Lambda_E}$ is unstable (see Problem 10.9). Any perturbation in a leads either to monotonic expansion (towards an expanding Lemaître universe) or monotonic contraction (towards a contracting Friedmann universe).

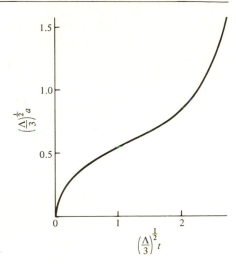

Fig. 10.16 *Radius of curvature of a typical nonempty Lemaître universe; $\Lambda = 1.17\,\Lambda_E$.*

EXERCISE 10. Show that (120) is a solution of (94) if $\Lambda = \Lambda_E$.

EXERCISE 11. Show that the equation of motion (94) may be written as

$$\left(\frac{da}{dt}\right)^2 + \left(1 - \frac{4GM}{3\pi}\frac{1}{a} - \frac{\Lambda}{3}a^2\right) = 0 \qquad [121]$$

The quantity $Q(a) = (1 - 4GM/3\pi a - \Lambda a^2/3)$ may be regarded as an effective potential for the motion of the universe. Make a rough plot of $Q(a)$ as a function of a. Show that $Q(a)$ has a maximum at $a_m = (2GM/\pi\Lambda)^{1/3}$. If the value of this maximum is $Q(a_m) < 0$, then (da/dt) is always positive and the universe expands monotonically. If $Q(a_m) > 0$, then (da/dt) vanishes at some value of a smaller than a_m; that is, a motion that begins at $a = 0$ reaches a turning point and then the universe recontracts. Show that $Q(a_m) < 0$ if $\Lambda > \Lambda_E$ and that $Q(a_m) > 0$ if $\Lambda < \Lambda_E$.

Finally, note that Table 10.3 contains no empty Lemaître models that are monotonic and have $\Lambda < 0$. We can understand the absence of such models if we recall that a negative cosmological constant corresponds to a positive effective mass density of the vacuum; as the volume of the universe grows so does the total effective mass, and ultimately this stops the expansion and initiates a recontraction. We can draw an interesting conclusion: the general Lemaître model with a nonzero mass density and $\Lambda < 0$ must necessarily be of the oscillating type. If this universe were monotonic, it would gradually approach a monotonic empty model, which, as we have seen, does not exist.

10.11 PROPAGATION OF LIGHT

Light signals propagate along lightlike worldlines, i.e., lines with $ds^2 = 0$. If we place the origin of coordinates at the position of the light source, then the light will move out radially ($d\theta = d\phi = 0$) and hence for both the closed and the open model we obtain

$$0 = a(\eta)[d\eta^2 - d\chi^2]$$

The equation of the worldline of the light signal is then

$$\chi = \eta - \eta_0 \qquad [122]$$

Suppose that two pulses of light are sent out: the first when the time parameter has the value η_0, the second when it has the value $\eta_0 + \Delta\eta_0$. The worldlines are, respectively,

$$\chi = \eta - \eta_0$$
$$\chi = \eta - (\eta_0 + \Delta\eta_0) \qquad [123]$$

The two light signals therefore arrive at any given point χ with a difference

$$\Delta\eta(\chi) = \Delta\eta_0 \qquad [124]$$

between their time parameters. This says that $\Delta\eta$ is a constant, independent of χ, and implies that if time is measured by η, then there is no redshift. However, atomic clocks in the galaxies do *not* measure η-time, they measure τ-time. Since $\Delta\tau = a\,\Delta\eta$, it follows from (124) that $\Delta\tau/a$ remains constant as the two pulses propagate. The time interval between the pulses therefore changes as a changes. For a continuous wave train, in which each peak may be regarded as a pulse, the time interval between pulses equals $\Delta\tau = 1/\nu$ and therefore

$$a\nu = \text{constant} \qquad [125]$$

In an expanding universe, a is increasing and hence ν must decrease; this gives us the redshift. (Incidentally, we have already given an alternative derivation of (125) by means of the simple argument following Eq. (29).)

According to (125), the frequency with which a wave arrives at χ is

$$\nu(\chi) = \nu(0)\,\frac{a(\eta_0)}{a(\eta)} = \nu(0)\,\frac{a(\eta - \chi)}{a(\eta)} \qquad [126]$$

where $\nu(0)$ is the emitted frequency.

If the distance between the emitter and receiver is small ($\chi << 1$), then we can approximate

$$\nu(\chi) \simeq \nu(0)\left[1 - \chi \frac{\dot{a}(\eta)}{a(\eta)}\right] \qquad [127]$$

i.e.,

$$\frac{\nu(\chi) - \nu(0)}{\nu(0)} \simeq -\chi \frac{\dot{a}(\eta)}{a(\eta)} \qquad [128]$$

In terms of the wavelength $\lambda = c/\nu$ this can be written

$$\frac{\lambda(\chi) - \lambda(0)}{\lambda(0)} \simeq + \chi \frac{\dot{a}(\eta)}{a(\eta)} \qquad [129]$$

The distance between emitter and receiver, at the instant η, is given by (see also Eq. (66))

$$l = \int dl = \int a(\eta)d\chi = a(\eta)\chi \qquad [130]$$

and Eq. (129) takes the form of Hubble's law (compare Eqs. (5) and (6)):

$$\frac{\lambda(\chi) - \lambda(0)}{\lambda(0)} = \frac{\dot{a}}{a^2} l \qquad [131]$$

From this we can identify the Hubble constant as

$$H = \frac{\dot{a}}{a^2} \qquad [132]$$

Since $\dot{a} = a\,da/dt$ (see Eq. (80)), we can also write this as

$$H = \frac{1}{a}\frac{da}{dt}$$

We will put the expression (132) to some use in the next section.

A curious feature of the Friedmann models and other models with a big bang is that the universe has a part that is visible to us and a part that is invisible. Consider a light signal sent out from the origin ($\chi = 0$) when the universe began ($\eta_0 = 0$). At the time characterized by η, this light signal will have reached a point $\chi = \eta$ (see Eq. (122)). Conversely, a light signal emitted from the point $\chi = \eta$ will reach the origin at this time, but any light signal emitted from a more distant point will take a longer time. For example, in the closed Friedmann model at the time

$$t = a_*\,(\eta - \sin \eta) \qquad [133]$$

only that part of the universe which is within a radial coordinate

$$\chi = \eta \tag{134}$$

or

$$r = a \sin \eta \tag{135}$$

will be visible from the origin. The surface defined by Eq. (134) or (135) is the *boundary of our visible universe.*

The actual distance to the boundary (135) is given by Eq. (60),

$$l = a \sin^{-1}(r/a) = a \eta \tag{136}$$

and the volume of the visible part of the universe is

$$2\pi^2 a^3 (\eta - \sin \eta \cos \eta)/\pi \tag{137}$$

EXERCISE 12. Derive Eq. (137).

The factor $(\eta - \sin \eta \cos \eta)/\pi$ gives the visible fraction of the total volume. This fraction gradually increases. At the moment of maximum expansion $(\eta = \pi/2)$, one-half of the total volume is visible; at the final moment of collapse $(\eta = \pi)$, the entire volume is visible.

In view of this, it is entirely wrong to suppose that during the expansion phase the receding galaxies will reach the boundary of the universe and disappear. On the contrary, the boundary of the universe recedes even faster than the galaxies and more and more of the universe is included inside the boundary. If we look at a far-away region of the universe, we see this region as it was a long time ago, when the light reaching us now began its journey. If we look at the boundary of the visible universe, we see the matter as it was at the initial instant, we see the primordial fireball. As the boundary expands, more and more of this fireball comes into our visible universe. Of course, at the boundary the redshift is infinite (see Eq. (126)) and hence the light will actually be too faint to be seen.

10.12 COMPARISON OF THEORY AND OBSERVATION

The observable cosmological parameters are the mass density, the Hubble constant, the deceleration parameter, and the age of the universe. Different cosmological models predict different relations between these parameters.

In this and the following section we adopt the convention that parameters with the subscript $_0$ (such as ρ_0, H_0, q_0, t_0) are evaluated at the present time; quantities without the subscript are evaluated at an arbitrary time.

If we substitute the expression (132) for the Hubble constant into the equation of motion, we obtain

$$3\left(H^2 + \frac{k}{a^2}\right) = 8\pi G\rho + \Lambda \tag{138}$$

where $k = +1$, $k = 0$, or $k = -1$ for the case of positive, zero, or negative curvature, respectively (see Eqs. (94), (105), and (111)). It follows from this that curvature of the universe is determined by H, ρ, and Λ:

$$\frac{8\pi G\rho}{3H^2} > 1 - \frac{\Lambda}{3H^2} \qquad \text{for positive curvature}$$

$$\frac{8\pi G\rho}{3H^2} < 1 - \frac{\Lambda}{3H^2} \qquad \text{for negative curvature}$$

$$\frac{8\pi G\rho}{3H^2} = 1 - \frac{\Lambda}{3H^2} \qquad \text{for zero curvature} \tag{139}$$

Since Λ is not a directly observable parameter, it is best to eliminate it. According to Eq. (20),

$$q = \frac{4\pi G\rho}{3H^2} - \frac{\Lambda}{3H^2} \tag{140}$$

This relation between q, ρ, and Λ was derived in Section 10.3 without making explicit use of Einstein's equations. It is obvious from the arguments given in Section 10.9 (see Eqs. (96)–(101)) that Eq. (140) can be obtained by differentiating the Einstein equation once with respect to t.

EXERCISE 13. Multiply Eq. (138) by a^2 and differentiate with respect to t. Show that the result can be put in the form (140).

By means of (140) we can eliminate Λ from (139) and obtain

$$4\pi G\rho/H^2 > 1 + q \qquad \text{for positive curvature}$$
$$4\pi G\rho/H^2 < 1 + q \qquad \text{for negative curvature}$$
$$4\pi G\rho/H^2 = 1 + q \qquad \text{for zero curvature} \tag{141}$$

From here on it is convenient to treat the Friedmann and Lemaître models separately. In the former models $\Lambda = 0$, and the situation is as follows:

(1) POSITIVE CURVATURE   According to Eq. (140), $q = 4\pi G\rho/3H^2$ and hence we can express the condition (141) either in terms of q or in

terms of ρ. Furthermore, we can replace the density by the density parameter $\sigma = 4\pi G\rho/3H^2$ defined in Eq. (34). This leads to

$$q > \tfrac{1}{2}, \qquad \sigma > \tfrac{1}{2} \tag{142}$$

It is also possible to obtain an inequality between the true age of the universe and the Hubble age. The inequality is

$$0 < t < \tfrac{2}{3}H^{-1} \tag{143}$$

EXERCISE 14. Show that

$$t = H^{-1}\frac{(\eta - \sin \eta)\sin \eta}{(1 - \cos \eta)^2} \tag{144}$$

and show that if η is in the range $0 < \eta < \pi$, which corresponds to an expanding universe, then t is in the range given by (143).

(II) NEGATIVE CURVATURE   The conditions on q and σ are

$$q < \tfrac{1}{2}, \qquad \sigma < \tfrac{1}{2} \tag{145}$$

and the age satisfies the inequalities

$$\tfrac{2}{3}H^{-1} < t < H^{-1} \tag{146}$$

EXERCISE 15. Derive (146).

(III) ZERO CURVATURE   In this last case

$$q = \tfrac{1}{2}, \qquad \sigma = \tfrac{1}{2} \tag{147}$$

and

$$t = \tfrac{2}{3}H^{-1} \tag{148}$$

This takes care of the Friedmann models; we next consider the general Lemaître models with $\Lambda \neq 0$.

(I) POSITIVE CURVATURE   In terms of the density parameter σ, the condition (141) reads

$$3\sigma > 1 + q \tag{149}$$

According to the discussion at the end of Section 10.10 the universe will expand monotonically if $\Lambda > \Lambda_E$, and it will eventually recontract if

$\Lambda < \Lambda_E$. Let us express the critical value Λ_E in terms of the observable parameters. By Eq. (138) we have

$$a^2 = \left(\frac{8\pi G\rho}{3} + \frac{\Lambda}{3} - H^2\right)^{-1} \tag{150}$$

and hence

$$\Lambda_E = \left(\frac{\pi}{2GM}\right)^2 = \left[\frac{\pi}{2G(2\pi^2 a^3\rho)}\right]^2$$

$$= \frac{1}{16\pi^2 G^2\rho^2}\left(\frac{8\pi G\rho}{3} + \frac{\Lambda}{3} - H^2\right)^3 \tag{151}$$

The condition $\Lambda > \Lambda_E$ for monotonic behavior then reads

$$(4\pi G\rho - 3H^2 q) > \frac{1}{16\pi^2 G^2\rho^2}\left[4\pi G\rho - (q+1)H^2\right]^3 \tag{152}$$

where the expression $\Lambda = 4\pi G\rho - 3H^2 q$ has been used for Λ (see Eq. (140)). In terms of the density parameter σ, the inequality (152) becomes

$$\sigma - q > \frac{1}{\sigma^2}\left[\sigma - \tfrac{1}{3}(q+1)\right]^3 \qquad \text{(monotonic)} \tag{153}$$

As this type of universe expands, it gradually approaches the empty Lemaître model (i) of Table 10.3.

The condition $\Lambda < \Lambda_E$ for oscillatory behavior is obtained by reversing the sign of the inequality in Eq. (153):

$$\sigma - q < \frac{1}{\sigma^2}\left[\sigma - \tfrac{1}{3}(q+1)\right]^3 \qquad \text{(oscillatory)} \tag{154}$$

Note that according to this inequality, a universe with a negative value of Λ, i.e., a negative value of $\sigma - q$, must oscillate. The absence of monotonic models with $\Lambda < 0$ has already been explained in Section 10.10.

If both sides of (153) are equal, then the universe gradually approaches an Einstein universe; however, the instability of this configuration will eventually result in either expansion or recontraction.

(II) NEGATIVE CURVATURE   The condition for this case is

$$3\sigma < 1 + q \tag{155}$$

The universe continues to expand if $\Lambda > 0$, recontracts if $\Lambda < 0$. These two conditions may be expressed as

$$\sigma - q > 0 \qquad \text{(monotonic)} \tag{156}$$

and

$$\sigma - q < 0 \qquad \text{(oscillatory)} \tag{157}$$

respectively.

(III) ZERO CURVATURE   In this case

$$3\sigma = 1 + q \tag{158}$$

The conditions for continual expansion and for recontraction are as in case (ii),

$$\sigma - q > 0 \qquad \text{(monotonic)} \tag{159}$$

$$\sigma - q < 0 \qquad \text{(oscillatory)} \tag{160}$$

Note that, in the Friedmann model, if any two of the four parameters ρ_0, q_0, H_0, t_0 are determined from observation, then the model fixes the remaining two and the evolution of the universe. Thus, cosmology may be described, in Sandage's words, as "a search for two numbers."[23] In the Lemaître models, there is an extra adjustable constant (Λ) and we must carry out a search for three numbers.

Table 10.4 summarizes the best available observational data on σ_0, q_0, H_0, and t_0 (see Sections 10.3, 10.4, 10.6).

TABLE 10.4 OBSERVED VALUES OF COSMOLOGICAL PARAMETERS

$$0.02 \leq \sigma_0 \leq 0.2$$
$$-2 \leq q_0 \leq 1.5$$
$$3 \leq H_0 \leq 12 \text{ cm sec}^{-1} \text{ pc}^{-1}$$
$$8 \leq t_0 \leq 18 \text{ billion years}$$

Let us first try to fit these parameters with a Friedmann model. Since $\sigma_0 < \frac{1}{2}$, the universe will have to be open (see Eq. (145)). We could only have a closed universe if the mass density is considerably larger than the upper limit given in Table 10.4.* But the existence of a large

---

* It should be pointed out that in the context of a Friedmann model, the observational data actually constrain σ_0 to a value below $\sigma_0 \simeq 0.05$. This constraint has to do with the age; a larger value of σ_0 would require an age in excess of 18 billion years.[24]

TABLE 10.5 TYPICAL MODELS OF OUR UNIVERSE

Model	Closed Friedmann	Open Friedmann	Closed Lemaître[a]	Open Lemaître[a]
Density parameter σ_0	1.0	0.035	0.15	0.035
Deceleration parameter q_0	1.0	0.035	−1.0	−0.5
Hubble constant H_0	5.5 cm sec^{-1} pc^{-1}	5.5 cm sec^{-1} pc^{-1}	10.0 cm sec^{-1} pc^{-1}	9.0 cm sec^{-1} pc^{-1}
Average mass density ρ_0	8.5×10^{-30} g/cm^3	4.0×10^{-31} g/cm^3	5.7×10^{-30} g/cm^3	1.1×10^{-30} g/cm^3
Cosmological constant Λ	0	0	3.6×10^{-35}/sec^2	1.4×10^{-35}/sec^2
Hubble age H_0^{-1}	1.8×10^{10} years	1.8×10^{10} years	0.98×10^{10} years	1.1×10^{10} years
True age t_0	1.1×10^{10} years	1.6×10^{10} years	1.1×10^{10} years	1.2×10^{10} years
Radius of curvature a_0	2.5×10^{10} light years	1.9×10^{10} light years	1.5×10^{10} light years	1.7×10^{10} light years
Time for completion of one expansion-contraction cycle	$2\pi a_* = 2.4 \times 10^{11}$ years	∞	∞	∞
Radius of curvature at maximum expansion	$2a_* = 7.7 \times 10^{10}$ light years	∞	∞	∞

[a] These Lemaître universes are *not* empty. However, the mass density is low and hence the future behavior will be essentially that of an empty Lemaître model (see Section 10.10). In the past the mass density was larger and hence the true age cannot be calculated from the formulas of Section 10.10; the true age has been calculated by numerical integration (see ref. 28).

amount of extra, unobserved matter in the universe can only be reconciled with the theory of deuterium production (see Section 10.6) if one makes some very artificial assumptions about this extra matter; for example, one might postulate that there is a large amount of mass in the form of black holes carefully hidden away in remote regions of intergalactic space. An open universe seems to give the most straightforward fit to the data.[24]

If we try to fit the observational parameters with a Lemaître model, then the large uncertainties in these parameters permit both open and closed universes.

Table 10.5 gives some examples of Friedmann and Lemaître universes. These models are consistent with the values listed in Table 10.4, except that the closed Friedmann model has an excessive mass density. There is obviously a large range of cosmological models that fit the observational data. Furthermore, it may be well to keep in mind that cosmology is traditionally afflicted by large uncertainties; for example, the value of the Hubble constant has been the subject of several drastic revisions—the value we now believe in is about ten times smaller than what Hubble believed in. It will be a while before we can predict with confidence the future of our universe.

10.13   THE BIG BANG AND THE EVOLUTION OF THE FIREBALL

At present, the universe is matter-dominated. This means that the main contribution to the energy density is due to massive particles and the contribution of the blackbody radiation is negligible. However, at an earlier age the universe was radiation-dominated. To see this, let us compare the densities of matter and of the blackbody radiation. The former is inversely proportional to a^3,

$$\rho_{matter} = \rho_0 \left(\frac{a_0}{a}\right)^3 \qquad [161]$$

The latter is proportional to the fourth power of the temperature,

$$\rho_{rad} = 4\sigma T^4 \qquad [162]$$

where $\sigma = 5.67 \times 10^{-5}$ gm/sec^3($^\circ$K)^4 is the Stefan-Boltzmann constant. Since $T \propto 1/a$, we obtain

$$\rho_{rad} = \rho_{rad,0} \left(\frac{a_0}{a}\right)^4 \qquad [163]$$

By comparing (161) and (163) we see that in the early universe, when a is small, the density of radiation dominates.

The transition from a radiation-dominated to a matter-dominated universe occurs when $\rho_{matter} = \rho_{rad}$, i.e., when

$$\rho_0 \left(\frac{a_0}{a}\right)^3 = \rho_{rad,0} \left(\frac{a_0}{a}\right)^4 \qquad [164]$$

The present density of radiation is $\rho_{rad,0} = 4.5 \times 10^{-34}$ g/cm³ (corresponding to $T = 2.7°$K in Eq. (162)). The present density of matter is somewhat uncertain; if we assume that the value given for the open Friedmann model in Table 10.5 is correct, then the domination of radiation comes to an end when

$$a/a_0 = 1.1 \times 10^{-3} \qquad [165]$$

With the values of a_0 and t_0 given in Table 10.5, Eqs. (108) and (109) imply that this happens at a Hubble age of $H^{-1} = 1.4 \times 10^{13}$ sec or a true age of $t = 5.3 \times 10^{13}$ sec. The temperature of the blackbody radiation was $\sim 2.4 \times 10^3$ °K at that time.

At earlier times the dynamical equation (107) of the Friedmann model becomes inapplicable. For the sake of simplicity we will completely neglect the density of matter. We can then use Eq. (105) and insert the density of radiation,

$$\frac{3}{a^4} (\dot{a}^2 - a^2) = 8\pi G \, \rho_{rad,0} \left(\frac{a_0}{a}\right)^4 \qquad [166]$$

This is the dynamical equation for an open universe filled with radiation. The solution of this equation is

$$a(\eta) = b_* \sinh \eta \qquad [167]$$

where

$$b_* = \sqrt{\frac{8\pi G}{3} \, a_0^4 \, \rho_{rad,0}} \qquad [168]$$

From Eq. (167) it follows that

$$t = b_*(\cosh \eta - 1) \qquad [169]$$

In the early universe $\eta << 1$. Therefore we can approximate Eqs. (167) and (169) as follows

$$a(\eta) \simeq b_* \eta \qquad [170]$$

$$t \simeq b_* \eta^2/2 \qquad [171]$$

and hence

$$a(t) \simeq \sqrt{2b_* t} \qquad [172]$$

EXERCISE 16. Show that with these approximations we also have

$$\rho_{rad} \simeq \frac{3}{32\pi G} \frac{1}{t^2} \qquad [173]$$

We can obtain some convenient expressions for the Hubble age by returning to our equation of motion (166). In the early universe we can neglect a^2 compared to $\dot{a}^2$ and hence Eq. (166) takes the form

$$3 \frac{\dot{a}^2}{a^4} = 8\pi G \, \rho_{rad} \qquad [174]$$

Since $\dot{a}/a^2 = H$, we can write the Hubble age as

$$H^{-1} = \left(\frac{3}{8\pi G} \frac{1}{\rho_{rad}} \right)^{1/2} \qquad [175]$$

Alternatively, we can use (162) to express ρ_{rad} in terms of the temperature,

$$H^{-1} = \left(\frac{3}{32\pi G \sigma T^4} \right)^{1/2} \qquad [176]$$

or we can use (173) to express ρ_{rad} in terms of the true age,

$$H^{-1} = 2t \qquad [177]$$

Hence, in the radiation-filled early universe the true age is simply one-half the Hubble age.

Note that although the above calculations were carried out for the case of an open universe, Eqs. (170)–(177) are equally valid in the case of a closed universe. The reason is that the dynamical equation for the open universe differs from Eq. (166) only in that $(\dot{a}^2 - a^2)$ is replaced by $(a^2 + \dot{a}^2)$. As long as we neglect a^2 compared to $\dot{a}^2$, this difference is of no consequence.

EXERCISE 17. Write down the dynamical equations for the closed universe filled with radiation. Solve for $a(\eta)$ and for $t(\eta)$. Derive the approximate results (170)–(177).

At very early times, the universe was so extremely hot that the typical thermal energies of the blackbody photons were large enough to permit the creation of electron-positron pairs. The required temperature is given by $kT \sim m_e c^2$, i.e., $T \sim 6 \times 10^9$ °K. According to (176), this temperature corresponds to an age of ~ 10 sec.

At higher temperatures, the electrons, positrons, and also neutrinos form a relativistic gas of leptons. Such a relativistic gas of particles has properties that are very similar to those of the gas of photons making up the blackbody radiation. The energy density of the gas is proportional to the fourth power of the temperature, but because there are more kinds of leptons than photons, the constant of proportionality is somewhat larger. Thus, the universe was lepton-dominated at very early times.

The dynamical equation for the lepton-dominated universe is essentially the same as for the radiation-dominated case. Only some of the constants of proportionality are slightly changed; we will ignore these changes. Using Eq. (176) we find that at an age of $\sim 10^{-4}$ sec the universe had a temperature of $\sim 10^{12}$ °K which is sufficient to permit the creation of muons and pions by thermal fluctuations. At slightly earlier times even the creation of protons, neutrons, and other baryons and antibaryons was possible and the universe was in a baryon-dominated stage.

We can therefore distinguish four eras in the chronology of the universe (see Fig. 17):

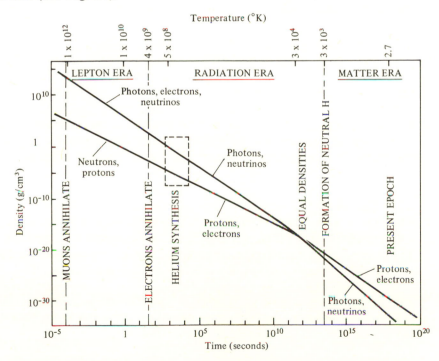

Fig. 10.17 *Eras in the evolution of the universe. (After Harrison,* Annual Review of Astronomy and Astrophysics, *1973.)*

(I) *THE HADRON ERA** ($0 < t < 10^{-4}$ sec; $T > 10^{12}$ °K)   During this era the universe contains particles and antiparticles of all known (and many unknown) kinds. The numbers of baryons and antibaryons in this primordial fireball are nearly equal, but there is a small excess of baryons (representing a small fraction of the whole). As the temperature drops towards 10^{12} °K, the baryons and antibaryons annihilate, leaving only the small excess of baryons which accounts for the number of baryons found in the present universe.

Incidentally, it has been suggested that the universe actually contains equal amounts of matter and antimatter so that perhaps some galaxies are made of matter and others of antimatter. Such a view is hard to reconcile with the big bang. In the dense early universe the particles and antiparticles are in close contact and, as the universe cools, they will tend to annihilate rather completely (radiation catastrophe). The resulting universe would then contain a smaller number of material particles and a larger number of photons than are observed in our universe. In order to avoid this difficulty one must suppose that matter and antimatter are not homogeneously mixed in the early universe, but separated into many small regions containing almost pure matter or pure antimatter. Onmès[26] has suggested that these conditions could arise as a result of phase separation so that the early universe has the character of an emulsion.

(II) *THE LEPTON ERA* ($10^{-4} < t < 10$ sec; $10^{12} > T > 4 \times 10^9$ °K) In this era only a much smaller number of particles survive. The remaining particles are photons, electrons, neutrinos, and the corresponding antiparticles. Besides, there are a few muons and pions and, of course, a residue of protons and neutrons. Even nowadays the neutrinos from this era should still be with us and fill our intergalactic space with an energy density amounting to 45% of the blackbody photon density. Since the interactions of neutrinos are extremely weak, there is not much hope for their detection.

(III) *THE RADIATION ERA* ($10 < t < 10^{13}$ sec; $4 \times 10^9 > T > 10^4$ °K) This era begins when electrons annihilate, leaving only a relatively small residue which balances the positive charge density of the protons. The density of the remaining electrons, and that of protons and neutrons, is negligible compared to the densities of photons and neutrinos.

During this era, at a temperature of $\sim 10^9$ °K, the neutrons fuse with the protons and form He^4 nuclei. About 27% (by mass) of the matter is made into helium and some very small amounts are made into deuterium, tritium, lithium, and other light nuclei.

---

* A hadron is a particle that participates in the strong interactions. All baryons and mesons (not including the muon) are hadrons.

(IV) *THE MATTER ERA* ($t > 10^{13}$ or 10^{14} sec; $T < 10^4$ or 10^3 °K)   The most recent era begins when the density of radiation drops below the density of matter. The exact birthdate of this era depends on the present mass density of the universe and is therefore somewhat uncertain. Near the beginning of this era electrons combine with protons and form neutral atoms of hydrogen and helium. This happens when the temperature drops to ~3,000 °K.

Once matter becomes neutral, the interaction between matter and radiation is drastically reduced and there is insufficient exchange of energy to maintain thermal equilibrium. Matter and radiation expand and cool separately, each with its own temperature. The temperature of matter varies as $T \propto 1/a^2$;* the temperature of radiation varies as $T \propto 1/a$. Hence, the temperature of matter ultimately drops below that of the radiation.

As long as matter and radiation are coupled, any small irregularities in the matter distribution are quickly smoothed out by the radiation pressure. But once matter is on its own, a region with a slight excess of mass will tend to grow by the gravitational attraction it exerts on its surroundings. By some such condensation process galaxies began to form at $t \sim 10^{16}$ sec. And within one of these galaxies the solar system formed at $t \sim 10^{17}$ sec after the big bang.

FURTHER READING

North, *The Measure of the Universe*, traces the history and conceptual foundations of modern cosmology; it includes discussions of alternative theories of gravitation and of metaphysical issues and is rather wordy but very readable.

Both Misner, Thorne, and Wheeler, *Gravitation*, and Peebles, *Physical Cosmology*, include outlines of the history of modern cosmology. Hubble, *The Realm of the Nebulae*, is a classic which relates the discovery of the extragalactic systems.

An elementary, but very nice, introduction to cosmology is Schatzmann, *The Structure of the Universe*.

Peebles, *Physical Cosmology;* Rees, Ruffini and Wheeler, *Black Holes, Gravitational Waves and Cosmology;* Sciama, *Modern Cosmology;* and Weinberg, *Gravitation and Cosmology*, provide excellent general descriptions of the observed properties of the universe. The first of these books contains the most informative discussion of the mass density; the second and third, of the radio sources and number counts; and the fourth, of the distance scale. Incidentally, Sciama's book is written with a minimum of mathematics and is therefore very readable.

Past and present difficulties in the measurement of H_0 and q_0 are simply described by Sandage in the article "Cosmology: A Search for Two Numbers"

---

* This corresponds to adiabatic expansion of a perfect gas.

(*Physics Today,* February, 1970). An account of Sandage's more recent work is contained in Misner, Thorne, and Wheeler, *Gravitation.*

The problem of deuterium production in the early universe and the implications for the mass density are presented in the article "What Can Deuterium Tell Us?" by Schramm and Wagoner (*Physics Today,* December, 1974).

The properties of the cosmic blackbody radiation are described in the books by Peebles, Rees et al., Sciama, and Weinberg (see above). Penzias, in the article "Cosmology and Microwave Astronomy" (*in* Reines, ed., *Cosmology, Fusion and Other Matters*), tells the story of the discovery.

Different determinations of the age of the universe are presented in the article "The Age of the Elements" by Schramm (*Scientific American,* January, 1974). The article "Nucleocosmochronology" by the same author (see ref. 14), covers some of the same topics in a more technical way.

The solution of Einstein's equations for an isotropic, homogeneous universe is of course given in many books on gravitational theory. The following contain a general discussion of open and closed models, with and without a cosmological constant: Misner, Thorne, and Wheeler, *Gravitation;* Robertson and Noonan, *Relativity and Cosmology;* Tolman, *Relativity, Thermodynamics, and Cosmology;* Weinberg, *Gravitation and Cosmology.*

Gott, Gunn, Schramm, and Tinsley, in "An Unbound Universe"?" (see ref. 24), make the case for the open Friedmann model; they review the data on the mass density and conclude that there is not enough mass to close the universe.

Details on the very early universe and the evolution of the fireball are given in the books by Rees et al., Weinberg (see above), and in the article "Standard Model of the Early Universe" by Harrison (see ref. 25).

Some interesting speculations regarding antimatter in the universe are contained in Alfvén, *Worlds—Antiworlds.*

Hawking and Ellis, *The Large Scale Structure of Space-Time,* uses the general singularity theorems to prove rigorously that the observed properties of our universe *demand* an initial singularity, at least if classical physics is applicable.

Ryan and Shepley, *Homogeneous Relativistic Cosmologies,* presents a mathematical discussion of diverse cosmological models, including some rather weird ones which have little to do with our universe. This book includes neat reviews of differential forms and of the concept of "singularity."

Our knowledge and ignorance of what happens at the singularity, at the beginning of the universe, is summarized in the article "Physical Processes Near Cosmological Singularities" by Novikov and Zel'dovich (*in* Goldberg, Layzer, and Phillips, eds., *Annual Review of Astronomy and Astrophysics,* vol. 11).

REFERENCES

1. C. M. Will and K. Nordvedt, Jr., Ap. J. **177,** 757 (1972).
2. K. Nordvedt, Jr., and K. M. Will, Ap. J. **177,** 775 (1972).
3. E. Hubble, *The Realm of the Nebulae* (Dover, New York, 1958), p. 186.
4. P. J. E. Peebles, *Physical Cosmology* (Princeton University Press, 1971), p. 37.

5. D. W. Sciama, *Modern Cosmology* (Cambridge University Press, 1973), Chapter 6.

6. M. Ryle, *in* L. Goldberg, D. Layzer, and J. G. Phillips, eds., *Annual Review of Astronomy and Astrophysics* (Annual Reviews Inc., Palo Alto, 1968), vol. 6.

7. M. Schmidt, Ap. J. **162**, 371 (1970); **176**, 273 (1972).

8. A. Sandage, Ap. J. **152**, L149 (1968).

9. S. Weinberg, *Gravitation and Cosmology* (Wiley, New York, 1972), p. 428.

10. A. Sandage and G. A. Tammann, Ap. J. **196**, 313 (1975); **197**, 265 (1975).

11. J. E. Gunn and J. B. Oke, Ap. J. **195**, 255 (1975).

12. J. E. Gunn, Ann. N.Y. Acad. Sci. **224**, 56 (1973).

13. R. E. Hills and J. N. Bahcall, Ann. N.Y. Acad. Sci. **224**, 58 (1973).

14. D. N. Schramm, *in* G. R. Burbidge, D. L. Layzer, and J. G. Phillips, eds., *Annual Review of Astronomy and Astrophysics* (Annual Reviews Inc., Palo Alto, 1974), vol. 12.

15. R. A. Alpher, H. A. Bethe, and G. Gamow, Phys. Rev. **73**, 803 (1948).

16. R. A. Alpher and R. C. Herman, Nature **162**, 774 (1948).

17. R. A. Penzias and R. W. Wilson, Ap. J. **142**, 419 (1965).

18. R. H. Dicke, P. J. E. Peebles, P. G. Roll, and D. T. Wilkinson, Ap. J. **142**, 414 (1965).

19. For a complete list of measurements and references, see ref. 9, p. 512. The limits in the infrared are due to D. P. Woody, J. C. Mather, N. S. Nishioka, and P. L. Richards, Phys. Rev. Lett. **34**, 1036 (1975).

20. See ref. 5, p. 197.

21. See ref. 9, p. 523.

22. See ref. 4, Chapter IV. We assume that the value of Hubble's constant is as given by Eq. (7).

23. A. R. Sandage, Physics Today, February 1970, p. 34.

24. J. R. Gott III, J. E. Gunn, D. N. Schramm, and B. M. Tinsley, Ap. J. **194**, 543 (1974).

25. E. R. Harrison, *in* L. Goldberg, D. Layzer, and J. G. Phillips, *Annual Review of Astronomy and Astrophysics* (Annual Reviews Inc., Palo Alto, 1973), vol. 11.

26. R. Omnès, Ann. N.Y. Acad. Sci. **224**, 339 (1973), and other references quoted therein.

27. D. N. Schramm and R. V. Wagoner, Physics Today, December 1974, p. 41.

28. S. Refsdal, R. Stabell, and F. G. de Lange, Mem. Roy. Astron. Soc. **71**, 143 (1967).

29. J. P. Ostriker and J. P. E. Peebles, Ap. J. **186**, 467 (1973).

30. J. P. Ostriker, J. P. E. Peebles, and A. Yahil, Ap. J. **193**, L1 (1974).

31. M. Rees, R. Ruffini, and J. A. Wheeler, *Black Holes, Gravitational Waves and Cosmology* (Gordon and Breach, New York, 1974).

PROBLEMS

1. Consider a local geodesic reference frame whose origin is comoving with an expanding Friedmann universe. A particle of mass m is at a distance l from

the origin. Use the equation of geodesic deviation (Eq. (7.54)) to show that the gravitational field of the universe exerts a (radial) tidal force

$$f = -mH_0^2 q_0 l$$

on this particle. Show that this force coincides with the Newtonian attractive force of the matter inside the sphere of radius l about the origin (see Eq. (16)).

2. Plot the value of the circumference of a circle as a function of the measured radius. Do this for the range $0 <$ (measured radius) $< \pi a$, for the uniformly curved three-geometries of both positive and negative curvature.

3. Plot the value of the square root of the area of a sphere as a function of the measured radius. Do this for the range $0 <$ (measured radius) $< \pi a$, for the uniformly curved three-geometries of both positive and negative curvature.

4. Expand the formulas for the volume of a sphere (Eqs. (63) and (71)) in powers of b/a; keep terms of order $(b/a)^2$. Consider a sphere with $b = 10^7$ light years in a universe with $a = 10^{10}$ light years. By what fraction does the volume deviate from $4\pi r^3/3$?

5. Find the true (measured) distance from the earth to the quasar QSO(OH 471) which has a redshift of $z = 3.40$. Assume that our universe is a closed Friedmann universe with the parameters of Table 10.5.

6. Show that the motion of a particle in a Lemaître universe is such that

$$ap = \text{constant}$$

Here p is the momentum of the particle as measured by an observer whose position coincides with that of the particle and whose velocity coincides with the local velocity of expansion of the universe (comoving observer). (Hint: Show that the $\mu = 1$ component of the geodesic equation is

$$\frac{d}{d\tau}\left(-a^2 \frac{d\chi}{d\tau}\right) = 0$$

The distance measured by the observer is $dl = a \, d\chi$ and $p \propto dl/d\tau$.)

7. Derive the result of the preceding problem by noting that in quantum mechanics a particle of momentum p is represented by a wave of wavelength $\lambda = h/p$ and that this wavelength partakes of the expansion of the universe.

8. At present, the velocity of the Andromeda galaxy relative to the comoving coordinates of the universe is ~ 90 km/sec. Use the result of Problem 6 and the parameters for the open Friedmann universe given in Table 10.5 to calculate the velocity of the Andromeda galaxy when the universe was $\frac{1}{10}$ of its present age. (This calculation ignores the gravitational attraction between our galaxy and Andromeda.)

9. Prove that the Einstein universe is unstable, i.e., prove that if a is slightly larger than $(\Lambda_E)^{-1/2}$ then $\ddot{a} > 0$ and that if a is slightly smaller than $(\Lambda_E)^{-1/2}$ then $\ddot{a} < 0$.

10. Find an expression, analogous to Eq. (137), for the volume of the visible

universe in the case of the *open* Friedmann model. Calculate the rate, in cubic light years per year, at which the volume of the *visible* universe is increasing. Also, calculate the rate at which the mass included in the visible universe is increasing. Use the values of the parameters given in Table 10.5 for the open Friedmann universe. (Hint: The mass in the visible universe is directly proportional to the volume.)

11. The closed Friedmann universe with the parameters given in Table 10.5 will begin to contract at an age of 1.2×10^{11} years. At this time redshifts of galaxies will begin to convert into blueshifts. Will all the redshifts suddenly turn to blueshifts? Will there be both redshifted galaxies and blueshifted galaxies? Is there any time at which we will see only blueshifted galaxies on the sky? For a universe aged $\sim 1.8 \times 10^{11}$ years, make a rough (qualitative) plot of the observed frequency shift of galaxies as a function of distance from the earth.

12. Consider universes with the following values of the observational parameters:
    (a) $\sigma_0 = 0.6$, $q_0 = 0.6$, $H_0 = 5.5$ cm sec^{-1}pc^{-1}
    (b) $\sigma_0 = 0.5$, $q_0 = 0.5$, $H_0 = 5.5$ cm sec^{-1}pc^{-1}
    (c) $\sigma_0 = 0.2$, $q_0 = 0.2$, $H_0 = 5.5$ cm sec^{-1}pc^{-1}
    Prepare a table giving the quantities listed in Table 10.5 for each of these universes.

13. Consider a universe with the following values of the observational parameters:

$$\sigma_0 = 0, \quad q_0 = -0.5, \quad H_0 = 9.0 \text{ cm sec}^{-1}\text{pc}^{-1}$$

Prepare a table listing the quantities that appear in Table 10.5. Compare your results with the open Lemaître universe of Table 10.5. Does the mass density in the latter make much of a difference?

Appendix

The Variational Principle in Field Theory and the Canonical Energy-Momentum Tensor

A.1 THE LAGRANGIAN EQUATIONS FOR A SYSTEM OF PARTICLES

The Lagrangian formalism for a system of fields is a generalization of the formalism for a system of particles. We therefore begin with a short review of the latter.

Suppose that we have a system of particles which can be described by a finite set of generalized coordinates. We write these coordinates as $q_i(t)$ where $i = 1, 2, 3, \ldots, N$. The Lagrangian formalism rests on the assumption that the dynamical equations of motion can be derived from Hamilton's variational principle as follows: We define the *action I* as the integral

$$I = \int_{t_1}^{t_2} L(q_i(t), \dot{q}_i(t)) dt \qquad [1]$$

where the Lagrangian $L(q_i, \dot{q}_i)$ is some given function of the coordinates q_i and the velocities $\dot{q}_i = dq_i/dt$. The equations of motion are then obtained from the requirement that the action remain stationary for infinitesimal variations of the functions $q_i(t)$.* These variations are arbitrary except for the constraint that they vanish at the times $t = t_1$ and $t = t_2$.

If the variation of the coordinate at the time t is $\delta q_i(t)$, then the velocities suffer a corresponding change $\delta \dot{q}_i(t) = (d/dt)\delta q_i$ and consequently the action changes by

$$\delta I = \int \sum_{i=1}^{N} \left(\frac{\partial L}{\partial q_i} \delta q_i + \frac{\partial L}{\partial \dot{q}_i} \delta \dot{q}_i \right) dt$$

$$= \int \sum_{i=1}^{N} \left(\frac{\partial L}{\partial q_i} \delta q_i + \frac{\partial L}{\partial \dot{q}_i} \frac{d}{dt} \delta q_i \right) dt \qquad [2]$$

The terms containing the time derivatives $(d/dt)\delta q_i(t)$ can be integrated by parts. Since the variations δq_i vanish at $t = t_1$ and $t = t_2$, we simply obtain

$$\delta I = \int \sum_{i=1}^{N} \left(\frac{\partial L}{\partial q_i} - \frac{d}{dt} \frac{\partial L}{\partial \dot{q}_i} \right) \delta q_i dt \qquad [3]$$

The action will be stationary for arbitrary choices of the functions $\delta q_i(t)$ if and only if all these functions have zero coefficients, i.e.,

$$\frac{d}{dt} \frac{\partial L}{\partial \dot{q}_i} - \frac{\partial L}{\partial q_i} = 0 \qquad i = 1, 2, \ldots, N \qquad [4]$$

These are the Euler-Lagrange equations for the system. Note that the number of these differential equations is N, the same as the number of degrees of freedom of the system.

The equations (4) can also be written as

$$\frac{d}{dt} \pi_i - \frac{\partial L}{\partial q_i} = 0 \qquad [5]$$

where

$$\pi_i = \frac{\partial L}{\partial \dot{q}_i} \qquad [6]$$

is the canonical momentum conjugate to q_i.

Note that in the integral (1) and in the Lagrange equations (4) the time t plays the role of independent variable. In Section 3.4 we derived Lagrangian equations using the proper time τ as the independent variable.

---

* We will not consider the possibility that L has an *explicit* time dependence (i.e., an extra time dependence in addition to the time dependence implied by $(q_i(t), \dot{q}_i(t))$. An explicit time dependence is only possible if external forces act on the system.

Although the use of τ is more elegant, we are now concerned with the development of the Lagrangian formalism for fields and for that purpose the use of t is more suitable.

The equations (4) have the consequence that the *Hamiltonian*,

$$H = \sum_{i=1}^{N} \left(\dot{q}_i \frac{\partial L}{\partial \dot{q}_i} \right) - L \tag{7}$$

is a constant of the motion. The proof is trivial:

$$\frac{dH}{dt} = \sum_{i=1}^{N} \left(\ddot{q}_i \frac{\partial L}{\partial \dot{q}_i} + \dot{q}_i \frac{d}{dt} \frac{\partial L}{\partial \dot{q}_i} \right) - \sum_{i=1}^{N} \left(\frac{\partial L}{\partial q_i} \dot{q}_i + \frac{\partial L}{\partial \dot{q}_i} \ddot{q}_i \right)$$

$$= \sum_{i=1}^{N} \dot{q}_i \left(\frac{d}{dt} \frac{\partial L}{\partial \dot{q}_i} - \frac{\partial L}{\partial q_i} \right) = 0 \tag{8}$$

By way of illustration, let us apply the above results to the very simple case of a system consisting of just one free particle. In this case it is easy to verify that the Lagrangian

$$L = -m \sqrt{1 - (\dot{x}^2 + \dot{y}^2 + \dot{z}^2)} \tag{9}$$

gives the correct equation of motion and the correct (relativistic) momenta p_x, p_y, p_z. Furthermore, the corresponding Hamiltonian H equals the total energy (rest mass energy plus kinetic energy) of the particle:

$$H = \frac{m}{\sqrt{1 - (\dot{x}^2 + \dot{y}^2 + \dot{z}^2)}} \tag{10}$$

EXERCISE 1. Evaluate Eqs. (4) and (6) for the Lagrangian L and also prove Eq. (10).

Note that the equality between H and energy hinges on making the correct choice for L. Obviously, there are many Lagrangians that will give the same equation of motion for a free particle. What distinguishes the Lagrangian (9) is that the canonical momentum coincides with the relativistic momentum.

EXERCISE 2. Consider the Lagrangian

$$L = \tfrac{1}{2}m(\dot{x}^2 + \dot{y}^2 + \dot{z}^2)$$

Evaluate Eqs. (4) and (6) for this L and compare with the results of Exercise 1.

We will always make the choice (9) for the Lagrangian of a free particle. Furthermore, we will insist that for a system of interacting particles the Lagrangian have the form (9) in the limiting case of large separation between the particles (no interaction). We can then make the following generalization: the total Hamiltonian of any system of inter-acting particles and fields equals the total energy of the particles and fields. But before we justify this statement we must develop the La-grangian formalism for fields.

The conservation law for the Hamiltonian is directly related to invariance under time translation, that is, $dH/dt = 0$ is a direct con-sequence of the absence of any explicit time dependence in the Lagrangian (L depends on time only implicitly, through its dependence on q_i and $\dot{q}_i$). Furthermore, momentum conservation is a direct consequence of invariance under space translations, i.e., $d\pi_i/dt = 0$ if L is independent of q_i ($\partial L/\partial q_i = 0$; see Eq. (5)). Thus, the conservation of energy and momentum is intimately connected with symmetry under time and space translations. The general connection between a (continuous) symmetry group of the Lagrangian and conservation laws is known as Noether's theorem.

A.2 THE LAGRANGIAN EQUATIONS FOR FIELDS

Suppose now that we have a "system" which consists of a field. For the sake of simplicity we will first deal with the case of a single one-com-ponent field (scalar field). Of course we are ultimately interested in the electromagnetic field (vector field) and in the gravitational field (tensor field). But the equations for these many-component fields will not be hard to discover if we first reach an understanding of the simple case of a one-component field.

A one-component field can be described by a function $\psi(\mathbf{x}, t)$ of space and time. At a given time, the function $\psi(\mathbf{x})$ gives the amplitude of the field at the point $\mathbf{x}$. This amplitude plays a role analogous to the set of generalized coordinates q_i: to describe the field system we must specify $\psi(\mathbf{x})$ for all $\mathbf{x}$; to describe the particle system we must specify q_i for all i. In fact, this correspondence goes beyond a mere analogy be-cause we can describe the field approximately by giving its values at a discrete set of points in space (a procedure often used in the numerical integration of field equations). Suppose that we divide all of space into cubical cells of volume ΔV and consider only the amplitudes of the field at the center of each cube. If the coordinates of these centers are $\mathbf{x}_1$, $\mathbf{x}_2$, $\mathbf{x}_3$, . . . , then we can replace the fields by a set of generalized co-ordinates

$$q_i = \psi(\mathbf{x}_i) \qquad i = 1, 2, 3, \ldots \qquad [11]$$

In this approximation the field equations are replaced by a set of Euler-Lagrange equations of the type (4). The exact field may be regarded as the limiting case, with $\Delta V \rightarrow \dot{0}$, of such a discrete set of generalized coordinates. Note that the approximation (11) involves a discrete infinity of degrees of freedom while the exact field has a continuous infinity of degrees of freedom.

The action integral for a field can now be written down by analogy with the particle case (see Table A.1). The Lagrangian appearing in (1) of course contains a *summation* over i; the Lagrangian for fields will therefore contain an *integration* over **x**. We will indicate this integration explicitly by writing

$$L = \int \mathscr{L}(\psi, \partial\psi/\partial t, \partial\psi/\partial x^k) d^3x \qquad [12]$$

where $\mathscr{L}(\psi, \partial\psi/\partial t, \partial\psi/\partial x^k)$ is the *Lagrangian density*. It is assumed that $\mathscr{L}$ is some expression constructed out of ψ and the *first* derivatives of ψ. The presence of ψ and of the time derivative $\partial\psi/\partial t$ is obvious from the analogy with the Lagrangian for particles. The presence of the space derivative $\partial\psi/\partial x^k$ is less obvious but can be understood in the context of the approximation (11): Ignoring y and z we have

$$\frac{\partial\psi}{\partial x} \cong \frac{q_{i+1} - q_i}{x_{i+1} - x_i} \propto q_{i+1} - q_i \qquad [13]$$

where x_i is the x-coordinate of the center of one cube, and x_{i+1} that of the center of the adjacent cube. Hence the presence of $\partial\psi/\partial x$ simply means that the Lagrangian contains some "potential" energy associated with a difference in the values of adjacent generalized coordinates.

The action integral is then

$$I = \int_{t_1}^{t_2} \int \mathscr{L}(\psi, \partial\psi/\partial t, \partial\psi/\partial x^k) d^3x\, dt \qquad [14]$$

TABLE A.1 CORRESPONDENCE BETWEEN A PARTICLE SYSTEM AND A FIELD SYSTEM

	Particle	**Field**
State of system described by	$q_i(t)$	$\psi(\mathbf{x}, t)$
Independent variables	i, t	$\mathbf{x}, t$
Lagrangian	$L = L(q_i, \dot{q}_i)$	$L = \int \mathscr{L}(\psi, \partial\psi/\partial t, \partial\psi/\partial x^k) d^3x$
Equation of motion	$\dfrac{d}{dt}\dfrac{\partial L}{\partial \dot{q}_i} - \dfrac{\partial L}{\partial q_i} = 0$	$\dfrac{\partial}{\partial x^\mu}\dfrac{\partial\mathscr{L}}{\partial\psi_{,\mu}} - \dfrac{\partial\mathscr{L}}{\partial\psi} = 0$
Hamiltonian (energy)	$H = \Sigma\, \dot{q}_i \dfrac{\partial L}{\partial \dot{q}_i} - L$	$H = \int \psi_{,0} \dfrac{\partial\mathscr{L}}{\partial\psi_{,0}} d^3x - L$

A variation $\delta\psi(\mathbf{x},t)$ in the field produces corresponding variations

$$\delta\left(\frac{\partial\psi}{\partial t}\right) = \frac{\partial}{\partial t}\,\delta\psi \tag{15}$$

and

$$\delta\left(\frac{\partial\psi}{\partial x^k}\right) = \frac{\partial}{\partial x^k}\,\delta\psi \tag{16}$$

in the derivatives. This leads to the following change in the action:

$$\delta I = \int_{t_1}^{t_2}\int\left[\frac{\partial\mathscr{L}}{\partial\psi}\,\delta\psi + \frac{\partial\mathscr{L}}{\partial(\partial\psi/\partial t)}\,\delta\left(\frac{\partial\psi}{\partial t}\right) + \frac{\partial\mathscr{L}}{\partial(\partial\psi/\partial x^k)}\,\delta\left(\frac{\partial\psi}{\partial x^k}\right)\right]d^3x\,dt$$

$$= \int_{t_1}^{t_2}\int\left[\frac{\partial\mathscr{L}}{\partial\psi}\,\delta\psi + \frac{\partial\mathscr{L}}{\partial(\partial\psi/\partial t)}\,\frac{\partial}{\partial t}\,\delta\psi + \frac{\partial\mathscr{L}}{\partial(\partial\psi/\partial x^k)}\,\frac{\partial}{\partial x^k}\,\delta\psi\right]d^3x\,dt \tag{17}$$

We integrate the term containing $(\partial/\partial t)\delta\psi$ by parts and impose the usual constraint that $\delta\psi = 0$ at $t = t_1$ and $t = t_2$. We also integrate the terms containing $(\partial/\partial x^k)\delta\psi$ by parts and make the assumption that $\delta\psi = 0$ at the limits of the x-, y-, and z-integrations. The result is

$$\delta I = \int_{t_1}^{t_2}\int\left[\frac{\partial\mathscr{L}}{\partial\psi} - \frac{\partial}{\partial t}\frac{\partial\mathscr{L}}{\partial(\partial\psi/\partial t)} - \frac{\partial}{\partial x^k}\frac{\partial\mathscr{L}}{\partial(\partial\psi/\partial x^k)}\right]\delta\psi\,d^3x\,dt \tag{18}$$

Since $\delta\psi$ is an arbitrary function, the action will be stationary if and only if the term in brackets is zero. With the Einstein summation convention and the comma notation for derivatives this can be written

$$\frac{\partial}{\partial x^\mu}\frac{\partial\mathscr{L}}{\partial\psi_{,\mu}} - \frac{\partial\mathscr{L}}{\partial\psi} = 0 \tag{19}$$

This partial differential equation is the field equation for the given Lagrangian. Note that it resembles Eq. (4), but contains both time and space derivatives (see Table A.1).

The Hamiltonian for a system of fields resembles Eq. (7), but the summation over i must be replaced by an integration over $\mathbf{x}$,

$$H = \int \psi_{,0}\frac{\partial\mathscr{L}}{\partial\psi_{,0}}\,d^3x - L$$

$$= \int\left(\psi_{,0}\frac{\partial\mathscr{L}}{\partial\psi_{,0}} - \mathscr{L}\right)d^3x \tag{20}$$

The field equations can be used to show that $dH/dt = 0$; we will give the proof of this in the next section. Taking the conservation of H

for granted, let us first inquire into the physical interpretation of this quantity. This physical interpretation has already been mentioned in the preceding section: the Hamiltonian of a closed system* is the total energy of the system.

To understand why this is so, consider a system of interacting particles and fields. The detailed mathematical treatment of interactions is beyond the scope of our discussion. Suffice it to say that the total Lagrangian, and hence also the total Hamiltonian, of an interacting system of particles and fields must contain some terms in which the particle and field variables appear together, in the form of a product. The total Hamiltonian will then consist of a sum of terms of the types (7), and (20), and more, and this total Hamiltonian will be conserved. Suppose that initially the particles are very widely separated and the fields absent. At a later time the particles come together, collide, produce strong fields, emit radiation, etc. (We may even suppose that an equal number of particles and antiparticles were present initially and that in the end they all annihilate, leaving nothing but fields!) Since the system initially consists of free particles, we know that the value of the Hamiltonian is initially equal to the energy (see Eq. (10)). We also know that both the Hamiltonian and the energy are conserved as the particles come together and generate fields. Hence we can conclude that the Hamiltonian and the energy are equal at all times. Thus, the identification of the Hamiltonian and the energy is justified by the conservation law. We may say: energy is that which is conserved.**

A.3   THE ENERGY-MOMENTUM TENSOR

We are now ready to construct the energy-momentum tensor. Since H is the energy, the integrand appearing in (20) can be regarded as the energy density,

$$t_0^0 = \psi_{,0} \frac{\partial \mathscr{L}}{\partial \psi_{,0}} - \mathscr{L} \qquad [21]$$

The complete energy-momentum tensor must then be such that its t_0^0 component agrees with (21). Obviously, the quantity

$$t_\mu^\nu = \psi_{,\mu} \frac{\partial \mathscr{L}}{\partial \psi_{,\nu}} - \delta_\mu^\nu \mathscr{L} \qquad [22]$$

* A closed system is a system not subject to any external forces.

** Incidentally, texts on classical mechanics often delight in pointing out examples in which the Hamiltonian differs from the energy. The reason for this apparent contradiction is that in mechanics one considers only the *mechanical energy* (energy that can be more or less directly associated with a particle), rather than the total energy. There is not always a simple connection between the mechanical Hamiltonian and the mechanical energy. Furthermore, the Hamiltonians used in mechanics often have an explicit time dependence.

satisfies this condition. In fact, arguments of the type given in Section 2.5 (see Exercise 2.21) show that $t_\mu{}^\nu$ is the *only* tensor that satisfies this condition. This tensor is called the *canonical energy-momentum tensor*. Next, we must prove that the conservation law

$$\frac{\partial}{\partial x^\nu} t_\mu{}^\nu = 0 \qquad [23]$$

is satisfied. The calculation is quite similar to that given in Eq. (8):

$$\frac{\partial}{\partial x^\nu} t_\mu{}^\nu = \psi_{,\mu,\nu} \frac{\partial \mathcal{L}}{\partial \psi_{,\nu}} + \psi_{,\mu} \frac{\partial}{\partial x^\nu} \frac{\partial \mathcal{L}}{\partial \psi_{,\nu}} - \frac{\partial \mathcal{L}}{\partial \psi} \psi_{,\mu} - \frac{\partial \mathcal{L}}{\partial \psi_{,\alpha}} \psi_{,\alpha,\mu} \qquad [24]$$

The first and last terms cancel, and the other two terms add to zero by virtue of the field equation (19).

The differential conservation law (23) implies the conservation of the energy (or Hamiltonian). We have

$$\frac{dH}{dt} = \frac{d}{dt} \int t_0{}^0 d^3x = \int \frac{\partial}{\partial t} t_0{}^0 d^3x$$

$$= - \int \frac{\partial}{\partial x^k} t_0{}^k d^3x \qquad [25]$$

Using Gauss's theorem, we can change the volume integral of the "divergence" $(\partial/\partial x^k) t_0{}^k$ into a surface integral. The surface must include all volume, i.e., the surface must be at infinity. Under the assumption that $t_0{}^k$ is exactly zero beyond some large distance or at least tends to zero faster than $1/r^2$, the surface integral vanishes and therefore H is constant.

By a similar argument one can show that the total momentum

$$P_k = \int t_k{}^0 d^3x \qquad [26]$$

is also constant.

EXERCISE 3. Prove this.

Before we consider some examples, we must issue a warning concerning the canonical energy-momentum tensor. It can happen that $t^{\mu\nu}$ is not symmetric in μ,ν (see, e.g., Eq. (36)). Since the gravitational field equations (3.48) are based on a symmetric energy-momentum tensor, this is a very serious defect.* But this defect can always be corrected by add-

---

* A lack of symmetry in the energy-momentum tensor also leads to troubles with the conservation of angular momentum.

ing to the tensor $t^{\mu\nu}$ of Eq. (20) an extra term $\bar{t}^{\mu\nu}$ chosen so that the final result is symmetric. This alteration of the energy-momentum tensor is permissible provided it leads to no change in the total energy (20) and momentum (26) and to no change in the conservation law (23). Mathematically this amounts to the requirements that $\int \bar{t}_0^0 d^3x = \int \bar{t}_k^0 d^3x = 0$ and $\partial_\nu \bar{t}_\mu{}^\nu = 0$. Thus, the extra term redistributes the energy (and momentum) in space, but does not change its total value. The freedom to add extra terms to $t^{\mu\nu}$ implies an ambiguity in the energy-momentum tensor. This ambiguity is a nuisance and it can only be eliminated by introducing some extra assumptions. For example, we might introduce the assumption that the energy-momentum tensor should be gauge invariant (see the case of electromagnetism discussed below.)

As a simple example of our results, let us consider the case of the one-component field ψ with a Lagrangian density

$$
\begin{aligned}
\mathscr{L} &= \tfrac{1}{2}(\psi_{,\alpha}\psi^{,\alpha} - m^2\psi^2) \\
&= \tfrac{1}{2}(\eta^{\alpha\beta}\psi_{,\alpha}\psi_{,\beta} - m^2\psi^2)
\end{aligned}
\tag{27}
$$

where m is a constant. The Lagrangian field equation is

$$
\frac{\partial}{\partial x^\mu}\,\eta^{\mu\alpha}\psi_{,\alpha} + m^2\psi = 0
$$

i.e.,

$$
\psi_{,\mu}{}^{,\mu} + m^2\psi = 0.
\tag{28}
$$

This is a famous equation of quantum mechanics — it is the Klein-Gordon equation for a free scalar field (of course we are now regarding ψ as a purely classical field rather than a quantum field).

The corresponding canonical energy-momentum tensor is

$$
\begin{aligned}
t_\mu{}^\nu &= \psi_{,\mu}\eta^{\nu\alpha}\psi_{,\alpha} - \delta_\mu{}^\nu\mathscr{L} \\
&= \psi_{,\mu}\psi^{,\nu} - \tfrac{1}{2}\delta_\mu{}^\nu(\psi_{,\alpha}\psi^{,\alpha} - m^2\psi^2)
\end{aligned}
\tag{29}
$$

EXERCISE 4. Show that the energy density is

$$
t_0^0 = \tfrac{1}{2}(\psi_{,0})^2 + \tfrac{1}{2}(\nabla\psi)^2 + \tfrac{1}{2}m^2\psi^2
\tag{30}
$$

Now we know all we need to know about the case of the one-component field and we are ready to write down the equations for a multi-component field. It is of course obvious that if instead of the one-component field ψ we have a field with several components (such as A_μ or $h_{\mu\nu}$), then when we perform the variation of the action, each component produces a term of the type shown in the brackets of Eq. (18). Since the different components have independent variations, each of these terms must vanish separately and we obtain as many field equations as there are components in the field. Thus, the equations for a four-component (vector) field are

$$\frac{\partial}{\partial x^\mu} \frac{\partial \mathscr{L}}{\partial A^\nu{}_{,\mu}} - \frac{\partial \mathscr{L}}{\partial A^\nu} = 0 \qquad \nu = 0, 1, 2, 3 \qquad [31]$$

and those for a sixteen-component (tensor) field are

$$\frac{\partial}{\partial x^\mu} \frac{\partial \mathscr{L}}{\partial h^{\alpha\beta}{}_{,\mu}} - \frac{\partial \mathscr{L}}{\partial h^{\alpha\beta}} = 0 \qquad \alpha, \beta = 0, 1, 2, 3 \qquad [32]$$

Equation (31) yields the field equations for a free electromagnetic field provided we take

$$\mathscr{L}_{em} = -\frac{1}{16\pi} (A_{\mu,\nu} - A_{\nu,\mu})(A^{\mu,\nu} - A^{\nu,\mu}) \qquad [33]$$

Eq. (32) yields the linear field equations for a free gravitational field provided we take*

$$\mathscr{L}_{(1)} = \frac{1}{4} (h_{\mu\nu,\lambda} h^{\mu\nu,\lambda} - 2h_{\mu\nu}{}^{,\mu} h^{\nu\lambda}{}_{,\lambda} + 2h_{\mu\nu}{}^{,\mu} h^{,\nu} - h_{,\nu} h^{,\nu}). \qquad [34]$$

EXERCISE 5. Show that $\mathscr{L}_{em}$ leads to the field equation (3.14) with $j^\nu = 0$.

EXERCISE 6. Show that $\mathscr{L}_{(1)}$ leads to the field equation (3.48) with $T^{\mu\nu} = 0$. (Hint: While working out the partial derivatives of $\mathscr{L}_{(1)}$ with respect to $h^{\alpha\beta}{}_{,\mu}$, you may treat $h^{\alpha\beta}$ as distinct from $h^{\beta\alpha}$, i.e., you need not worry about the symmetry of the tensor field).

The canonical energy-momentum tensors can be calculated from the appropriate generalization of Eq. (20):

$$t_{(em)\mu}{}^\nu = A^\alpha{}_{,\mu} \frac{\partial \mathscr{L}_{em}}{\partial A^\alpha{}_{,\nu}} - \delta_\mu^\nu \mathscr{L}_{em} \qquad [35]$$

$$t_{(1)\mu}{}^\nu = h^{\alpha\beta}{}_{,\mu} \frac{\partial \mathscr{L}_{(1)}}{\partial h^{\alpha\beta}{}_{,\nu}} - \delta_\mu^\nu \mathscr{L}_{(1)} \qquad [36]$$

EXERCISE 7. Use (33) and (35) to show that

$$t_{(em)\mu}{}^\nu = -\frac{1}{4\pi} \left(-A^\alpha{}_{,\mu} F^\nu{}_\alpha - \frac{1}{4} \delta_\mu^\nu F^{\alpha\beta} F_{\alpha\beta} \right) \qquad [37]$$

---

* The subscript (1) on $\mathscr{L}$ indicates that we are dealing with the linear approximation.

where

$$F^{\alpha\beta} = A^{\alpha,\beta} - A^{\beta,\alpha}$$

EXERCISE 8. Use (34) and (36) to find an expression for $t_{(1)\mu}{}^{\nu}$. Show that if the gauge condition $h^{\mu\nu}{}_{,\nu} = \tfrac{1}{2}h^{,\mu}$ is inserted in your expression, then the result is

$$t_{(1)\mu}{}^{\nu} = \frac{1}{4}\,[\,2h^{\alpha\beta}{}_{,\mu}h_{\alpha\beta}{}^{,\nu} - h_{,\mu}h^{,\nu} - \delta_{\mu}{}^{\nu}(h^{\alpha\beta}{}_{,\lambda}h_{\alpha\beta}{}^{,\lambda} - \tfrac{1}{2}\,h_{,\lambda}h^{,\lambda})\,]. \qquad [38]$$

Eq. (38) resembles Eq. (3.62). To see that these equations are actually identical, one need only express Eq. (38) in terms of $\varphi^{\alpha\beta}$ by means of $h^{\alpha\beta} = \varphi^{\alpha\beta} - \tfrac{1}{2}\,\eta^{\alpha\beta}\varphi$.

Eq. (37) resembles Eq. (2.117), but is *not identical* to it. In fact, the canonical energy-momentum tensor (37) suffers from the defect mentioned previously: it is not symmetric. This defect is easily repaired by adding an extra term

$$-\frac{1}{4\pi}\frac{\partial}{\partial x^{\alpha}}\,(A_{\mu}F^{\nu\alpha}) \qquad [39]$$

to $t_{(em)\mu}{}^{\nu}$. This extra term has a divergence that is identically zero,

$$\frac{\partial}{\partial x^{\nu}}\frac{\partial}{\partial x^{\alpha}}\,(A_{\mu}F^{\nu\alpha}) = 0 \qquad [40]$$

and hence has no effect on the conservation law.

EXERCISE 9. Prove Eq. (40). (Hint: The differential operator $\partial^2/\partial x^{\nu}\partial x^{\alpha}$ is symmetric in $\alpha\nu$, while $F^{\nu\alpha}$ is antisymmetric.)

Furthermore, if the $\nu = 0$ components of the extra term are integrated over all volume, the result is zero —

$$\int \frac{\partial}{\partial x^{\alpha}}\,(A_{\mu}F^{0\alpha})d^3x = 0 \qquad [41]$$

and hence the values of the energy and momentum in Eqs. (20) and (26) are not altered.

EXERCISE 10. Prove Eq. (41), assuming that the fields vanish at infinity. (Hint: Use $F^{00} = 0$ and use Gauss's theorem.)

Since $(\partial/\partial x^{\alpha})F^{\nu\alpha} = 0$ for a free electromagnetic field, we can also write the expression (39) as

$$-\frac{1}{4\pi} A_{\mu,\alpha} F^{\nu\alpha} \qquad\qquad\qquad [42]$$

which when added to the right side of Eq. (37) gives

$$-\frac{1}{4\pi}\left(-A^{\alpha}{}_{,\mu}F^{\nu}{}_{\alpha} - \frac{1}{4}\delta_{\mu}{}^{\nu}F^{\alpha\beta}F_{\alpha\beta}\right) - \frac{1}{4\pi} A_{\mu,\alpha}F^{\nu\alpha} \qquad [43]$$

This is easily seen to be identical to (2.117).

One might ask whether the choice of the extra term in Eq. (39) is unique. Could we have used a different extra term to symmetrize the energy-momentum tensor? The choice of the extra term is determined by gauge invariance. If the energy-momentum tensor is to serve as an unambiguous source of gravitational fields, it must obviously be left invariant by the electromagnetic gauge transformation of Eq. (3.19). The tensor (43) has this property; any other choice does not.

Incidentally, the gravitational energy-momentum tensor (38) is *not* invariant under the gravitational gauge transformation of Eq. (3.49). But this does no harm. When we use this term as a source of gravitation in Eq. (3.64), both the right and left sides of this equation have gauge-dependent terms, but the complete equation does not—the gauge-dependent terms cancel and there is no ambiguity in the final result.

A.4 THE VARIATIONAL PRINCIPLE FOR EINSTEIN'S EQUATIONS

We have seen in the preceding section that the approximate linear field equations for gravitation can be obtained from the variational principle with the Lagrangian $\mathscr{L}_{(1)}$ of Eq. (34). The exact nonlinear Einstein equations can also be obtained from a variational principle. The corresponding Lagrangian may be taken as

$$\mathscr{L} = \frac{1}{\kappa^2} R \sqrt{-g} \qquad\qquad\qquad [44]$$

where R is the curvature scalar and where $\kappa^2 = 16\pi G/c^4$. This Lagrangian contains not only first derivatives of the fields $g_{\mu\nu}$, but also second derivatives. In general, the presence of second derivatives leads to an Euler-Lagrange equation somewhat more complicated than (19). However, it turns out (see below) that in the special case of the Lagrangian (44), the terms contributed by the second derivatives cancel identically. Hence, the Lagrangian (44) effectively only depends on first derivatives.

To see how the second derivatives are eliminated, we begin with the identity (see Problem 6.15)

$$R \sqrt{-g} = \frac{\partial}{\partial x^\alpha} [g^{\mu\nu} \sqrt{-g} (-\Gamma^\alpha_{\mu\nu} + \delta_\mu{}^\alpha \Gamma^\beta_{\nu\beta})]$$

$$- g^{\mu\nu} \sqrt{-g} (\Gamma^\beta_{\mu\alpha}\Gamma^\alpha_{\nu\beta} - \Gamma^\alpha_{\mu\nu}\Gamma^\beta_{\alpha\beta}) \quad [45]$$

Second derivatives of $g_{\mu\nu}$ only appear in the first term (with square brackets) in Eq. (45); note that this term is a divergence. The contribution of this term to the action

$$I = \kappa^{-2} \int R \sqrt{-g} \, d^3x dt \qquad [46]$$

is of the form

$$\int \frac{\partial}{\partial x^\alpha} [\quad] \, d^3x dt \qquad [47]$$

This is the (four-dimensional) volume integral of a divergence. Hence, by integration by parts (47) can be converted into a surface integral over the boundary of the volume. Since the variational principle assumes that the variation $\delta g_{\mu\nu}$ vanishes on the boundary (see Eqs. (17), (18)), it follows that (47) gives no contribution to the variation of the action. We can, therefore, ignore the first term in (45) and regard

$$\mathcal{L} = - \kappa^{-2} g^{\mu\nu} \sqrt{-g} (\Gamma^\beta_{\mu\alpha}\Gamma^\alpha_{\nu\beta} - \Gamma^\alpha_{\mu\nu}\Gamma^\beta_{\alpha\beta}) \qquad [48]$$

as the effective Lagrangian. This Lagrangian contains first derivatives only.

It is a straight forward, but tedious, exercise to check that the Einstein equations do indeed follow from (48). We can reach this conclusion by a shortcut if we begin with the observation that the Euler-Lagrange equation will necessarily be a tensor equation. This can best be seen by writing the variation of the action in the form

$$\delta \int \kappa^{-2} R \sqrt{-g} \, d^3x dt$$

$$= \int \left[\frac{1}{\sqrt{-g}} \left(\frac{\partial \mathcal{L}}{\partial g_{\mu\nu}} - \frac{\partial}{\partial x^\alpha} \frac{\partial \mathcal{L}}{\partial g_{\mu\nu,\alpha}} \right) \delta g_{\mu\nu} \right] \sqrt{-g} \, d^3x dt \quad [49]$$

Since R is a scalar, the quantity in brackets on the right side must also be a scalar. But this quantity consists of the product of an (arbitrary) tensor $\delta g_{\mu\nu}$ with

$$\frac{1}{\sqrt{-g}} \left(\frac{\partial \mathscr{L}}{\partial g_{\mu\nu}} - \frac{\partial}{\partial x^{\alpha}} \frac{\partial \mathscr{L}}{\partial g_{\mu\nu,\alpha}} \right) \qquad [50]$$

Hence the quantity (50) must be a tensor and, therefore, the differential equation obtained by setting (50) equal to zero is a tensor equation.

We now want to check that the Einstein equations coincide with the Euler-Lagrange equations derived from (48). Since we are dealing with *tensor* equations, we only need to check this in some special, convenient coordinates. The tensor character of the equations guarantees that if agreement obtains in some special coordinates, then it also obtains in general coordinates. As our special coordinates, we take local geodesic coordinates (at one point). In these coordinates, the field equation will then only contain second order derivatives linearly. To single out these terms we write

$$g_{\mu\nu} = \eta_{\mu\nu} + \kappa h_{\mu\nu} \qquad [51]$$

and

$$\Gamma^{\alpha}{}_{\mu\nu} = \frac{\kappa}{2} \eta^{\alpha\beta} (h_{\nu\beta,\mu} + h_{\beta\mu,\nu} - h_{\mu\nu,\beta}) + \cdots \qquad [52]$$

and express the Lagrangian (48) as

$$\mathscr{L} = -\tfrac{1}{4}\eta^{\mu\nu} \left[(h_{\alpha}{}^{\beta}{}_{,\mu} + h^{\beta}{}_{\mu,\alpha} - h_{\mu\alpha}{}^{,\beta})(h_{\beta}{}^{\alpha}{}_{,\nu} + h^{\alpha}{}_{\nu,\beta} - h_{\nu\beta}{}^{,\alpha}) \right.$$
$$\left. -\tfrac{1}{4}(h_{\nu}{}^{\alpha}{}_{,\mu} + h^{\alpha}{}_{\mu,\nu} - h_{\mu\nu,}{}^{\alpha})(h^{\beta}{}_{\beta,\alpha}) \right] + \cdots \qquad [53]$$

where the dots stand for extra terms which are of no interest to us. Multiplying out the terms in the parentheses, and using the symmetry of $h_{\mu\nu}$ in μ,ν, we readily obtain

$$\mathscr{L} = \tfrac{1}{4}(h_{\alpha\beta,\mu} \, h^{\alpha\beta,\mu} - h_{\mu\alpha,\beta} \, h^{\beta\alpha,\mu} - h_{\alpha\beta,\mu} \, h^{\alpha\mu,\beta} + 2 h_{\mu\alpha}{}^{,\mu} \, h^{,\alpha}$$
$$- h_{,\alpha} \, h^{,\alpha}) + \cdots \qquad [54]$$

This Lagrangian agrees with $\mathscr{L}_{(1)}$ (see Eq. (34)) except that the term $-2h_{\mu\nu}{}^{,\mu} \, h^{\nu\lambda}{}_{,\lambda}$ which appears in the latter, has been replaced by a term $(-h_{\mu\alpha,\beta} \, h^{\beta\alpha,\mu} - h_{\alpha\beta,\mu} \, h^{\alpha\mu,\beta})$. But it is easy to check that this change does not make any difference at all for the differential equations generated by these Lagrangians.

EXERCISE 11. Show that both the above-mentioned terms give exactly the same contribution to the Euler-Lagrange equations.

We have, therefore, established that in local geodesic coordinates the Euler-Lagrange equations obtained from (48) agree with the usual equations of the linear approximation. A coordinate transformation from geodesic to general coordinates then tells us that the Euler-Lagrange equations always coincide with the Einstein equations (*in vacuo*).*

Incidentally, according to Eq. (46) we may say that the Einstein equations represent the condition for an extremum in the (four-dimensional) volume integral of the curvature. This circumstance justifies Whittaker's well-known remark: "Gravitation simply represents a continual effort of the universe to straighten itself out."

FURTHER READING

The last chapter of Goldstein, *Classical Mechanics,* gives a very nice introduction to the Lagrangian and Hamiltonian equations for fields with emphasis on the analogy with the corresponding equations for particles.

Wentzel, *Quantum Theory of Fields,* contains an excellent discussion of the Lagrangian methods in field theory. Although this book deals with quantum fields, any results that do not involve commutation relations are of course equally valid in the classical case. The canonical energy-momentum tensor is derived by Wentzel, and also by Landau and Lifshitz in *The Classical Theory of Fields.*

---

* The Einstein equations in the presence of matter can of course also be obtained from a variational principle, but we will not deal with this generalization here.

Answers to Problems

CHAPTER 1

1. $Q_{xx} = Q_{yy} = 3.3 \times 10^{42}$ g cm^2, $Q_{zz} = -6.6 \times 10^{42}$ g cm^2
   $g_{quad} \simeq 5 \times 10^{-3} g_{mono}$
2. $J_2 = 0.0011$
3. 1.7×10^{-3}
4. 1.1×10^{-5}; 1 rev/12 days
7. 1.7×10^{-3} rad; 1.7×10^{-12} rad
8. -3.9×10^{-25}, -4.2×10^{-10}, -1.3×10^{-6}, -4.4×10^{-2}
9. 1.2×10^9 cm
10. Spring: 77 cm. Neap: 29 cm
11. $\simeq 1 \times 10^{-10}$
13. 3×10^5 kW
14. 1800 ″/(century)^2, 1.9×10^{18} erg/sec, 4.9×10^{19} erg/sec

CHAPTER 2

2. $v > 0.56$, from New York to Camden

7. $\dfrac{2v_o}{a_o} (1 - v_o^2)^{1/2}$, $\dfrac{v_o}{a_o} (1 - 4v_o^2)^{1/2} + \dfrac{1}{2a_o} \sin^{-1} 2v_o$

9. $A'^{\mu}(x') = \left(\dfrac{-v}{\sqrt{1-v^2}}, \dfrac{1}{\sqrt{1-v^2}}, 1, 0 \right)$

$$\times \sin\left[\left(\dfrac{5-3v}{\sqrt{1-v^2}} \right) x'^1 - \left(\dfrac{3-5v}{\sqrt{1-v^2}} \right) x'^0 \right]$$

CHAPTER 3

9. $\theta = -\dfrac{2GM_\odot}{bv^2}(1+v^2)$

12. $\theta = -\dfrac{4GM}{bc^2} \pm \dfrac{4GS_z}{b^2c^3}$ (with $S_x = S_y = 0$); $\pm 5 \times 10^{-6}$ ''

CHAPTER 4

2. (a) $\epsilon'^{\mu\nu} = \epsilon_1^{\mu\nu}$ (b) $A'_I = A_I$ (c) $t'^{00}_{(1)} = \dfrac{1-v}{1+v} t^{00}_{(1)}$

4. $h = \text{constant} - \dfrac{\kappa}{4g} A_{II}\omega^2(1 - 2\cos^2\phi)R^2\sin^2\theta\cos\omega t$ (for a wave incident along the z-axis)
5. $\approx 10^{-19}$ rad/sec
6. $\approx 10^{-19}$ erg/sec
8. 2.0×10^9 erg/sec, -1.1×10^{-18} cm/sec
9. -6.1×10^{-14} sec/sec, -1.5×10^{-12} sec/sec, -4.6×10^{-3} sec/sec
10. $\approx 7 \times 10^{-14}$ erg; $r \approx 0$
11. $\approx 3 \times 10^{33}$ erg

CHAPTER 5

2. $\dfrac{\delta\nu}{\nu} = \dfrac{1}{2}\dfrac{\delta G}{G} + \dfrac{3}{2}\dfrac{\delta\alpha}{\alpha}$
5. $(1 - 1.1 \times 10^{-6})$

6. $\omega \simeq \omega_0\left(1 + \dfrac{GM}{Rc^2} - \dfrac{3GM}{2rc^2} + \dfrac{\Omega^2 R^2}{2}\right)$

7. -1.0×10^{-2} sec
8. $\Delta\nu/\nu = \pm 1.7 \times 10^{-6}$

CHAPTER 6

1. $\Gamma^1_{22} = -\sin\theta\cos\theta, \Gamma^2_{12} = \cot\theta$
   $R^1_{212} = \sin^2\theta, R^2_{121} = 1$
   $R_{11} = -1, R_{22} = -\sin^2\theta, R = -2/a^2$
2. $\Gamma^\mu_{\alpha\beta} = 0, R^\alpha_{\beta\mu\nu} = R_{\beta\mu} = R = 0$
3. $R^0_{202} = \ddot{a}a, R^2_{020} = -\ddot{a}/a$

4. (a) $g_{11} = (R + r\sin\theta)^2$, $g_{22} = r^2$, $g_{12} = g_{21} = 0$
   (b) (0,1)   (c) (0, R/r)

CHAPTER 7

6. 4000, 16, 120

CHAPTER 8

1. $[r(1 - 2GM/r)^{1/2} + 2GM\tanh^{-1}(1 - 2GM/r)^{1/2}]_{r_1 = L_1/2\pi}^{r_2 = L_2/2\pi}$; 1.0×10^5 cm

2. $\left[\dfrac{4\pi}{3}(1 - 2GM/r)^{1/2}\left(r^3 + \dfrac{5GMr^2}{2} + \dfrac{15G^2M^2r}{2}\right)\right.$
   $\left. + 5G^3M^3\tanh^{-1}(1 - 2GM/r)^{1/2}\right]_{r_1}^{r_2}$; 1.4×10^{28} cm^3

4. (a) $dr/dt = 1 - 2GM/r$ (b) $2(r_1 - r_2) + 4GM\log\dfrac{r_1 - 2GM}{r_2 - 2GM}$

   (c) $(1 - 2GM/r_1)^{1/2}\left[2(r_1 - r_2) + 4GM\log\dfrac{r_1 - 2GM}{r_2 - 2GM}\right]$

5. 0.89 °/year, 300 °/year, 4×10^{10} °/year

6. $\dfrac{(dt/d\tau)_{max} - (dt/d\tau)_{min}}{(dt/d\tau)_{max}} \simeq \dfrac{GM}{ac^2}\left(\dfrac{1}{1 - \epsilon} - \dfrac{1}{1 + \epsilon} + \dfrac{1}{2}\dfrac{1 + \epsilon}{1 - \epsilon} - \dfrac{1}{2}\dfrac{1 - \epsilon}{1 + \epsilon}\right)$
   $= 5.1 \times 10^{-6}$ for $M = 1.5M_\odot$, $a = 1.7 \times 10^{11}$ cm

CHAPTER 9

1. 5×10^{11} dyne
3. (b) $8r_S$
8. (a) 2.2×10^8 cm   (b) 1.2×10^6 cm

CHAPTER 10

4. $\dfrac{4\pi}{3}b^3\left(1 \pm \dfrac{1}{30}\dfrac{b^2}{a^2}\right)$; $\pm 3.3 \times 10^{-8}$
5. 1.6×10^{10} light years
8. 650 km/sec
10. $2\pi a^3(-\eta + \sinh\eta\cosh\eta)$, 8.3×10^{24} cubic light year/year, 2.8×10^{48} g/year

BIBLIOGRAPHY

Adler, R., M. Bazin, and M. Schiffer, *Introduction to General Relativity* (Mc-Graw-Hill, New York, 1965).

Alfvén, H., *Worlds-Antiworlds* (W. H. Freeman, San Francisco, 1966).

Anderson, J. L., *Principles of Relativity Physics* (Academic Press, New York, 1967).

Annual Review of Astronomy and Astrophysics (Annual Reviews, Palo Alto).

Beer, A., *Vistas in Astronomy* (Pergamon Press, New York).

Bergmann, P. G., *Introduction to the Theory of Relativity* (Prentice-Hall, Englewood Cliffs, 1942).

Bohm, D., *The Special Theory of Relativity* (W. A. Benjamin, New York, 1965).

Burbidge, G., and M. Burbidge, *Quasi-Stellar Objects* (W. H. Freeman, San Francisco, 1967).

Chiu, H., and W. F. Hoffman, eds., *Gravitation and Relativity* (W. A. Benjamin, New York, 1964).

Clancy, E. P., *The Tides, Pulse of the Earth* (Doubleday, New York, 1969).

Cohen, I. B., ed., *Isaac Newton's Papers and Letters on Natural Philosophy* (Harvard University Press, Cambridge, Mass., 1958).

Danby, J. M., *Fundamentals of Celestial Mechanics* (Macmillan, New York, 1962).

Defant, A., *Ebb and Flow* (University of Michigan Press, Ann Arbor, 1958).

———, *Physical Oceanography*, vol. 2 (Pergamon Press, New York, 1961).

Deser, S., and K. W. Ford, eds. *Lectures on Particles and Fields* (Prentice-Hall, Englewood Cliffs, 1965).

Deser, S., M. Grisaru, and H. Pendleton, eds., *Lectures on Elementary Particles and Quantum Field Theory* (MIT Press, Cambridge, Mass., 1970).

DeWitt, C., and B. DeWitt, eds., *Black Holes* (Gordon and Breach, New York, 1973).

———, *Relativity, Groups and Topology* (Gordon and Breach, New York, 1964).

Dicke, R. H., *Gravitation and the Universe* (American Philosophical Society, Philadelphia, 1970).

———, *The Theoretical Significance of Experimental Relativity* (Gordon and Breach, New York, 1964).

Eddington, A. S., *Space, Time, and Gravitation* (Cambridge University Press, Cambridge, England, 1920).

———, *The Mathematical Theory of Relativity* (Cambridge University Press, Cambridge, England, 1923).

Einstein, A., *The Meaning of Relativity,* 5th ed. (Princeton University Press, Princeton, 1955).

Eisenhart, L. P., *Riemannian Geometry* (Princeton University Press, Princeton, 1925).

Fock, V., *The Theory of Space, Time, and Gravitation,* trans. by N. Kemmer (Pergamon Press, Oxford, England, 1964).

French, A. P., *Special Relativity* (W. W. Norton, New York, 1968).

Galilei, G., *Le Opere di Galileo Galilei,* edited by A. Favaro (Edizione Nationale, Florence, 1898).

Goldberg, L., ed., *Annual Review of Astronomy and Astrophysics* (Annual Reviews, Palo Alto).

Goldstein, H., *Classical Mechanics* (Addison-Wesley, Reading, Mass., 1959).

Grosser, M., *The Discovery of Neptune* (Harvard University Press, Cambridge, Mass., 1962).

Harrison, B. K., K. S. Thorne, M. Wakano, and J. A. Wheeler, *Gravitation Theory and Gravitational Collapse* (University of Chicago Press, Chicago, 1965).

Hawking, S. W., and G. F. R. Ellis, *The Large Scale Structure of Space-Time* (Cambridge University Press, Cambridge, England, 1973).

Hoffman, B., *Albert Einstein, Creator and Rebel* (Viking Press, New York, 1972).

Hoyle, F., *Frontiers of Astronomy* (Harper and Brothers, New York, 1955).

———, and J. V. Narlikar, *Action at a Distance in Physics and Cosmology* (W. H. Freeman, San Francisco, 1974).

Hubble, E., *The Realm of the Nebulae* (Dover Publications, New York, 1958).

Infeld, L., and J. Plebanski, *Motion and Relativity* (Pergamon Press, New York, 1960).

Jackson, J. D., *Classical Electrodynamics* (John Wiley and Sons, New York, 1972).

Jammer, M., *Concepts of Mass* (Harvard University Press, Cambridge, Mass., 1961).

Jordan, P., *Schwerkraft und Weltall* (Vieweg, Braunschweig, 1951).

Kittel, C., W. D. Knight, and M. A. Ruderman, *Mechanics* (McGraw-Hill, New York, 1962).

Klauder, J. R., ed., *Magic without Magic: John Archibald Wheeler* (W. H. Freeman, San Francisco, 1972).

Koestler, A., *The Sleepwalkers* (Grosset and Dunlap, New York, 1963).

Koyré, A., *The Astronomical Revolution*, trans. by R. E. W. Maddison (Cornell University Press, Ithaca, 1973).

———, *From the Closed World to the Infinite Universe* (The Johns Hopkins Press, Baltimore, 1957).

———, *Newtonian Studies* (Harvard University Press, Cambridge, Mass., 1965).

Landau, L. D., and E. M. Lifshitz, *The Classical Theory of Fields* (Addison-Wesley, Reading, Mass., and Pergamon Press, Oxford, England, 1962).

Levi-Civita, T., *The Absolute Differential Calculus* (Blackie and Son, London, 1929).

Lightman, A. P., W. H. Press, R. H. Price, and S. A. Teukolsky, *Problem Book in Relativity and Gravitation* (Princeton University Press, Princeton, 1975).

Lorentz, H. A., A. Einstein, H. Minkowski, and H. Weyl, *The Principle of Relativity* (Dover Publications, New York, 1924).

Mach, E., *Die Mechanik*, 9th ed. (Brockhaus, Wiesbaden, 1933).

———, *The Science of Mechanics, A Critical and Historical Account of Its Development* (Open Court, La Salle, Ill., 1960).

MacMillan, W. D., *The Theory of the Potential* (Dover Publications, New York, 1958).

Marsden, B. G., and A. G. W. Cameron, eds., *The Earth-Moon System* (Plenum Press, New York, 1966).

McVittie, G. C., ed., *Problems of Extra-Galactic Research* (The MacMillan Company, New York, 1962).

Misner, C. W., K. S. Thorne, and J. A. Wheeler, *Gravitation* (W. H. Freeman, San Francisco, 1973).

Møller, C., *The Theory of Relativity* (Clarendon Press, Oxford, England, 1952).

———, ed., *Evidence for Gravitational Theories* (Academic Press, New York and London, 1962).

More, L. T., *Isaac Newton* (Charles Scribner's Sons, New York, 1934).

Newton, I., *Papers and Letters on Natural Philosophy,* edited by I. B. Cohen (Harvard University Press, Cambridge, Mass., 1958).

———, *Principia,* edited by F. Cajori (University of California Press, Berkeley, 1962).

Newton, R. R., *Ancient Astronomical Observations and the Accelerations of the Earth and Moon* (Johns Hopkins Press, Baltimore, 1970).

North, J. D., *Isaac Newton* (Oxford University Press, Oxford, England, 1967).

———, *The Measure of the Universe: A History of Modern Cosmology* (Oxford University Press, Oxford, England, 1965).

O'Raifertaigh, L., ed., *General Relativity: Papers in Honor of J. L. Synge* (Clarendon Press, Oxford, England, 1972).

Pauli, W., *Theory of Relativity* (Pergamon Press, London, 1958).

Peebles, P. J. E., *Physical Cosmology* (Princeton University Press, Princeton, 1971).

Rees, M., R. Ruffini, and J. A. Wheeler, *Black Holes, Gravitational Waves, and Cosmology* (Gordon and Breach, New York, 1974).

Reines, F., ed., *Cosmology, Fusion and Other Matters* (Colorado Associated University Press, Boulder, 1972).

Robertson, H. P., and T. W. Noonan, *Relativity and Cosmology* (Saunders, Philadelphia, 1968).

Ruffini, R., and H. Gursky, eds. *Neutron Stars, Black Holes, and Binary X-Ray Sources* (Reidel, Amsterdam, 1975).

Ryan, M. P., Jr., and L. C. Shepley, *Homogeneous Relativistic Cosmologies* (Princeton University Press, Princeton, 1975).

Salam, A., and E. P. Wigner, eds., *Aspects of Quantum Theory* (Cambridge University Press, Cambridge, England, 1972).

Schatzmann, E. L., *The Structure of the Universe* (McGraw-Hill, New York, 1968).

Schilpp, P. A., ed., *Albert Einstein: Philosopher-Scientist* (Harper and Brothers, New York, 1959).

Schwartz, M., *Principles of Electrodynamics* (McGraw-Hill, New York, 1972).

Schweber, S. S., *An Introduction to Relativistic Quantum Field Theory* (Harper and Row, New York, 1961).

Sciama, D. W., *Modern Cosmology* (Cambridge University Press, Cambridge, England, 1973).

————, *The Physical Foundations of General Relativity* (Doubleday, Garden City, N.Y., 1969).

————, *The Unity of the Universe* (Doubleday, Garden City, N.Y., 1961).

Sokolnikoff, I. S., *Tensor Analysis* (John Wiley and Sons, New York, 1951).

Streater, R. F., and A. S. Wightman, *PCT, Spin, Statistics, and All That* (W. A. Benjamin, New York, 1964).

Struve, O., *Stellar Evolution* (Princeton University Press, Princeton, 1950).

Synge, J. L., *Relativity: The General Theory* (North-Holland, Amsterdam, 1971).

————, *Relativity: The Special Theory* (North-Holland, Amsterdam, 1965).

————, *Talking About Relativity* (North-Holland, Amsterdam, 1970).

————, and A. Schild, *Tensor Calculus* (University of Toronto Press, Toronto, 1956).

Taylor, E. F., and J. A. Wheeler, *Spacetime Physics* (W. H. Freeman, San Francisco, 1963).

Tolman, R. C., *Relativity, Thermodynamics, and Cosmology* (Clarendon Press, Oxford, England, 1934).

Weinberg, S., *Gravitation and Cosmology: Principles and Applications of the General Theory of Relativity* (John Wiley and Sons, New York, 1972).

Wentzel, G., *Quantum Theory of Fields* (Interscience, New York, 1949).

Weyl, H., *Space-Time-Matter* (Dover Publications, New York, 1922).

Wheeler, J. A., *Einsteins Vision* (Springer-Verlag, Berlin, 1968).

————, *Geometrodynamics* (Academic Press, New York, 1962).

Whittaker, E., *A History of the Theories of Aether and Electricity* (Harper and Brothers, New York, 1960).

Wigner, E. P., *Symmetries and Reflections* (Indiana University Press, Bloomington and London, 1967).

Witten, L., ed., *Gravitation: An Introduction to Current Research* (John Wiley and Sons, New York, 1962).

Zwicky, F., and E. Herzog, *Catalogue of Galaxies and of Clusters of Galaxies* (California Institute of Technology, Pasadena, 1963).

INDEX

FUNDAMENTAL CONSTANTS[a]

Speed of light $\qquad c = 3.00 \times 10^{10}$ cm/sec

Planck's constant $\qquad \hbar = 1.05 \times 10^{-27}$ erg sec

$\qquad = 6.58 \times 10^{-22}$ MeV sec

Gravitational constant $\qquad G = 6.67 \times 10^{-8}$ cm^3 g^{-1} sec^{-2}

$\qquad \kappa = (16\pi G/c^4)^{1/2} = 2.04 \times 10^{-24}$ sec (cm g)$^{-1/2}$

Proton charge $\qquad e = 4.80 \times 10^{-10}$ esu

Fine structure constant $\qquad \alpha = e^2/\hbar c = 1/137.0$

Electron mass $\qquad m_e = 0.911 \times 10^{-27}$ g

$\qquad = 0.511$ MeV/c^2

Proton mass $\qquad M_p = 1.67 \times 10^{-24}$ g

$\qquad = 938$ MeV/c^2

Neutron mass $\qquad M_n = M_p + 2.31 \times 10^{-27}$ g

$\qquad = M_p + 1.29$ MeV/c^2

Compton wavelength $\qquad \lambdabar_C = \hbar/m_e c = 3.86 \times 10^{-11}$ cm

Bohr radius $\qquad a_0 = \hbar^2/m_e e^2 = 0.529 \times 10^{-8}$ cm

Rydberg constant $\qquad R_\infty = \tfrac{1}{2} m_e c^2 \alpha^2 = 13.6$ eV

Boltzmann constant $\qquad k = 1.38 \times 10^{-16}$ erg/°K

$\qquad = 8.62 \times 10^{-5}$ eV/°K

Stefan-Boltzmann constant $\qquad \sigma = \pi^2 k^4/60 h^3 c^2 = 5.67 \times 10^{-5}$ g sec^{-3} °K^{-4}

CONVERSION CONSTANTS

1 year (y) $\qquad = 3.16 \times 10^7$ sec

1 astronomical unit (A.U.) $\qquad = 1.50 \times 10^{13}$ cm

1 light year (ly) $\qquad = 0.946 \times 10^{18}$ cm

1 parsec (pc) $\qquad = 3.26$ light years

$\qquad = 3.08 \times 10^{18}$ cm

1 second of arc (″) $\qquad = 4.85 \times 10^{-6}$ radians

1 electron volt (eV) $\qquad = 1.60 \times 10^{-12}$ erg

[a] The values of all constants have been rounded off to three significant figures; this is convenient and (usually) sufficient.